T0133961

Statistical Treatment of Analytical Data

Zeev B. Alfassi

Department of Nuclear Engineering, Ben-Gurion University, Israel

Zvi Boger

OPTIMAL – Industrial Neural Systems, Ltd, Beer Sheva, Israel

Yigal Ronen

Department of Nuclear Engineering, Ben-Gurion University, Israel

Blackwell
Science

CRC Press

© 2005 by Blackwell Science Ltd
A Blackwell Publishing Company

Editorial offices:
Blackwell Publishing Ltd, 9600 Garsington Road, Oxford OX4 2DQ, UK
 Tel: +44 (0)1865 776868
Blackwell Publishing Asia Pty Ltd, 550 Swanston Street, Carlton, Victoria 3053, Australia
 Tel: +61 (0)3 8359 1011

ISBN 0-632-05367-4

Published in the USA and Canada (only) by
CRC Press LLC, 2000 Corporate Blvd., N.W., Boca Raton, FL 33431, USA
Orders from the USA and Canada (only) to
CRC Press LLC

USA and Canada only:
ISBN 0-8493-2436-X

This book contains information obtained from authentic and highly regarded sources. Reprinted
material is quoted with permission, and sources are indicated. Reasonable efforts have been made to
publish reliable data and information, but the author and the publisher cannot assume responsibility for
the validity of all materials or for the consequences of their use.

Trademark notice: Product or corporate names may be trademarks or registered trademarks, and are
used only for identification and explanation, without intent to infringe.

First published 2005

Library of Congress Cataloging-in-Publication Data:
A catalog record for this title is available from the Library of Congress

British Library Cataloguing-in-Publication Data:
A catalogue record for this title is available from the British Library

Set in 10/12 Times
by Kolam Information Services Pvt. Ltd, Pondicherry, India
Printed and bound in India
by Gopsons Papers Ltd, Noida

The publisher's policy is to use permanent paper from mills that operate a sustainable forestry policy,
and which has been manufactured from pulp processed using acid-free and elementary chlorine-free
practices. Furthermore, the publisher ensures that the text paper and cover board used have met
acceptable environmental accreditation standards.

For further information on Blackwell Publishing, visit our website:
www.blackwellpublishing.com

Contents

Preface

Chapters 1–11 were written by Zeev B. Alfassi, chapter 12 was written by Yigal Ronen, and chapter 13 was written by Zvi Boger.

Zeev B. Alfassi
Zvi Boger
Yigal Ronen

1 Introduction

1.1 Statistics and quality assurance, control and assessment

The appraisal of quality has a considerable impact on analytical laboratories. Laboratories have to manage the quality of their services and to convince clients that the advocated level of quality is attained and maintained. Increasingly accreditation is demanded or used as evidence of reliability. At present there are American and European standards (ISO 25 and EN45001) that describe how a laboratory ought to be organized in order to manage the quality of its results. These standards form the basis for accreditation of analytical labs. Terms used frequently are *quality assurance* and *quality control*. Quality assurance is a wider term which includes both quality control and quality assessment.

Quality control of analytical data (QCAD) was defined by the ISO Committee as: 'The set of procedures undertaken by the laboratory for continuous monitoring of operations and results in order to decide whether the results are reliable enough to be released'. QCAD primarily monitors the batch-wise accuracy of results on quality control materials, and precision on independent replicate analysis of 'test materials'. Quality assessment was defined (Taylor 1987) as 'those procedures and activities utilized to verify that the quality control system is operating within acceptable limits and to evaluate the data'.

The standards of quality assurance (American ISO 25; European EN 45001) were written for laboratories that do analyses of a routine nature and give criteria for the implementation of a quality system which ensures an output with performance characteristics stated by the laboratory. *An important aspect of the quality assurance system is the full documentation of the whole analysis process.* It is essential to have well designed and clear worksheets. On the worksheets both the raw data and the calculated results of the analyses should be written. Proper worksheets reduce the chances of computing error and enable reconstruction of the test if it appears that a problem has occurred. The quality assurance system (or Standard) also treats the problems of personnel, equipment, materials and chemicals. The most important item is the methodology of the analysis. Quality control is not meaningful unless the methodology used has been validated properly. Validation of a methodology means the proof of suitability of this methodology to provide useful analytical data. A method is validated when the performance characteristics of the method are adequate and when it has been established that the measurement is under statistical control and produces accurate results.

'Statistical control' is defined as 'A phenomenon will be said to be "statistically controlled" when, through the use of past experience, we can predict, at least within limits, how the phenomenon may be expected to vary in the future. Here it is understood that prediction means that we can state at least approximately, the probability that the observed phenomenon will fall within the given limits.'

The quality assurance systems required for accreditation of analytical laboratories are very important and are dealt with in several recent books (Kateman & Buydens 1987; Guennzler 1994; Funk *et al.* 1995; Pritchard 1997). However, these systems are well beyond the scope of this book, which will be devoted mainly to *quality assessment* of analytical data.

The quality of chemical analysis is usually evaluated on the basis of its uncertainty compared to the requirements of the users of the analysis. If the analytical results are consistent and have small uncertainty compared to the requirements, e.g. minimum or maximum concentration of special elements in the sample and its tolerances, the analytical data are considered to be of adequate quality. When the results are excessively variable or the uncertainty is larger than the needs, the analytical results are of low or inadequate quality. Thus, the evaluation of the quality of analysis results is a relative determination. What is high quality for one sample could be unacceptable for another. *A quantitative measurement is always an estimate of the real value of the measure and involves some level of uncertainty.* The limits of the uncertainty must be known within a stated probability, otherwise no use can be made of the measurement. *Measurement must be done in such a way that could provide this statistical predictability.*

Statistics is an integral part of quality assessment of analytical results, e.g. to calculate the precision of the measurements and to find if two sets of measurements are equivalent or not (in other words if two different methods give the same result for one sample).

Precise and accurate, which are synonyms in everyday language, have distinctly different meaning in analytical chemistry methodology. There are precise methods, which means that repeated experiments give very close results which are inaccurate since the measured value is not equal to the true value, due to systematic error in the system. For example, the deuterium content of a H_2O/D_2O mixture used to be determined by the addition of $LiAlH_4$, which reduces the water to hydrogen gas. The gas is transferred and measured by a mass spectrometer. However, it was found that although the method is precise, it is inaccurate since there is an isotope effect in the formation of the hydrogen.

Figure 1.1 explains simply the difference between precision and accuracy. *Statistics deals mainly with precision*, while accuracy can be studied by comparison with known standards. In this case, statistics play a role in analyzing whether the results are the same or not.

Old books dealt with only statistical methods. However the trend in the last decade is to include other mathematical methods that are used in analytical chemistry. Many analytical chemists are using computer programs to compute analytically areas of the various peaks in a spectrum or a chromatogram (in a spectrum the intensity of the signal is plotted vs. the wavelength or the mass [in mass spectra], while in the chromatogram it is plotted as a function of the time of the separation process). Another example is the use of the Fourier Transform either in 'Fourier Transform Spectroscopy' (mainly FTIR and FT-NMR, but recently also other spectroscopies) or in smoothing of experimental curves. The combination of statistics

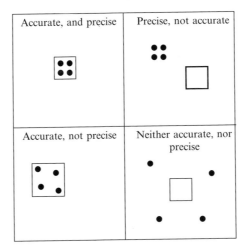

Fig. 1.1 Illustration of the meaning of accuracy and precision.

and other mathematical methods in chemistry is often referred to as chemometrics. However due to the large part played by statistics, and since many are 'afraid' of the general term of chemometrics, we prefer the title of Statistical and Mathematical Methods. These methods can be used as a black box, but it is important for educated analysts to understand the basic theory in order to take advantages of the full possibilities of these techniques and to choose intelligently the parameters as well as recognizing the limitation of these methods. It is clear that the choice of the mathematical tools is subjective, hence some methods are not included in this book because the authors feel that they are less important. Including the other methods would make this book too large.

1.2 References

Funk, W., Damman, V. & Donnevert, G. 1995, *Quality Assurance in Analytical Chemistry*, VCH, Weinheim.

Guennzler, H. 1994, *Accreditation and Quality Assurance in Analytical Chemistry*, Springer, Berlin.

Kateman, G. & Buydens, L. 1987, *Quality Control in Analytical Chemistry*, John Wiley, New York.

Pritchard, E. 1997, *Quality in the Analytical Chemistry Lab*, John Wiley, Chichester, UK.

Taylor, G. K. 1987, *Quality Assurance of Chemical Measurements*, John Wiley, Chichester, UK.

2 Statistical measures of experimental data

2.1 Mean and standard deviation

One of the best ways to assess the reliability of the precision of a measurement is to repeat the measurement several times and examine the different values obtained. Ideally, all the repeating measurements should give the same value, but in reality the results deviate from each other. Ideally, for a more precise result many replicate measurements should be done, however cost and time usually limit the number of replicate measurements possible. Statistics treats each result of a measurement as an item or individual and all the measurements as the *sample*. All possible measurements, including those which were not done, are called the *population*.

The basic parameters that characterize a population are the *mean*, μ, and the *standard deviation*, σ. In order to determine the *true* μ and σ, the entire population should be measured, which is usually impossible to do. In practice, measurement of several items is done, which constitutes a sample. Estimates of the mean and the standard deviation are calculated and denoted by $\bar{x}$ and s, respectively. The values of $\bar{x}$ and s are used to calculate confidence intervals, comparison of precisions and significance of apparent discrepancies. The mean, $\bar{x}$, and the standard deviation, s, of the values x_1, x_2, $\ldots, x_n$ obtained from n measurements is given by the equations:

$$\bar{x} = \frac{x_1 + x_2 + \ldots + x_n}{n} \tag{2.1a}$$

$$s = \sqrt{\left(\frac{(x_1 - \bar{x})^2 + (x_2 - \bar{x})^2 + \ldots + (x_n - \bar{x})^2}{n - 1}\right)} \tag{2.2a}$$

These equation can be written in a shorter way using the Σ notation:

$$\bar{x} = \frac{\sum\limits_{i=1}^{n} x_i}{n} \tag{2.1b}$$

$$s = \sqrt{\left(\frac{\sum\limits_{i=1}^{n}(x_i - \bar{x})^2}{n - 1}\right)} = \sqrt{\left(\frac{\sum\limits_{i=1}^{n} x_i^2}{n - 1} - \frac{n(\bar{x})^2}{n - 1}\right)} = \sqrt{\left(\frac{\sum\limits_{i=1}^{n} x_i^2}{n - 1} - \frac{\left(\sum\limits_{i=1}^{n} x_i\right)^2}{n(n - 1)}\right)} \tag{2.2b}$$

In some older books the use of the term 'average' instead of 'mean' (Youden 1994), can be found, but the common term nowadays is 'mean'. There are different kinds of 'means' (Woan 2000) (e.g. arithmetic mean, harmonic mean), but if not explicitly written the 'mean' is meant to be the arithmetic mean as defined by Equation (2.1).

There are several reasons why the arithmetic mean and not the other ones is chosen. The main reason is because it is the simplest one:

Arithmetic mean: $\quad \bar{x}_a = \dfrac{1}{n}(x_1 + x_2 + \ldots + x_n)$

Geometric mean: $\quad \bar{x}_g = (x_1 \times x_2 \times x_3 \times \ldots \times x_n)^{1/n}$

Harmonic mean: $\quad \bar{x}_h = n\left(\dfrac{1}{x_1} + \dfrac{1}{x_2} + \ldots + \dfrac{1}{x_n}\right)^{-1}$

Another reason to choose the arithmetic mean is that it fulfils the least squares criterion (Cantrell 2000), i.e. $\bar{x}_a$ fulfils the requirement:

$$\sum_{j=1}^{n} (x_j - \bar{x}_a)^2 = \text{minimum}$$

The names of these means come from the corresponding sequences. If we have an odd number of consecutive terms of a geometric sequence, then the middle term is given by the geometric mean of all these terms. The same is true for the arithmetic mean (in the case of an arithmetic sequence) and for the harmonic mean (in the case of an harmonic sequence). From now on we will use only the arithmetic mean and will refer to it in the general form:

$$\bar{x} = \bar{x}_a = \frac{x_1 + x_2 + \ldots + x_n}{n} \tag{2.1c}$$

The mean of the sum of squares of the deviation of the observed data from the mean is called the *variance*:

$$V = \frac{(x_1 - \bar{x})^2 + (x_2 - \bar{x})^2 + \ldots + (x_n - \bar{x})^2}{n - 1} \tag{2.3}$$

The division by $(n - 1)$ and not by n is done because we do not know the true value of $\bar{x}$, i.e. μ, and instead we used the calculated value of $\bar{x}$. For the calculation of $\bar{x}$, we use one degree of freedom (one unit of information), and this is the reason that we divide by $(n - 1)$ (the number of degrees of freedom, i.e. the number of free units of information which were left).

The dimension of *the variance*, V, is the square of the dimension of our observation and in order to get the same dimension we take the square root of V, which is called the *standard deviation, s*. In many cases the variance is not denoted by V, but is written as s^2.

$$s = \sqrt{\left(\frac{(x_1 - \bar{x})^2 + (x_2 - \bar{x})^2 + \ldots + (x_n - \bar{x})^2}{n - 1}\right)} = \sqrt{\left(\frac{n\Sigma x_i^2 - (\Sigma x_i)^2}{n(n - 1)}\right)} \tag{2.2c}$$

The values of $\bar{x}$ and s can be calculated using a computer program or a calculator. It is important to note that all scientific calculators have two keys, one depicted as σ_n and the other one as σ_{n-1}. Equation (2.2) fits the key σ_{n-1}. The other key uses n instead of $(n-1)$ in Equation (2.2). The key σ_{n-1} gives the standard deviation of our sample, but not of the whole population, which can be obtained by doing an infinite number of repeated measurements. In other words, σ_n is the standard deviation if the true mean μ is known. Otherwise, one degree of freedom is lost on the calculation of $\bar{x}$. For a small number of repetitions, the equation with $(n-1)$ gives a better estimate of the true σ, which is unknown. The mean $\bar{x}$ is a better estimate for the true value than one measurement alone. The standard deviation σ (or its estimate s) represents the dispersion of the measured values around the mean. The standard deviation has the same dimension as that of the measured values, x_i. Often, analysts prefer to use a dimensionless quantity to describe the dispersion of the results. In this case they use the *relative standard deviation* as a ratio (SV) (also called the *coefficient of variation*, CV) or as a percentage (RSD):

$$SV = s/\bar{x} \tag{2.4}$$

$$RSD = CV \times 100 \tag{2.5}$$

When calculating small absolute standard deviations using a calculator, sometimes considerable errors are caused by rounding, due the limited number of digits used. In order to overcome this problem, and in order to simplify the punching on the calculator, it is worth subtracting a constant number from all the data points, so that x_i will be not large numbers but rather of the same magnitude as their differences. The standard deviation will be unchanged but *the subtracted constant should be added to the mean*. In other words, if we have n data points, $x_1, \ldots, x_n$, which are large numbers, it is better to key into the calculator $(x_1 - c)$, $(x_2 - c), \ldots, (x_n - c)$ such that $(x_i - c)$ are no longer large numbers. The real mean of x_i is $\bar{x}_i = c + (\overline{x_i - c})$ and the standard deviation remains the same, $s(x_i) = s(x_i - c)$. Thus for calculating the mean and standard deviation of 50.81, 50.85, 50.92, 50.96, 50.83, we can subtract the constant 50.8, key 0.01, 0.05, 0.12, 0.16, 0.03 and obtain $\bar{x} = 0.074$ and $s = 0.06348$. The real mean is $50.8 + 0.074 = 50.874$ and s remains the same i.e. 0.06348. We could subtract only 50, key 0.81, 0.85, 0.92, 0.96, 0.83 and will obtain $\bar{x} = 0.874$ and $s = 0.06348$. The real mean is $50 + 0.874 = 50.874$ and s is 0.06348 as before.

Usually we choose the constant c as the smallest integer number of our data, so that the smallest number of $(x_i - c)$ is less than one. For example, if the data points are 92.45, 93.16, 91.82, 95.43, 94.37, we subtract 91 from all the data points, and calculate the mean and standard deviation of 1.45, 2.16, 0.82, 4.43, 3.37. The calculator will give $\bar{x} = 2.446$ and $s = 1.4584$. Adding 91 to the obtained mean, we get $\bar{x} = 93.446$ and $s = 1.4584$. Some will find it more easy to subtract just 90.

2.1.1 Significant figures

At this stage it is important to emphasize the importance of significant figures, especially nowadays when all calculations are made with calculators or computers, which yield results with many digits. Since our original data were given with two digits after the decimal point, any additional digits are meaningless. Consequently in the previous example there is no point giving $\bar{x} = 93.446$; we should round it off to $\bar{x} = 93.45$ and similarly $s = 1.46$. Usually the number of significant figures does not refer to the number of decimal digits but to the total number of figures. Thus, for example, the number 92.45 has four significant figures. This means that our precision of the measurement is 10^{-4}. In this case a result should not be given as 25.3 but rather as 25.30, in order to emphasize the precision of the measurement. *The mean of values should have the same number of significant figures as the values themselves.* However, the standard deviation, which is usually smaller, should have the same number of decimal digits as the measurements themselves, rather than the same number of significant figures. Thus, in our example we use for s only three significant figures i.e. $s = 1.46$, since the important factor is the decimal digits.

2.1.2 Frequency tables

When large numbers of measurements are made (on the same aliquot if it is not consumed by the measurement, or on different aliquots of the same sample or on different samples), some values are obtained more than once. Sometimes, instead of discrete values, a range of values is chosen as one value. In both cases it is simpler to concentrate the data in a *frequency table* – a table that gives the number of times (named frequency) each value was obtained. For example, the concentration of salt in drinking water was measured each day for a whole year. The results are given in Table 2.1 (given to two significant figures).

In this case the mean and the standard deviation are calculated by the equations:

$$\bar{x} = \frac{f_1 x_1 + f_2 x_2 + \ldots + f_n x_n}{f_1 + f_2 + \ldots + f_n} \quad \Rightarrow \quad \bar{x} = \frac{\sum\limits_{i=1}^{n} f_i x_i}{\sum\limits_{i=1}^{n} f_i} \tag{2.6}$$

Table 2.1 Concentration of salt in drinking water measured each day for one year.

Concentration (mg/ℓ) x_i	Numbers of days f_i
3.5	18
3.6	22
3.7	25
3.8	35
3.9	46
4.0	55
4.1	45
4.2	40
4.3	32
4.4	27
4.5	20

$$s = \sqrt{\left(\frac{f_1(x_1 - \bar{x})^2 + f_2(x_2 - \bar{x})^2 + \ldots + f_n(x_n - \bar{x})^2}{f_1 + f_2 + \ldots + f_n - 1}\right)} \quad \Rightarrow \quad s = \sqrt{\left(\frac{\sum\limits_{i=1}^{n} f_i(x_i - \bar{x})^2}{\left(\sum\limits_{i=1}^{n} f_i\right) - 1}\right)}$$

(2.7)

The summation is carried out over all the various values of x_i (n different values) and the total number of measurements is:

$$f_1 + f_2 + \ldots + f_n = \sum_{i=1}^{n} f_i$$

(2.8)

Most scientific calculators can calculate the mean value and the standard deviation from frequency tables. In our example the following results will be obtained:

$$\bar{x} = 4.0, \quad s = 0.3$$

(*remember to use the n−1 key*). The units of both $\bar{x}$ and s are the same as each sample, i.e. mg/ℓ. In short the concentration of the salt is written as 4.0 ± 0.3 mg/ℓ.

2.2   Graphical distributions of the data – bar charts or histograms

The standard deviation gives a measure of the spread of the results around the mean value. However, it does not indicate the shape of the spread.

Frequency tables and, even more so, drawing them as a rod diagram or as a histogram give a clearer picture of the spread of the measurement. A histogram describes the real situation better than bar charts since the real values are not discrete values of only two significant digits, and 3.7 mg/ℓ stands, for example, for the range 3.650 01 to 3.750 00. If the table were to three rather than two significant digits, there would be many more columns in the histogram. Increasing the number of measurements and the number of significant figures will lead to a continuous distribution.

Most spreadsheet data programs, such as Lotus 1-2-3, Quattro Pro, Excel or Origin can draw the frequency table in the form of column charts or histograms. Figures 2.1 and 2.2 are, for example, the result of the use of the chart wizard of Excel on the data of Table 2.1.

2.3   Propagation of errors (uncertainties)

In some cases we are interested in a value of a variable, which cannot be determined directly but can be calculated from several measurements of different properties. Thus for the measurement of the area of a rectangle we need to measure both its length L and the width W. The area A, is given by:

$$A = L \times W$$

For the volume of a box, V, we need in addition to measure its height, H:

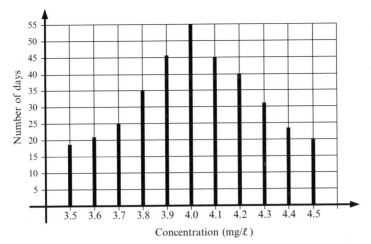

Fig. 2.1 Rod chart.

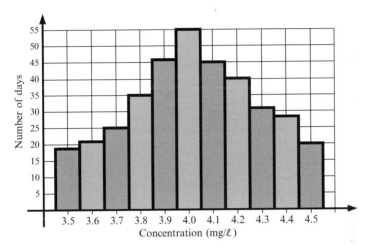

Fig. 2.2 Histogram.

$$V = L \times W \times H$$

How do the uncertainties (possible errors) in the estimation of L and W affect the resulting uncertainty in the value of the area, A? One way to calculate the possible error in A is to take the highest values of L and W, calculate from them the obtained A and compare it with the average value and the minimal values. Thus:

$$\overline{L}, \ \overline{W} \ \Rightarrow \ \overline{A} = \overline{L} \times \overline{W}$$

$$\overline{L} + \Delta L, \ \overline{W} + \Delta W \ \Rightarrow \ \overline{A} + \Delta A = (\overline{L} + \Delta L) \times (\overline{W} + \Delta W)$$
$$= \overline{L}\,\overline{W} + \overline{L} \times \Delta W + \overline{W} \times AL + \Delta W \times \Delta L$$

$$\overline{L} - \Delta L, \ \overline{W} - \Delta W \ \Rightarrow \ \overline{A} - \Delta A = (\overline{L} - \Delta L) \times (\overline{W} - \Delta W)$$
$$= \overline{L}\,\overline{W} - (\overline{L} \times \Delta W + \overline{W} \times AL) + \Delta W \times \Delta L$$

If we assume that the last term $(\Delta W \times \Delta L)$ can be neglected, due to the fact that the product of two small terms will lead to a smaller term, we can see that both directions will lead to the same value of ΔA:

$$\Delta A = \bar{L} \times \Delta W + \bar{W} \times \Delta L \tag{2.9}$$

The same equation will be obtained by calculus:

$$dA = \frac{\partial A}{\partial L} \, d\ell + \frac{\partial A}{\partial W} \, dW \;\Rightarrow\; dA = W \, d\ell + \ell \, dW \tag{2.10}$$

In the general case, where y was measured from the separate quantities x, z, etc., we can write:

$$y = f(x, z, \ldots) \tag{2.11}$$

The mean of y is calculated from the mean values of the different quantities:

$$\bar{y} = f(\bar{x}, \bar{z}, \ldots) \tag{2.12}$$

The different values of y can be written as:

$$y_i - \bar{y} \cong (x_i - \bar{x}) \frac{\partial y}{\partial x} + (z_i - \bar{z}) \frac{\partial y}{\partial z} + \ldots \tag{2.13}$$

and the variance σ_y^2 is:

$$\sigma_y^2 = \lim_{N \to \infty} \frac{1}{N} \sum (y_i - \bar{y})^2 \tag{2.14}$$

The variance, σ_y^2, can be expressed in terms of the variance of the separate measured quantities σ_x^2, σ_z^2, etc:

$$\sigma_y^2 \cong \lim_{N \to \infty} \frac{1}{N} \sum \left[(x_i - \bar{x}) \frac{\partial y}{\partial x} + (z_i - \bar{z}) \frac{\partial y}{\partial z} + \ldots \right]^2$$

$$\sigma_y^2 = \lim_{N \to \infty} \frac{1}{N} \sum \left[(x_i - \bar{x})^2 \left(\frac{\partial y}{\partial x} \right)^2 + (z_i - \bar{z})^2 \left(\frac{\partial y}{\partial z} \right)^2 \right.$$

$$\left. + 2(x_i - \bar{x})(z_i - \bar{z}) \left(\frac{\partial y}{\partial x} \right) \left(\frac{\partial y}{\partial z} \right) + \ldots \right]$$

$$\sigma_y^2 = \left(\frac{\partial y}{\partial x} \right)^2 \lim_{N \to \infty} \frac{1}{N} \sum (x_i - \bar{x})^2 + \left(\frac{\partial y}{\partial z} \right)^2 \lim_{N \to \infty} \frac{1}{N} \sum (z_i - \bar{z})^2$$

$$+ 2 \frac{\partial y}{\partial x} \frac{\partial y}{\partial z} \lim_{N \to \infty} \frac{1}{N} \sum (x_i - \bar{x})(z_i - \bar{z}) + \ldots$$

The first two sums are σ_x^2 and σ_z^2 respectively:

$$\sigma_x^2 = \lim_{N \to \infty} \frac{1}{N} \sum (x_i - \bar{x})^2 \qquad \sigma_z^2 = \lim_{N \to \infty} \frac{1}{N} \sum (z_i - \bar{z})^2 \tag{2.15}$$

Similarly the third sum can be defined as σ_{xz}^2:

$$\sigma_{xz}^2 = \lim_{N \to \infty} \frac{1}{N} \sum (x_i - \bar{x})(z_i - \bar{z}) \tag{2.16}$$

With this definition the approximation for the standard deviation of y is given by:

$$\sigma_y^2 \simeq \sigma_x^2 \left(\frac{\partial y}{\partial x}\right)^2 + \sigma_z^2 \left(\frac{\partial y}{\partial z}\right)^2 + 2\sigma_{xz}^2 \left(\frac{\partial y}{\partial x}\right)\left(\frac{\partial y}{\partial z}\right) + \ldots \tag{2.17}$$

The first two terms, which will presumably dominate, are averages of squares of deviation. The third term is the average of cross terms. If the functions in x and z are uncorrelated, the cross term will be small and will vanish in the limit of a large number of random observations. Thus for uncorrelated variables:

$$\sigma_y^2 \simeq \sigma_x^2 \left(\frac{\partial y}{\partial x}\right)^2 + \sigma_z^2 \left(\frac{\partial y}{\partial z}\right)^2 \tag{2.18}$$

Where there are more than two variables, each variable will contribute a similar term.

Addition and subtraction: If y is given by a linear combination of x and z, i.e. $y = ax \pm bz$, then:

$$\frac{\partial y}{\partial x} = a \qquad \frac{\partial y}{\partial z} = \pm b$$

Hence:

$$y = ax \pm bz \implies \sigma_y^2 = a^2\sigma_x^2 + b^2\sigma_y^2 \pm 2ab\sigma_{xz}^2 \tag{2.19}$$

In most cases the errors in x and z are uncorrelated and the mixed covariance, σ_{xz}, is equal to zero. However if the errors are correlated, σ_y^2 might vanish due to compensations by the covariance σ_{xz}^2. Usually we write:

$$\sigma_y^2 = a^2\sigma_x^2 + b^2\sigma_z^2 \tag{2.20}$$

Multiplication and division: If y is given by $y = axz$ or $y = a(x/z)$ then:

$$y = axz \implies \frac{\partial y}{\partial x} = az \qquad \frac{\partial y}{\partial z} = ax$$

Thus for multiplication:

$$\sigma_y^2 = a^2 z^2 \sigma_x^2 + a^2 x^2 \sigma_z^2 \tag{2.21}$$

Dividing by $y^2 = a^2 x^2 z^2$ leads to:

$$\frac{\sigma_y^2}{y^2} = \frac{\sigma_x^2}{x^2} + \frac{\sigma_z^2}{z^2} \tag{2.22}$$

$$y = a\left(\frac{x}{z}\right) \implies \frac{\partial y}{\partial x} = \frac{a}{z} \qquad \frac{\partial y}{\partial z} = -\frac{ax}{z^2}$$

$$y = a\left(\frac{x}{z}\right) \implies \sigma_y^2 = \left(\frac{a^2}{z^2}\right)\sigma_x^2 + \left(\frac{a^2 x^2}{z^4}\right)\sigma_z^2$$

Dividing by $y^2 = a^2 x^2 / z^2$ will lead to the same equation for the relative standard deviation as in multiplication (Equation (2.22)):

$$\frac{\sigma_y^2}{y^2} = \frac{\sigma_x^2}{x^2} + \frac{\sigma_z^2}{z^2}$$

Powers: If the function is $y = a\,x^b$, the derivative is:

$$\frac{\partial y}{\partial x} = b\,a\,x^{b-1} = b\,\frac{ax^b}{x} = b\,\frac{y}{x}$$

According to Equation (2.18), the relative σ_y:

$$\sigma_y^2 = b^2\,\frac{y^2}{x^2}\,\sigma_x^2 \;\Rightarrow\; \frac{\sigma_y}{\sigma_x} = b\,\frac{y}{x} \quad \text{or} \quad \frac{\sigma_y}{y} = b\,\frac{\sigma_x}{x} \tag{2.23}$$

Exponential: If the function is $y = a\,e^{bx}$, the derivative is:

$$\frac{\partial y}{\partial x} = a\,b\,e^{bx} \;\Rightarrow\; \frac{\partial y}{\partial x} = by$$

Hence the variance of y and the relative σ_y are:

$$\sigma_y^2 = b^2\,y^2\,\sigma_x^2 \;\Rightarrow\; \frac{\sigma_y}{\sigma_x} = by \;\Rightarrow\; \frac{\sigma_y}{y} = b\,\sigma_x \tag{2.24}$$

If the base is not e but a constant c, we can write:

$$c^{bx} = \left(e^{\ln c}\right)^{bx} = e^{b\ln c\,x}$$

and using Equation (2.24) gives:

$$y = a\,c^{bx} \;\Rightarrow\; \frac{\sigma_y}{y} = b\,\ln c\,\sigma_x \tag{2.25}$$

Logarithm: If the function is $y = a\ln b\,x$, the derivative is:

$$\frac{\partial y}{\partial x} = \frac{a}{x} \;\Rightarrow\; \sigma_y^2 = \frac{a^2}{x^2}\,\sigma_x^2 \;\Rightarrow\; \sigma_y = a\,\frac{\sigma_x}{x} \tag{2.26}$$

Looking at the functional form $y = a\ln b + \ln x^a$, we can see that the first term, which is a constant, has no influence and that σ_y for $y = x^a$ and $y = \ln x^a$ is the same.

In both cases,

$$\sigma_y = a\,\frac{\sigma_x}{x}$$

2.4   References

Cantrell, C. D. 2000, *Modern Mathematical Methods for Physicists and Engineers*, Cambridge University Press, p. 41.

Woan, G. 2000, *The Cambridge Handbook of Physics Formulas*, Cambridge University Press, p. 27.

Youden, W. J. 1994, *Experimentation and Measurement*, NIST Special Publication 672, US Department of Commerce.

3 Distribution functions

3.1 Confidence limit of the mean

The mean of a sample of measurements, $\bar{x}$, provides an estimate of the true value, μ, the quantity we are trying to measure. However, it is quite unlikely that $\bar{x}$ is exactly equal to μ, and an important question is to find a range of values in which we are certain that the true value lies. This range depends on the measured mean but also on the distribution of the various x_i, on the number of measurements done *and on the question of how certain we want to be*. The more certain we want to be, the larger the range we have to take. The larger the number of experiments done, the closer $\bar{x}$ is to μ, and a smaller range has to be taken for the same percentage of certainty. Usually statistics tables refer not to the number of repeated experiments but to the *degrees of freedom* (usually given as v or df). In the previous chapter it was seen that in the calculation of the standard deviation, the number of degrees of freedom is $(n-1)$, where n is the number of repeated measurements. The number of degrees of freedom refers to the number of independent variables (units of information). When calculating the standard deviation n terms of $(x_i - \bar{x})^2$ are used, but only $(n-1)$ terms are independent, since the nth term can be calculated from the equation:

$$\sum_{i=1}^{n} (x_i - \bar{x}) = 0$$

i.e. from the definition of the mean. The standard deviation tells us if the spread of the data is large or small or, in other words, if the distribution function is wide or narrow. But from the few data we usually have, this cannot be deduced from the distribution function itself. Various distribution functions are *assumed* in the use of statistics. The width of the distribution function that is assumed comes from the confidence limit of the calculated mean. The narrower the distribution function, the better the results and the smaller the confidence limit (range).

3.2 Measurements and distribution functions

In some measurements only discrete results are available, e.g. the reaction occurs or does not occur. Also in the case of a colored product, although we can characterize the product by a continuous variable, the wavelength of maximum absorption (peak of absorption), we will usually refer to the color by discrete values i.e. the names: black, white, green, etc. In most analytical chemistry measurements the value of the result is of a continuous nature. Yet in many cases we will gather the various results into several groups, making the results discrete as each group has a unique value and the number of groups is limited (albeit by us).

Let us assume that we have k groups of balls of varying sizes. Each group contains balls of a range of sizes and there is no overlapping between the groups, i.e according to the size of the ball we know unequivocally to which group it belongs. If the ith group has a radius in the range a_i to b_i and there are n_i balls in the group, then the total number of balls is

$$\sum_{i=1}^{k} n_i$$

The probability that in choosing a ball at random its size will be between a_i and b_i is

$$\left(n_i \Big/ \sum_{i-1}^{k} n_i \right)$$

The plot of n_i vs. i gives us the distribution function of the size of the balls as a histogram rather than as a rod chart. The sum of all the rectangles in the histogram is $\sum n_i$. If we plot the ratio

$$\left(n_i \Big/ \sum_{i-1}^{k} n_i \right) \text{ vs. } i$$

we will get the *probability function* for a ball chosen at random to have a radius in the range a_i to b_i. The difference between these two plots (either n_i or the ratio $n_i / \sum n_i$) is the multiplication of the first one by the factor $(1/\sum n_i)$ in order to change the area under the function curve (histogram) from $\sum n_i$ to 1, since the *total probability must be unity*. This multiplication is usually called *normalization*. A normalized distribution function is one which has an area under its curve (the integral from $-\infty$ to ∞ for a continuous distribution function) of unity, and hence can be used as a *probability function*.

If the distribution function of an experiment is known, it can be used to calculate the confidence limits of the mean. This confidence limit depends on the *certainty* with which we want to know it. If, for example, we want to know the confidence limit for 95% certainty (5% uncertainty), we will take 0.025 from both sides of the maximum of the distribution function area. This area of $0.05 = (2 \times 0.025)$ represents 5% of the total area under the distribution (probability) function curve. For common distribution functions the values which fit different areas (uncertainties) are tabulated, since the integrals cannot be calculated analytically.

3.3   Mathematical presentation of distribution and probability functions

A probability function is a function that assigns a probability of occurrence to each outcome of an experiment or a measurement. In the case of discrete outcomes, a *probability function (PF)* $f(x)$ is defined as that which gives the probability of getting an outcome equal to x. In the case of continuous outcomes, a *probability density function (PDF)* $f(x)\,dx$ is defined as that which gives the probability that an outcome will lie in the interval between x and $(x + dx)$.

The probability must have a real non-negative value not greater than 1, so in both cases of discrete and continuous outcomes:

$$0 \leq f(x) \leq 1$$

Since in any experiment we have some outcomes (results), the probabilities must sum to unity. Thus:

$$\text{for PF} \quad \sum_i f(x_i) = 1 \quad \text{and for PDF} \quad \int_{-\infty}^{\infty} f(x)\, dx = 1 \tag{3.1}$$

The summation or integration is done for all allowed values of x. For a PDF of common distribution functions, $f(x)$ can be expressed analytically, however the integration must be done numerically. The numerical integration is tabulated for the *cumulative probability function* (*CPF*) $F(x)$ defined as:

$$F(x) = \int_{-\infty}^{x} f(u)\, du$$

i.e. $F(x)$ is the area under the probability density function $f(u)$ from $-\infty$ to x.

The probability that the outcome of a measurement will lie between the limits a_1 and a_2 is given by $F(a_2) - F(a_1)$, i.e.

$$p(a_1 \leq x \leq a_2) = \int_{a_1}^{a_2} f(u)\, du = \int_{-\infty}^{a_2} f(u)\, du - \int_{-\infty}^{a_1} f(u)\, du = F(a_2) - F(a_1)$$

The mean and the variance are connected to the distribution function by definition of the *moments of the distribution function*.

The *kth moment* of a distribution about zero (the origin) which is assigned by $E(x^k)$ is defined by the equations:

$$E(x^k) = \sum_i u_i^k\, f(u_i) \quad \text{for PF (discrete outcomes)}$$

$$E(x^k) = \int_{-\infty}^{\infty} u^k\, f(u)\, du \quad \text{for PDF (continuous outcomes)}$$

The *mean* of a distribution (known also as its *expectation value*) is defined as the first moment about zero:

$$\mu = E(x) = \sum_i u_i\, f(u_i) \quad \text{for PF} \quad \text{and} \quad \mu = E(x) = \int_{-\infty}^{\infty} u\, f(u)\, du \quad \text{for PDF} \tag{3.2}$$

where u_i are all the possible values of the random variable x.

When we talk about the mean of a sample (a portion of the whole population) we call it $\bar{x}$ and similarly the standard deviation is designated by s. When we refer to the whole population and similarly to the distribution function we use μ and σ for the mean and the standard deviation, respectively.

The moment about zero is a special case of the moment about the mean (center), where the center is chosen as zero. The definition of the *kth moment* of distribution about the center (the mean) is defined as:

$$E\left[(x - \mu)^k\right] = \sum (u_i - \mu)^k \, f(u_i) \quad \text{for } \mathbf{PF} \text{ (discrete probability)}$$

$$E\left[(x - \mu)^k\right] = \int (u - \mu)^k \, f(u) \, du \quad \text{for } \mathbf{PDF} \text{ (continuous probability)}$$

(3.3)

where $\mu = E(x)$

The *variance* (the square of the standard deviation) is defined as the second central moment (the second moment about the mean):

$$\sigma^2 = V(x) = E\left[(x - \mu)^2\right] = \sum (u_i - \mu)^2 \, f(u_i) \quad \text{or} \quad \int (u_i - \mu)^2 \, f(u) \, d(u) \quad (3.4)$$

The variance, σ^2, can be transformed to a moment about zero using the following treatment:

$$\sigma^2 = \sum (x_i - \mu)^2 \, f(x_i) = \sum \left(x_i^2 - 2\mu x_i + \mu^2\right) f(x_i)$$
$$= \sum x_i^2 \, f(x_i) - 2\mu \sum x_i \, f(x_i) + \mu^2 \sum f(x_i)$$

As $\sum x_i \, f(x_i)$ is defined as μ, and since the distribution function is normalized, $\sum f(x_i) = 1$:

$$\sigma^2 = \sum x_i^2 \, f(x_i) - 2\mu^2 + \mu^2 \quad \Rightarrow \quad \sigma^2 = \sum x_i^2 \, f(x_i) - \mu^2 \quad (3.5a)$$

In the moments notation (exception value) we can write:

$$\sigma^2 = E\left(x^2\right) - [E(x)]^2 \quad \Rightarrow \quad \sigma^2 = E\left(x^2\right) - \mu^2 \quad (3.5b)$$

In order to see an example for these definitions let us look at the throwing of a die. This is a discrete probability function, which for the six possible outcomes 1, 2, 3, 4, 5, 6 has a value of $1/6$, and for all other numbers the probability is zero.

$$\mu = \sum (u_i) \, f(u_i) = 1 \times \frac{1}{6} + 2 \times \frac{1}{6} + 3 \times \frac{1}{6} + 4 \times \frac{1}{6} + 5 \times \frac{1}{6} + 6 \times \frac{1}{6} = \frac{21}{6} = 3.5$$

$$\sigma^2 = \sum (u_i - \mu)^2 \, f(u_i) = \sum (u_i - 3.5)^2 \, f(u_i)$$

$$\sigma^2 = (-2.5)^2 \times \frac{1}{6} + (-1.5)^2 \times \frac{1}{6} + (-0.5)^2 \times \frac{1}{6} + 0.5^2 \times \frac{1}{6} + 1.5^2 \times \frac{1}{6} + 2.5^2 \times \frac{1}{6}$$

$$= \frac{17.5}{6} = 2.9167$$

$$\sigma = \sqrt{2.9167} \quad \Rightarrow \quad \sigma = 1.7078$$

In most analytical chemical experiments we have only continuous distribution functions, yet in some cases discrete distribution functions are important. For this reason we will learn about the distributions of both types. In most cases the 'true' distribution function corresponding to our measurements is not known. The number of measurements done is usually too small to find the real distribution function. Here we will learn about some theoretical distribution functions and will use them as an approximation for the real situation. From the systematic point of view it might be more appropriate to start with discrete probability functions, however as continuous functions are more common in analytic chemistry, we will start with them.

3.4 Continuous distribution functions

3.4.1 The normal distribution function

The most common distribution function is the *normal* one, a distribution which fits most natural phenomena, although not all of them, when the sample is large enough and the random errors are sufficiently small. Most natural phenomena are symmetric about the mean, forming a bell-shape plot. For a function to be symmetrical about the mean μ, it should involve either $|x - \mu|$ or $(x - \mu)^n$, where n is an even integer, the simplest case being $n = 2$. The most appropriate form of mathematical function should be proportional to $\exp\left[-c|x - \mu|\right]$ or $\exp\left[-c(x - \mu)^2\right]$. The inclusion of c is to allow for different widths of the distribution. After fixing c, the proportionality constant, A, can be determined by normalizing the distribution function, i.e requiring that $\int_{-\infty}^{\infty} f(x)\, dx = 1$. For the normal distribution, c is chosen as $c = 1/(2\sigma^2)$. Substituting $u = (x - \mu)/\sigma$, the normal distribution function becomes

$$f(u) = A\ \exp\left(-u^2/2\right)$$

This substitution of x by $(x - \mu)$ leads all normal distribution functions to be symmetric around the same number – the origin. The division of $(x - \mu)$ by σ leads all normal distribution functions to have the same width. Using $u = (x - \mu)/\sigma$ leads to $f(u)$ being the *standard normal distribution function*. The variable u is called the *standard variable*:

$$u = \frac{x - \mu}{\sigma} \tag{3.6a}$$

The value of the integral

$$\int_{-\infty}^{\infty} \exp\left(-u^2/2\right)\, du$$

can be found by the following method. The function is symmetrical about zero and hence the area under the function from $-\infty$ to 0 is equal to the area from 0 to ∞, hence $\int_{-\infty}^{\infty} = 2\int_{0}^{\infty}$

Let $I = \int_{0}^{\infty} \exp\left(-x^2/2\right)\, dx$, and by using the same notation for another variable, $I = \int_{0}^{\infty} \exp\left(-y^2\right)\, dy$, multiplication of the two integrals yields:

$$I^2 = \int_{0}^{\infty} \exp\left(-x^2/2\right)\, dx \times \int_{0}^{\infty} \exp\left(-y^2/2\right)\, dy$$

As x and y are independent variables and since $\exp\left(-x^2/2\right) \times \exp\left(-y^2/2\right) = \exp\left[-(x^2 + y^2)/2\right]$

$$I^2 = \int_0^\infty \int_0^\infty \exp\left[-(x^2 + y^2)/2\right] dx\, dy$$

Let us change the cartesian coordinates to polar ones. In polar coordinates, $x^2 + y^2 = r^2$. The range of the variables x and y from 0 to ∞ means the first quadrant and hence r is going from 0 to infinity and θ from 0 to $\pi/2$. In polar coordinates, $dx\, dy = r\, dr\, d\theta$, and hence:

$$I^2 = \int_0^{\pi/2} \int_0^\infty \exp\left(-r^2/2\right) r\, dr\, d\theta$$

Calculating the inner integral yields:

$$\int_0^\infty \exp\left(-r^2/2\right) r\, dr = \left[-\exp\left(-r^2/2\right)\right]_0^\infty = -(0 - 1) = 1$$

Hence:

$$I^2 = \int_0^{\pi/2} d\theta = \frac{\pi}{2} \quad \Rightarrow \quad I = \frac{1}{2}\sqrt{(2\pi)}$$

From this integral we can find the normalization factor of the distribution function:

$$\int_{-\infty}^\infty f(u)\, du = 2A\, I \quad \Rightarrow \quad \int_{-\infty}^\infty f(u)\, du = A\sqrt{(2\pi)} = 1$$

The normalization constant is $A = 1/\sqrt{(2\pi)}$ and the *standard normal distribution function* is:

$$f(u) = \frac{1}{\sqrt{(2\pi)}}\, \exp\left(-u^2/2\right) \tag{3.7}$$

When changing from the standard variable u to the original variable x, the normalization constant also changes since $u = (x - \mu)/\sigma$ leads to $dx = \sigma\, du$ and hence

$$\int_{-\infty}^\infty f(x)\, dx = \int_{-\infty}^\infty f(u)\, \sigma\, du = \sigma \int_{-\infty}^\infty f(u)\, du = \sigma\sqrt{(2\pi)}$$

So the normalization constant is $1/(\sigma\sqrt{2\pi})$.
The *normal distribution function* for the variable x is:

$$f(x) = \frac{1}{\sigma\sqrt{(2\pi)}} \exp\left[-(x - \mu)^2/2\sigma^2\right] = \frac{1}{\sigma\sqrt{(2\pi)}} \exp\left[-\frac{1}{2}\left(\frac{x - \mu}{\sigma}\right)^2\right] \tag{3.8}$$

It can be proven that for $f(u)$ the mean is zero and the variance is equal to 1.

$$\mu(u) = \int_{-\infty}^{\infty} u\, f(u)\, du = \int_{-\infty}^{\infty} u\, \frac{\exp(-u^2/2)}{\sqrt{(2\pi)}}\, du = \frac{1}{\sqrt{(2\pi)}} \left[-\exp\left(\frac{-u^2}{2}\right) \right]_{-\infty}^{\infty}$$

$$= \frac{1}{\sqrt{(2\pi)}}(0-0) = 0$$

Using Equation (3.5):

$$\sigma^2 = \int_{-\infty}^{\infty} u^2\, f(u)\, dx - \mu^2 = 1 - 0 \quad \Rightarrow \quad \sigma^2(u) = 1$$

For f(x), the mean is μ and the variance is equal to σ^2. The function f(x) is a probability function, since $\int_{-\infty}^{\infty} f(x)\, dx = 1$, which means that f(x) dx is the probability in a random experiment to receive a result which lies in the range x and (x + dx).

The cumulative probability function $\phi(x)$ is defined as the integral of the probability function from $-\infty$ to x:

$$\phi(x) = \int_{-\infty}^{x} f(y)\, dy \tag{3.9}$$

For the normal distribution, the integral $\int_{-\infty}^{x} \exp(-y^2/2)\, dy$ cannot be calculated analytically, however it can be calculated numerically and has been tabulated. Table 3.1 gives the function $\phi(z)$, where z is the standard variable:

$$z = \frac{x - \mu}{\sigma} \tag{3.6b}$$

In using this table for calculating probabilities, it should always be remembered that *the variable should be first transformed to the standard variable*. Some examples of calculations using this table follows.

Given a population where the monthly expenditures of a family are *distributed normally*, with a mean of $2000 and a standard deviation of $800, calculate the percentage of the families who:

(1) Spend monthly less than $3000.
(2) Spend monthly more than $3500.
(3) Spend monthly less than $1000.
(4) Spend monthly between $1500 and $2800.

Solution: The first step is to transform the nominal variable (x) to the standard one (z), using Equation (3.6):

$$x = 3000 \quad \Rightarrow \quad z = \frac{3000 - 2000}{800} = 1.25$$

$$x = 3500 \quad \Rightarrow \quad z = 1.875$$

$$x = 1000 \quad \Rightarrow \quad z = \frac{1000 - 2000}{800} = -1.25$$

$$x = 1500 \quad \Rightarrow \quad z = -0.625 \qquad x = 2800 \quad \Rightarrow \quad z = 1$$

$\phi(z)$ is found from Table 3.1.

Table 3.1 Cumulative distribution function for the standard normal distribution.

z	0	1	2	3	4	5	6	7	8	9
0.0	0.500	0.504	0.508	0.512	0.516	0.520	0.524	0.528	0.532	0.536
0.1	0.540	0.544	0.548	0.552	0.556	0.560	0.564	0.568	0.571	0.575
0.2	0.579	0.583	0.587	0.591	0.595	0.599	0.603	0.606	0.610	0.614
0.3	0.618	0.622	0.625	0.629	0.633	0.637	0.641	0.644	0.648	0.652
0.4	0.655	0.659	0.663	0.666	0.670	0.674	0.677	0.681	0.684	0.688
0.5	0.692	0.695	0.699	0.702	0.705	0.709	0.712	0.716	0.719	0.722
0.6	0.726	0.729	0.732	0.736	0.739	0.742	0.745	0.749	0.752	0.755
0.7	0.758	0.761	0.764	0.767	0.770	0.773	0.776	0.779	0.782	0.785
0.8	0.788	0.791	0.794	0.797	0.800	0.802	0.806	0.809	0.811	0.813
0.9	0.816	0.819	0.821	0.824	0.826	0.829	0.832	0.834	0.837	0.839
1.0	0.841	0.844	0.846	0.848	0.851	0.853	0.855	0.858	0.860	0.862
1.1	0.864	0.866	0.869	0.871	0.873	0.875	0.877	0.879	0.881	0.883
1.2	0.885	0.887	0.889	0.891	0.893	0.894	0.896	0.898	0.900	0.902
1.3	0.903	0.905	0.907	0.908	0.910	0.911	0.913	0.915	0.916	0.918
1.4	0.919	0.921	0.922	0.924	0.925	0.926	0.928	0.929	0.931	0.932
1.5	0.933	0.935	0.936	0.937	0.938	0.939	0.941	0.942	0.943	0.944
1.6	0.945	0.946	0.947	0.948	0.9495	0.9505	0.9515	0.9525	0.9535	0.9545
1.7	0.9554	0.9564	0.9573	0.9582	0.9591	0.9599	0.9608	0.9616	0.9625	0.9633
1.8	0.9641	0.9650	0.9657	0.9664	0.9671	0.9678	0.9686	0.9693	0.9699	0.9706
1.9	0.9713	0.9719	0.9726	0.9732	0.9738	0.9744	0.9750	0.9756	0.9762	0.9767
2.0	0.9773	0.9778	0.9783	0.9788	0.9793	0.9798	0.9803	0.9808	0.9812	0.9817
2.1	0.9821	0.9826	0.9830	0.9834	0.9838	0.9842	0.9846	0.9850	0.9854	0.9857
2.2	0.9861	0.9865	0.9868	0.9871	0.9875	0.9878	0.9881	0.9884	0.9887	0.9890
2.3	0.9893	0.9896	0.9898	0.9901	0.9904	0.9906	0.9909	0.9911	0.9913	0.9916
2.4	0.9918	0.9920	0.9922	0.9925	0.9927	0.9929	0.9931	0.9932	0.9934	0.9936
2.5	0.9938	0.9940	0.9941	0.9943	0.9945	0.9946	0.9948	0.9949	0.9951	0.9952
2.6	0.9954	0.9955	0.9956	0.9957	0.9959	0.9960	0.9961	0.9962	0.9963	0.9964
2.7	0.9965	0.9966	0.9967	0.9968	0.9969	0.9970	0.9971	0.9972	0.9973	0.9974
2.8	0.9974	0.9975	0.9976	0.9977	0.9977	0.9978	0.9979	0.9979	0.9980	0.9981
2.9	0.9981	0.9982	0.9983	0.9983	0.9984	0.9984	0.9985	0.9985	0.9986	0.9986
3.0	0.9987	0.9987	0.9987	0.9988	0.9988	0.9989	0.9989	0.9989	0.9990	0.9990

Family (1): For $z = 1.25$, we look at row 1.2 and then at the number in column 5 (which is our third digit) and find 0.894. In some statistics tables the decimal point appears only in the column headed by zero, so in the column headed by 5 we would find 894, but we have to write 0. to the left of the number found as the probability is always less than one, i.e. 0.894. Finding $\phi(1.25) = 0.894$ means that the integral $\int_{-\infty}^{1.25} f(z)\, dz = 0.894$. It means that 0.894 from all the population lies in the range $[-\infty,\ 1.25]$. Hence 89.4% ($= 0.894 \times 100$) spends monthly less than \$3000.

Family (2): $z = 1.875$ cannot be found directly in Table 3.1, since the table gives only $\phi(1.87)$ and $\phi(1.88)$ in row 1.8 in the columns headed by 7 and 8. Since $\phi(1.87) = 0.9693$ and $\phi(1.88) = 0.9699$, then by interpolation $\phi(1.875) = 0.9696$.

Thus 96.96% ($= 0.9696 \times 100$) spends less than $3500 and hence 0.04% ($100 - 96.96$) of the population spends monthly more than $3500.

Notice that directly from the table we get only 'less than'. In order to obtain 'more than', the tabulated value has to be subtracted from the whole (1.0 in probability or 100% in population) population.

Family (3): For negative values of z we cannot find $\phi(z)$ in Table 3.1. Some books give tables which include negative values of z, but most books give a table with $\phi(z)$ only for positive values of z. In this case, $\phi(z)$ for negative values of z is calculated by the relation:

$$\phi(-z) = 1 - \phi(z) \tag{3.10}$$

This equation is explained by Fig. 3.1, where z has a positive value. Since the probability function is symmetrical about the zero, Equation (3.10) must hold.

We found earlier that $f(1.25) = 0.894$ and thus $\phi(-1.25) = 1 - 0.894 = 0.106$. Hence 0.106 or 10.6 % of the population spends less than $1000.

When we are interested in the probability of having an expenditure between z_1 and z_2 we subtract the cumulative probability of those values (Fig. 3.2).

Hence the fraction of the population spending between $1500 and $2800 is:

$$\phi(1) - \phi(-0.625) = 0.841 - (1 - 0.734) = 0.841 - 0.266 = 0.575$$

i.e. 57.5 % of the population spend in this range.

In some statistics books, instead of tabulating the cumulative probability, tabulations of the normal curve area are given which take the lower boundary of the integral as 0 and not as $-\infty$, as can be seen in Table 3.2. In most cases the table gives a figure to indicate the boundaries of the integration. It is advisable to check if $\phi(0) = 0.5$, which means that the lower boundary is $-\infty$, or $\phi(0) = 0$, which means that the lower boundary is 0.

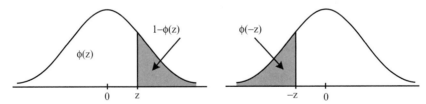

Fig. 3.1

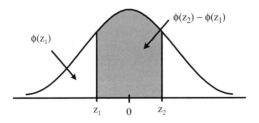

Fig. 3.2

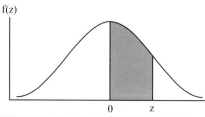

f(z)

Table 3.2 Normal curve areas.

z	0.00	0.01	0.02	0.03	0.04	0.05	0.06	0.07	0.08	0.09
0.0	0.0000	0.0040	0.0080	0.0120	0.0160	0.0199	0.0239	0.0279	0.0319	0.0359
0.1	0.0398	0.0438	0.0478	0.0517	0.0557	0.0596	0.0636	0.0675	0.0714	0.0753
0.2	0.0793	0.0832	0.0871	0.0910	0.0948	0.0987	0.1026	0.1064	0.1103	0.1141
0.3	0.1179	0.1217	0.1255	0.1293	0.1331	0.1368	0.1406	0.1443	0.1480	0.1517
0.4	0.1554	0.1591	0.1628	0.1664	0.1700	0.1736	0.1772	0.1808	0.1844	0.1879
0.5	0.1915	0.1950	0.1985	0.2019	0.2054	0.2088	0.2123	0.2157	0.2190	0.2224
0.6	0.2257	0.2291	0.2324	0.2357	0.2389	0.2422	0.2454	0.2486	0.2517	0.2549
0.7	0.2580	0.2611	0.2642	0.2673	0.2707	0.2734	0.2764	0.2794	0.2823	0.2852
0.8	0.2881	0.2910	0.2939	0.2967	0.2995	0.3023	0.3051	0.3078	0.3106	0.3133
0.9	0.3159	0.3186	0.3212	0.3238	0.3264	0.3289	0.3315	0.3340	0.3365	0.3389
1.0	0.3413	0.3437	0.3461	0.3485	0.3508	0.3531	0.3554	0.3577	0.3599	0.3621
1.1	0.3643	0.3665	0.3686	0.3708	0.3729	0.3749	0.3770	0.3790	0.3810	0.3830
1.2	0.3849	0.3869	0.3888	0.3907	0.3925	0.3944	0.3962	0.3980	0.3997	0.4015
1.3	0.4032	0.4049	0.4066	0.4082	0.4099	0.4115	0.4131	0.4147	0.4162	0.4177
1.4	0.4192	0.4207	0.4222	0.4236	0.4251	0.4265	0.4279	0.4292	0.4306	0.4319
1.5	0.4332	0.4345	0.4357	0.4370	0.4382	0.4394	0.4406	0.4418	0.4429	0.4441
1.6	0.4452	0.4463	0.4474	0.4484	0.4495	0.4505	0.4515	0.4525	0.4535	0.4545
1.7	0.4554	0.4564	0.4573	0.4582	0.4591	0.4599	0.4608	0.4616	0.4625	0.4633
1.8	0.4641	0.4649	0.4656	0.4664	0.4671	0.4678	0.4686	0.4693	0.4699	0.4706
1.9	0.4713	0.4719	0.4726	0.4732	0.4738	0.4744	0.4750	0.4756	0.4761	0.4767
2.0	0.4772	0.4778	0.4783	0.4788	0.4793	0.4798	0.4803	0.4808	0.4812	0.4817
2.1	0.4821	0.4826	0.4830	0.4834	0.4838	0.4842	0.4846	0.4850	0.4854	0.4857
2.2	0.4861	0.4864	0.4868	0.4871	0.4875	0.4878	0.4881	0.4884	0.4887	0.4890
2.3	0.4893	0.4896	0.4898	0.4901	0.4904	0.4906	0.4909	0.4911	0.4913	0.4916
2.4	0.4918	0.4920	0.4922	0.4925	0.4927	0.4929	0.4931	0.4932	0.4934	0.4936
2.5	0.4938	0.4940	0.4941	0.4943	0.4945	0.4946	0.4948	0.4949	0.4951	0.4952
2.6	0.4953	0.4955	0.4956	0.4957	0.4959	0.4960	0.4961	0.4962	0.4963	0.4964
2.7	0.4965	0.4966	0.4967	0.4968	0.4969	0.4970	0.4971	0.4972	0.4973	0.4974
2.8	0.4974	0.4975	0.4976	0.4977	0.4977	0.4978	0.4979	0.4979	0.4980	0.4981
2.9	0.4981	0.4982	0.4982	0.4983	0.4984	0.4984	0.4985	0.4985	0.4986	0.4986
3.0	0.4987	0.4987	0.4987	0.4988	0.4988	0.4989	0.4989	0.4989	0.4990	0.4990

Not all random variables are distributed normally. There are other probability distribution functions. Many of them are derived from the *standard normal distribution* function which is called $N(0, 1)$ i.e. normal distribution with mean $= \mu = 0$ and variance $= \sigma^2 = 1$. The most common distributions, other the normal one, are the chi-square (χ^2) distribution, the t (student's) distribution and the F distribution.

3.4.2 Chi-square (χ^2) distribution

The chi-square distribution function is derived from the normal distribution in the following way.

It can be shown (Howell 1984) that if Z_1, Z_2, Z_3, ..., Z_n are independent variables which are normally distributed with a mean of zero and variance of 1, i.e. all the variables have a *standard normal distribution*, then the sum of the squares, i.e. $y = Z_1^2 + Z_2^2 + Z_3^2 + ... + Z_n^2$ has a χ^2 distribution with n degrees of freedom. More explicitly, we have a large population of species for which we measure different independent properties $(Z_1, Z_2, ..., Z_n)$, where Z is the collection of the values of one property for the whole population, and each of these different properties is distributed in a standard normal distribution. For example, for a large population if we measure the height (Z_1), the weight (Z_2), the chest width (Z_3), the shoe size (Z_4), etc. and if each of these properties is independent and is distributed by a *standard normal distribution*, then $y = Z_1^2 + Z_2^2 + ... + Z_n^2$ has a χ^2 distribution with n degrees of freedom. Mathematicians will write it briefly as: if Z_1, Z_2, ..., Z_n are N(0, 1) and independent, then $y = Z_1^2 + Z_2^2 + Z_3^2 + ... + Z_n^2$ is $\chi^2(n)$. *N(0, 1) means that it is normally distributed with $\mu = 1$ and $\sigma^2 = 1$, i.e. standard normal distribution.*

The chi-square (written as χ^2) distribution function is a very important distribution function in analytical chemistry. This distribution function is a one-parameter distribution function, which is a special case of the gamma distribution function that has two positive parameters α and β. If β is taken as $\beta = 2$, the gamma distribution becomes the chi-square distribution function. The gamma distribution got this name because it includes the gamma function used in advanced calculus (Hogg & Craig 1971). The gamma function is the definite integral defined by:

$$\Gamma(\alpha) = \frac{1}{\beta^\alpha} \int_0^\infty x^{\alpha-1} \exp(-x/\beta)\, dx \quad \text{for } \alpha > 0,\ \beta > 0 \tag{3.11}$$

This definition of the gamma function is actually a more extended definition of the usual definition of the gamma function in advanced calculus. One important aspect of the gamma function is its connection to the factorial operation. The common definition of the gamma function is by taking $\beta = 1$.

For $\beta = 1$ and when α is a positive integer, then $\Gamma(\alpha)$ is $(\alpha - 1)$! ($\alpha - 1$ factorial). Since $\Gamma(1) = (1)$ and for $\alpha > 1$ the function $\Gamma(\alpha)$ fulfilled the equation (Hogg & Craig 1971):

$$\Gamma(\alpha) = (\alpha - 1)\, \Gamma(\alpha - 1) \tag{3.12a}$$

For non-integers values of α, $\Gamma(\alpha)$ can be approximated numerically by the equation:

$$\Gamma(\alpha) = (\alpha+5.5)^{\alpha+0.5}\, \frac{e^{-(\alpha+5.5)}}{\alpha} \left(m_0 + \frac{m_1}{\alpha+1} + \frac{m_2}{\alpha+2} + \frac{m_3}{\alpha+3} + \frac{m_4}{\alpha+4} + \frac{m_5}{\alpha+5} + \frac{m_6}{\alpha+6} \right)$$

where m_i ($i = 0, 1, ..., 6$) are constants (Hogg & Craig 1971).
From Equation (3.11), the definition of the gamma function, we can write:

$$\frac{\int_0^\infty x^{\alpha-1} \exp(-x/\beta)}{\Gamma(\alpha)\, \beta^\alpha} = 1$$

As the denominator $\Gamma(\alpha)\,\beta^{\alpha}$ is independent of x, we can write the integral sign for the whole left side:

$$\int_0^\infty \frac{x^{\alpha-1}\exp(-x/\beta)}{\Gamma(\alpha)\,\beta^{\alpha}}\,dx = 1 \tag{3.12b}$$

Since the integral is equal to 1, we can treat the integrand as a probability distribution function. If we define the probability distribution function as the integrand for $x > 0$, and equal to zero for $x \le 0$:

Gamma probability distribution: $f(x) = \begin{cases} \frac{x^{\alpha-1}\exp(-x/\beta)}{\Gamma(\alpha)\,\beta^{\alpha}} & \text{for } x > 0 \\ 0 & \text{for } x \le 0 \end{cases}$ \hfill (3.13)

If we take $\beta = 2$ and equate α to $n/2$, then we obtain the chi-square probability distribution function, χ^2:

χ^2 probability function: $f(x) = \begin{cases} \frac{x^{(n/2)-1}\exp(-x/2)}{\Gamma(n/2)\,2^{(n/2)}} & \text{for } x > 0 \\ 0 & \text{for } x \le 0 \end{cases}$ \hfill (3.14)

The reason for changing α to $n/2$ is that the parameter $n = 2\alpha$ is the mean of the distribution, as we will see later. Let us calculate the mean and the standard deviation of the χ^2 probability distribution using the first and second moments. It is sufficient to calculate the integral between the limits 0 and ∞, i.e. $\int_0^\infty$, as the probability function is zero for negative variables:

$$\mu = E(x) = \int_0^\infty x\,f(x)\,dx = \int_0^\infty \frac{x^{(n/2)}\exp(-x/2)}{\Gamma(n/2)\,2^{(n/2)}}\,dx$$

Substituting $x/2 = t$, $x = 2t$, $dx = 2dt$:

$$\mu = E(x) = \frac{2}{\Gamma(n/2)}\int_0^\infty t^{(n/2)}\exp(-t)\,dt$$

The integral is the definition of $\Gamma(n/2 + 1)$. Since Equation (3.12a) yields $\Gamma(\alpha) = (\alpha - 1)\,\Gamma(\alpha - 1)$, then $\Gamma(n/2 + 1)/\Gamma(n/2) = n/2$ and hence:

$$\mu = E(x) = n \tag{3.15}$$

In this distribution, n is called also the number of degrees of freedom.

In the same way $E(x^2) = 2(n/2 + 1)\,n = n\,(n + 2)$ and from Equation (3.5):

$$\sigma^2 = E(x^2) - \mu^2 \quad \Rightarrow \quad \sigma^2 = n^2 + 2n - n^2 \quad \Rightarrow \quad \sigma^2 = 2n \tag{3.16}$$

In summary, for the χ^2 probability distribution the mean is equal to n and the variance (square of standard deviation) is equal to $2n$.

The shape of the χ^2 probability distribution (density) function is different for each value of n, which is usually called the number of degrees of freedom and designated by ν. Thus there is a family of chi-square distributions. The following figure (Fig. 3.3) gives the shape of the χ^2 probability distribution function for various n, the number of degrees of freedom. The probability distribution function

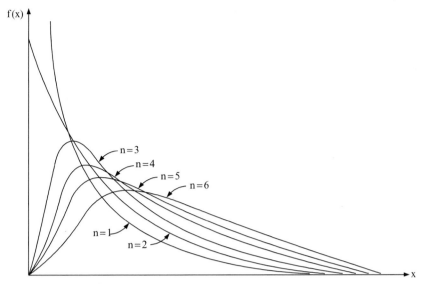

Fig. 3.3 Distribution of χ^2 for various degrees of freedom.

curves are positively skewed. The curve shifts to the right as n (degrees of freedom $=$ df $= \nu$) increases.

The cumulative distribution function of the χ^2 distribution cannot be calculated analytically and it is tabulated in handbooks and in statistics books. Actually two different methods can be found in various tables, as is explained with the help of the following graph:

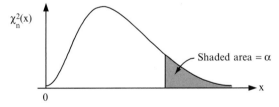

There are tables which give the real cumulative area, i.e. cumulative probability function (the unshaded area), as we saw in the standard normal distribution function. However, most tables give the shaded area, i.e. one minus the cumulative probability distribution. The value of the shaded area, which is designated usually by α is given in the headings of the columns. The value of x required in order to obtain this value of α for a given n is given in the body of the table. When the shaded area is given, the table is usually called 'Critical values of $\chi^2_{n,\alpha}$ for χ^2 distribution', as can be seen in Table 3.3. This name will be explained later when statistical tests are discussed.

The χ^2 table is arranged differently from the standard normal distribution table, the N (0, 1) table. While in the N (0, 1) table x is given in the headings (top row and

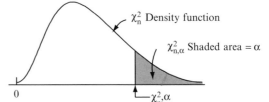

Table 3.3 Values of $\chi^2_{n,\,\alpha}$ for the chi-square distribution.

α	0.995	0.99	0.975	0.95	0.90	0.10	0.05	0.025	0.01	0.005
$n=v$										
1	0.000	0.000	0.001	0.004	0.016	2.706	3.843	5.025	6.637	7.882
2	0.010	0.020	0.051	0.103	0.211	4.605	5.992	7.378	9.210	10.597
3	0.072	0.115	0.216	0.352	0.584	6.251	7.815	9.348	11.344	12.837
4	0.207	0.297	0.484	0.711	1.064	7.779	9.488	11.143	13.277	14.860
5	0.412	0.554	0.831	1.145	1.610	9.236	11.070	12.832	15.085	16.748
6	0.646	0.827	1.237	1.635	2.204	10.645	12.592	14.440	16.812	18.548
7	0.989	1.239	1.690	2.167	2.833	12.017	14.067	16.012	18.474	20.276
8	1.344	1.646	2.180	2.733	3.490	13.362	15.507	17.534	20.090	21.954
9	1.735	2.088	2.700	3.325	4.168	14.684	16.919	19.022	21.665	23.585
10	2.156	2.558	3.247	3.940	4.865	15.987	18.307	20.483	23.209	25.188
11	2.603	3.053	3.816	4.575	5.578	17.275	19.675	21.920	24.724	26.755
12	3.047	3.571	4.404	5.226	6.304	18.549	21.026	23.337	26.217	28.300
13	3.565	4.107	5.009	5.892	7.041	19.812	22.362	24.735	27.687	29.817
14	4.075	4.660	5.629	6.571	7.790	21.064	23.685	26.119	29.141	31.319
15	4.600	5.229	6.262	7.261	8.547	22.307	24.996	27.488	30.577	32.799
16	5.142	5.812	6.908	7.962	9.312	23.542	26.296	28.842	32.000	34.267
17	5.967	6.407	7.564	8.682	10.085	24.769	27.587	30.190	33.408	35.716
18	6.265	7.015	8.231	9.390	10.865	25.989	28.869	31.526	34.805	37.156
19	6.843	7.632	8.906	10.117	11.651	27.203	30.143	32.852	36.190	38.580
20	7.434	8.260	9.591	10.851	12.443	28.412	31.410	34.170	37.566	39.997
21	8.033	8.897	10.283	11.591	13.240	29.615	32.670	35.478	38.930	41.399
22	8.693	9.542	10.982	12.338	14.042	30.813	33.924	36.781	40.289	42.796
23	9.260	10.195	11.688	13.090	14.848	32.007	35.172	38.075	41.637	44.179
24	9.886	10.856	12.401	13.848	15.659	33.196	36.415	39.364	42.980	45.558
25	10.519	11.523	13.120	14.611	16.473	34.381	37.652	40.646	44.313	46.925
26	11.160	12.198	13.844	15.379	17.292	35.563	38.885	41.923	45.642	48.290
27	11.807	12.878	14.573	16.151	18.114	36.741	40.113	43.194	46.962	49.642
28	12.461	13.565	15.308	16.928	18.939	37.916	41.337	44.461	48.278	50.993
29	13.120	14.256	16.147	17.708	19.768	39.087	42.557	45.772	49.586	52.333
30	13.787	14.954	16.791	18.493	20.599	40.256	43.773	46.979	50.892	53.672

left column) and the cumulative probability is given in the body of the table, in the $\chi^2(n)$ table the cumulative probability, or α, is given in the top row (the heading), while x is given in the body of the table. Table 3.3 can be used also for the calculation of x in the interval $[x_1,\ x_2]$. For example, if we know that our sample is χ^2 (11) (i.e. x has a chi-square distribution with 11 degrees of freedom), then the probability of finding x in the interval $3.816 \le x < 21.92$ can be found by looking at the row for $n = 11$. The value 3.816 fits to $\alpha = 0.975$ and 21.92 fits to $\alpha = 0.025$, thus the probability of $3.816 \le x \le 21.92$ is $(0.975 - 0.025) = 0.95$. Similarly if we want to know x so that $\alpha = 0.05$ for χ^2 (5) distribution, the answer from Table 3.3 is $x = 11.070$.

3.4.3 Student's(t) distribution

The normal distribution assumed a fixed μ and σ^2. However this is not the real situation when the number of observations, n, is small. For a small number of observations, which are done several times, i.e. a collection of items where each one of them is a small collection of individual items, it was found that it was not the individual observations that were distributed normally but the mean of each collection. To be more explicit, when we have a collection of groups, each of which has a mean $(\bar{x})_i$, then the mean $(\bar{x})_i$ is distributed normally and the square of the standard deviation (variance) $s_{\bar{x}}^2$ follows a χ^2 distribution. In this case the individual observations follow the *Student's t-distribution*, which in a similar way to the normal distribution is symmetrical about zero but is wider and lower than the normal one (the area under all distributions must equal 1).

This strange name of 'Student's distribution' comes from the historical fact that its originator (W.S. Grosset) was not allowed by his employer (Guinness Breweries) to publish and hence he published his statistical research under the pseudonym 'Student'. It is known by the names *t*-distribution or student's distribution or even *t*-student distribution.

If the random variables z and x are independent and z is $N(0, 1)$ while x is $\chi^2(n)$ (i.e. z has standard normal distribution and x has chi-square distribution with df $= n$), then the random variable :

$$T = \frac{z}{\sqrt{(x/n)}} \tag{3.17}$$

has t distribution, which is defined by the probability distribution function:

$$f(t) = \frac{\Gamma\left(\frac{n+1}{2}\right)}{\sqrt{(n\pi)} \cdot \Gamma\left(\frac{n}{2}\right)} \left(1 + \frac{t^2}{n}\right)^{-\frac{n+1}{2}} \qquad \text{for } -\infty < t < \infty$$

where n is the number degrees of freedom of the χ^2 distribution. We saw that the usual notation for a variable with standard normal distribution is z. The usual notation for a variable with t-statistic is either t or T.

For a random sample $x_1, x_2, \ldots, x_n$ (each sample consists of several items and each has its mean $\bar{x}$ and its standard deviation s) which are drawn from a $N(\mu, \sigma^2)$ distribution, it can be shown that the variable $(\bar{x} - \mu)/(\sigma/\sqrt{n})$ is exactly distributed as $N(0, 1)$ for any n. However, when σ in the denominator is replaced by the standard deviation of the sample, s, then the new variable is no longer standard normal distribution but rather the t-distribution. The new variable is called the T-statistic.

$$T = \frac{\bar{x} - \mu}{s/\sqrt{n}} \tag{3.18}$$

Since the t-distribution function fits small samples better than the normal distribution, it is very important in analytical chemistry where the number of experiments is usually small.

Mathematicians abbreviate the sentence 'Variable x has a t-distribution with n degrees of freedom (df)' by writing $x \sim T_n$. The t probability distribution function (density function) has a different shape for each value of n, as can be seen in Fig. 3.4.

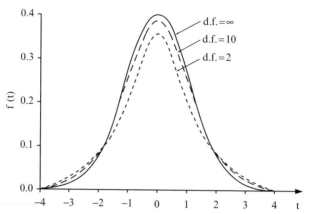

Fig. 3.4 Student's t probability distribution function curves for $n = 2, 10$ and ∞.

As n increases, the t-distribution function becomes more peaked and a lower fraction is present in its tails. For $n \to \infty$, the t-distribution function approaches the standard normal distribution, since the sample standard deviation s is approaching the true standard deviation σ.

The cumulative t probability distribution function is tabulated in Table 3.4 in a similar mode to the tabulation of χ^2 in Table 3.3, i.e. the area in the tail (one minus the cumulative probability) is given in the top row (headings) while the numbers in the body of the table are the values of t. The left column, n, is the number of degrees of freedom. For example, for a variable with t-distribution with df $= 7$, we look at the row with $n = 7$. If we want to know the value of the variable for which the cumulative probability is 0.95, we have to look at right tail probability of $1 - 0.95 = 0.05$. Looking at column 0.05 we find at row 7 that the value of the variable is $x = 1.895$. Similarly for T_{13} with probability 0.975, we find $x = 2.160$.

3.4.4   F-distribution

The t-distribution was defined in the previous section as the distribution of a variable that is the ratio of two variables, the nominator having normal distribution $N(0, 1)$ and the denominator having χ_n^2 distribution. If both the nominator and the denominator have chi-square distributions, the resulting F variable can be shown to have a distribution called the F-distribution. The F *variable* (also called F *statistic*) is defined as:

$$F = \frac{U/n_1}{V/n_2} \tag{3.19}$$

where U and V are independent random variables possessing χ^2 distribution, with degrees of freedom n_1 and n_2 respectively.

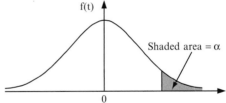

Table 3.4 Critical values $t_{n,\alpha}$ for the student's t-distribution ($n = \nu =$ degrees of freedom).

α / n	0.10	0.05	0.025	0.01	0.005	0.001	0.0005
1	3.078	6.314	12.706	31.821	63.657	318.31	636.62
2	1.886	2.920	4.303	6.965	9.925	22.326	31.598
3	1.638	2.353	3.182	4.541	5.841	10.213	12.924
4	1.533	2.132	2.776	3.747	4.604	7.173	8.610
5	1.476	2.015	2.541	3.365	4.032	5.893	6.869
6	1.440	1.943	2.447	3.143	3.707	5.208	5.959
7	1.415	1.895	2.365	2.998	3.499	4.785	5.408
8	1.397	1.860	2.306	2.896	3.355	4.501	5.041
9	1.383	1.833	2.262	2.821	3.250	4.297	4.781
10	1.372	1.812	2.228	2.764	3.169	4.144	4.587
11	1.363	1.796	2.201	2.718	3.106	4.025	4.437
12	1.356	1.782	2.179	2.681	3.055	3.930	4.318
13	1.350	1.771	2.160	2.650	3.012	3.852	4.221
14	1.345	1.761	2.145	2.624	2.977	3.787	4.140
15	1.341	1.753	2.131	2.602	2.947	3.733	4.073
16	1.337	1.746	2.120	2.583	2.921	3.686	4.015
17	1.333	1.740	2.110	2.567	2.898	3.646	3.965
18	1.330	1.734	2.101	2.552	2.878	3.610	3.922
19	1.328	1.729	2.093	2.539	2.861	3.579	3.883
20	1.325	1.725	2.086	2.528	2.845	3.552	3.850
21	1.323	1.721	2.080	2.518	2.831	3.527	3.819
22	1.321	1.717	2.074	2.508	2.819	3.505	3.792
23	1.319	1.714	2.069	2.500	2.807	3.485	3.767
24	1.318	1.711	2.064	2.492	2.797	3.467	3.745
25	1.316	1.708	2.060	2.485	2.787	3.450	3.725
26	1.315	1.706	2.056	2.479	2.779	3.435	3.707
27	1.314	1.703	2.052	2.473	2.771	3.421	3.690
28	1.313	1.701	2.048	2.467	2.763	3.408	3.674
29	1.311	1.699	2.045	2.462	2.756	3.396	3.659
30	1.310	1.697	2.042	2.457	2.750	3.385	3.646
40	1.303	1.684	2.021	2.423	2.704	3.307	3.551
60	1.296	1.671	2.000	2.390	2.660	3.232	3.460
120	1.289	1.658	1.980	2.358	2.617	3.160	3.373
∞	1.282	1.645	1.960	2.326	2.576	3.090	3.291

The statistic (variable) F has the f probability distribution function:

$$f(F) = \frac{n_1^{n_1/2}\, n_2^{n_2/2}\, \Gamma\left(\frac{n_1+n_2}{2}\right)}{\Gamma\left(\frac{n_1}{2}\right)\Gamma\left(\frac{n_2}{2}\right)} F^{\frac{n_1-2}{2}} (n_2 + n_1 F)^{-\frac{n_1+n_2}{2}} \tag{3.20}$$

The name of this distribution is after Sir Ronald Fisher and in some books it is referred to as the Fisher F-distribution.

Since the F-distribution function has two parameters, n_1 and n_2, its cumulative probability distribution function cannot be given in a two-dimensional table, since it requires three numbers in order to get the fourth one. The four numbers are n_1, n_2, F and the cumulative probability. The problem is solved by having a separate table for each cumulative probability, or rather for the integral of the right tail which is equal to one minus the cumulative probability. Table 3.5 is an example for this distribution for $\alpha = 0.01$ (i.e. cumulative probability $= 0.99$).

The top row (heading) and the left column give the values of n_1 and n_2, respectively, and the numbers in the body of the table are the values of F which yield $\alpha = 0.01$. Thus, for example, for $n_1 = 7$ and $n_2 = 10$, F should be 5.20 in order to get $\alpha = 0.01$.

When writing the two degrees of freedom the first one is the df of the numerator and the second is the df of the denominator. Thus $F_{7,11}$ means that n_1 the df of the numerator is 7 and the df of the denominator is 11. Figure 3.5 shows the F distribution for two sets of n_1 and n_2. As can be seen, F is always positive and f_{n_1, n_2} is right skewed (which means that more than 0.5 of the total integral is to the right of the maximum probability).

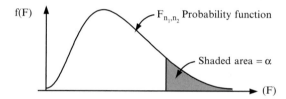

Table 3.5 Critical values $f_{n_1, n_2, \alpha}$ for the F distribution (for $\alpha = 0.01$).

				Degrees of freedom of the numerator (n_1)									
	1	2	3	4	5	6	7	8	9	10	12	15	20
1	4052.0	4999.5	5403.0	5625.0	5764.0	5859.0	5928.0	5982.0	6022.0	6056.0	6106.0	6157.0	6209.0
2	98.50	99.00	99.17	99.25	99.30	99.33	99.36	99.37	99.39	99.40	99.42	99.43	99.45
3	34.12	30.82	29.46	28.71	28.24	27.91	27.67	27.49	27.35	27.23	27.05	26.87	26.69
4	21.20	18.00	16.69	15.98	15.52	15.21	14.98	14.80	14.66	14.55	14.37	14.20	14.02
5	16.26	13.27	12.06	11.39	10.97	10.67	10.46	10.29	10.16	10.05	9.89	9.72	9.55
6	13.75	10.92	9.78	9.15	8.75	8.47	8.26	8.10	7.98	7.87	7.72	7.56	7.40
7	12.25	9.55	9.45	7.85	7.46	7.19	6.99	6.84	6.72	6.62	6.47	6.31	6.16
8	11.26	8.65	7.59	7.01	6.63	6.37	6.18	6.03	5.91	5.81	5.67	5.52	5.36
9	10.56	9.02	6.99	6.42	6.06	5.80	5.61	5.47	5.35	5.26	5.11	4.96	4.81
10	10.04	7.56	6.55	5.99	5.64	5.39	5.20	5.06	4.94	4.85	4.71	4.56	4.41
11	9.65	7.21	6.22	5.67	5.32	5.07	4.89	4.74	4.63	4.54	4.40	4.25	4.10
12	9.33	6.93	5.95	5.41	5.06	4.82	4.64	4.50	4.39	4.30	4.16	4.01	3.86
13	9.07	6.70	5.74	5.21	4.96	4.62	4.44	4.30	4.19	4.10	3.96	3.82	3.66
14	8.86	6.51	5.56	5.04	4.69	4.46	4.28	4.14	4.03	3.94	3.80	3.66	3.51
15	8.68	6.36	5.42	4.89	4.36	4.32	4.14	4.00	3.89	3.80	3.67	3.52	3.37
16	8.53	6.23	5.29	4.77	4.44	4.20	4.03	3.89	3.78	3.69	3.55	3.41	3.26
17	8.40	6.11	5.18	4.67	4.34	4.10	3.93	3.79	3.68	3.59	3.46	3.31	3.16
18	8.29	6.01	5.09	4.58	4.25	4.01	3.84	3.71	3.60	3.51	3.37	3.23	3.08
19	8.18	5.93	5.01	4.50	4.17	3.94	3.77	3.63	3.52	3.43	3.30	3.15	3.00
20	8.10	5.85	4.94	4.43	4.10	3.87	3.70	3.56	3.46	3.37	3.23	3.09	2.94

Degrees of freedom of the denominator (n_2)

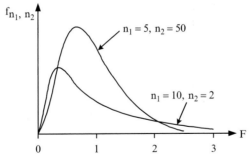

Fig. 3.5 The *f*-distribution function of the statistic *F*.

If the order of the two dfs is changed there is a correlation which is given by the relation:

$$f_{n_1, n_2, 1-\alpha} = \frac{1}{f_{n_2, n_1, 1-\alpha}}$$ (3.21)

where $f_{n_1, n_2, \alpha}$ is the critical value of F for df $= n_1$, n_2 and the right tail cumulative probability is α.

A relation exists between the t-distribution and the F-distribution. Using the definition of the T variable:

$$T_n = \frac{z}{\sqrt{(u/n)}}$$

where z and u are independent random variables distributed as N(0, 1) and χ_n^2, respectively. Squaring this equation we obtain:

$$T_n^2 = \frac{z^2}{(u/n)} \quad \Rightarrow \quad T_n^2 = \frac{z^2/1}{(u/n)}$$

We learned that z^2 has χ_1^2 distribution and hence T_n^2 is the ratio of two χ^2 distributed independent functions. Hence:

$$T_n^2 \sim F_{1, n}$$

i.e. the distribution function of T_n^2 is the same as that of F with $n_1 = 1$ and $n_2 = n$. For the critical value it can be shown that:

$$t_{n, \alpha/2}^2 = f_{1, n, \alpha}$$

3.4.5 *Exponential distribution*

The exponential distribution function has several applications. One example is in studying the length of life of radioactive material. It can be seen as a special case of the gamma distribution function, where $\alpha = 1$.

$$\text{Exponential distribution: } f(x) = \begin{cases} \frac{1}{\beta} \exp(-x/\beta) & for \ x > 0 \\ 0 & for \ x \le 0 \end{cases}$$ (3.22)

For radioactive material, β is the mean life which is equal to the reciprocal value of the decay constant or $\beta = t_{1/2}/\ln 2$, where $t_{1/2}$ is the half-life of the radioactive material.

The exponential distribution has been found also to be an appropriate model for calculating the probability that a piece of equipment will function properly for a total of x time units before it fails, especially if failure is due principally to external causes rather than internal wear. Since the exponential distribution function can be integrated analytically there is no need for tabulation. Let us look at the following example. An electronic instrument was found by experience to last on average 5 years before any breakdown (i.e. the mean of the times before first breakdown is 5 years). The first breakdown follows an exponential distribution. Let us look at two questions: (1) If the guarantee for the instrument is for one year, what fraction of customers will need their instrument repaired during the guarantee period? (2) What will be the fraction if the guarantee period is two years? Since the mean life is 5 years then $\beta = 5$ and the distribution function is $f(x) = \exp(-x/5)/5$. The failure during n years is:

$$\int_0^n \frac{\exp(-x/5)}{5}\, dx = \int_0^{n/5} \exp(-t)\, dt = 1 - \exp(-n/5)$$

Thus for one year ($n = 1$) the fraction of failures is 0.181, i.e. 18.1% of the customers will come for repair, while for two years 33.0% of the instruments will be brought back.

3.5 Discrete distribution functions

3.5.1 Binomial distribution functions

This distribution function deals with cases where each experiment has only two possible outcomes, which are referred to as 'a success' and 'a failure'. The 'success' and 'failure' do not have the common meanings of success and failure; we can designate them as we wish. For example, throwing a die we can call 5 or 6 'a success'. In this case, 1 or 2 or 3 or 4 will be 'failure'. If the probability for the outcome 'success' is constant, p, for each single trial, then for our example $p = 2/6 = 1/3$ and the outcome of 'failure' will also be constant for each trial and equal to $(1 - p)$. For n experiments, the probability that the first k experiments will be 'successes' is p^k and the probability that the next $(n - k)$ experiments will be 'failures' is $(1 - p)^{n-k}$. Thus, the probability that the first k experiments will be 'successes' and the following $(n - k)$ experiments will be 'failures' is $p^k \times (1 - p)^{n-k}$.

The same probability will also be true for any combination of k 'successes' (not especially the first experiments) and $(n - k)$ 'failures'. The number of possible combinations of k 'successes' in n experiments is given by the binomial coefficient written as:

$$C_n^k = \binom{n}{k} = \frac{n!}{k!\,(n - k)!} \tag{3.23}$$

where $n!$, termed n factorial, is given by $n! = 1 \times 2 \times 3 \times \ldots \times n$ and $0! = 1$. So the probability of obtaining k successes in n experiments is given by:

$$\frac{n!}{k!\,(n-k)!}\, p^k (1-p)^{n-k}$$

The binomial distribution function for obtaining x 'successes' in n experiments $(n \geq x,\ n,\ x = 0,\ 1,\ 2\ldots)$ is:

$$f(x) = \frac{n!}{x!\,(n-x)!}\, p^x (1-p)^{n-x} \tag{3.24}$$

Using the binomial theorem:

$$(a+b)^n = \sum_{k=0}^{n} \frac{n!}{k!\,(n-k)!}\, a^k\, b^{n-k}$$

it is easy to see that

$$\sum_{x=0}^{n} f(x) = [p + (1-p)]^n = 1^n = 1$$

proving that the binomial distribution function is a normalized one and hence it is a probability function. Using the distribution function, we can calculate the mean and standard deviation:

$$\mu = E(x) = \sum_{x=0}^{n} x\, f(x) = \sum_{x=0}^{n} x \left(\frac{n!}{x!\,(n-x)!} \right) p^x\, (1-p)^{n-x} \tag{3.25}$$

The first term with $x = 0$ is equal to zero because of the multiplication by x. Thus:

$$\mu = \sum_{x=1}^{n} \frac{x\, n!}{x!\,(n-x)!}\, p^x\, (1-p)^{n-x}$$

$$\frac{x!}{x} = \frac{x\,(x-1)\,(x-2)\,\ldots\,(1)}{x} = (x-1)\,(x-2)\,\ldots(1) = (x-1)!$$

hence

$$\mu = \sum_{x=1}^{n} \frac{n!}{(x-1)!\,(n-x)!}\, p^x\, (1-p)^{n-x}$$

Substituting $x = y + 1$, will lead to the sum from $y = 0$ to $y = n - 1$:

$$\mu = \sum_{y=0}^{n-1} \frac{n!}{y!\,(n-1-y)!}\, p^{y+1}\, (1-p)^{n-1-y}$$

Substituting $n! = n\,(n-1)!$ and $p^{y+1} = p\, p^y$ and taking pn as a factor from the sum yields:

$$\mu = pn \sum_{y=0}^{n-1} \frac{(n-1)!}{y!\,(n-1-y)!}\, p^y\, (1-p)^{n-1-y}$$

According to the binomial theorem, the sum is equal to $[p + (1 - p)]^{n-1} = 1^{n-1} = 1$.
Hence:

$$\mu = pn \tag{3.26}$$

Using Equation (3.5), $\sigma^2 = \sum x^2 f(x) - \mu^2$, we can obtain σ^2 by calculating the second central moment of the probability function,

$$\sigma^2 + \mu^2 = \sum_{x=0}^{n} x^2 f(x) = \sum_{x=0}^{n} x^2 \frac{n!}{(n-x)! \, x!} \, p^x \, (1-p)^{n-x}$$

Substituting $x^2 = x^2 - x + x = x(x-1) + x$ leads to:

$$\sigma^2 + \mu^2 = \sum_{x=0}^{n} \frac{x(x-1)n!}{(n-x)! \, x!} \, p^x \, (1-p)^{n-x} + \sum_{x=0}^{n} \frac{x \, n!}{x! \, (n-x)!} \, p^x \, (1-p)^{n-x}$$

The right sum is equal to μ by definition. In the left sum the first two terms are equal to zero since either $x = 0$ or $x - 1 = 0$. Hence discarding these two terms:

$$\sigma^2 + \mu^2 - \mu = \sum_{x=2}^{n} \frac{n! \, x \, (x-1)}{(n-x)! \, x!} \, p^x \, (1-p)^{n-x} = \sum_{x=2}^{n} \frac{n!}{(n-x)! \, (x-2)!} \, p^x \, (1-p)^{n-x}$$

in the last step we use $x! = x \, (x-1) \, (x-2)!$ and hence $x!/[x(x-1)] = (x-2)!$
Substituting $x = y + 2$ we will have a sum over y from 0 to $(n-2)$:

$$\sigma^2 + \mu^2 - \mu = \sum_{y=0}^{n-2} \frac{n!}{(n-2-y)! \, y!} \, p^{y+2} \, (1-p)^{n-2-y}$$

Substituting $n! = n(n-1)(n-2)!$, $p^{y+2} = p^2 \, p^y$ and factoring $n(n-1) \, p^2$ from the sum yields:

$$\sigma^2 + \mu^2 - \mu = n(n-1) \, p^2 \sum_{y=0}^{n-2} \frac{(n-2)!}{(n-2-y)! \, y!} \, p^y \, (1-p)^{n-2-y}$$

According to Newton's binomial theorem the sum is equal to: $[p + (1-p)]^{n-2} = 1^{n-2} = 1$. Hence:

$$\sigma^2 + \mu^2 - \mu = n \, (n-1) \, p^2$$

Substituting $\mu = pn$ yields:

$$\sigma^2 = np - np^2 \quad \Rightarrow \quad \sigma^2 = np \, (1-p) \tag{3.27}$$

3.5.2 Approximation of the binomial distribution by the normal distribution

When n is large the binomial distribution can be approximated by the normal distribution with:

$$\mu = np \quad \text{and} \quad \sigma^2 = np \, (1-p) \tag{3.28}$$

The requirement that n is large enough for this approximation, is that both np and $n(1 - p)$ are equal to or larger than 5. It means that n will be equal to or larger than the larger value of $5/p$ and $5/(1 - p)$. The value of x represents the number of successes in the binomial experiment. Since in the binomial distribution x has a discrete value, while the variable of the normal distribution is continuous, we should replace x either by $(x + 0.5)$ or $(x - 0.5)$ depending on the nature of the problem. Thus for example, from 30 students chosen at random in a school that has 67% female students, calculating the probability that between 20 and 25 will be females, we have to use $20 - 0.5 = 19.5$ for the lower boundary and $25 + 0.5 = 25.5$ as the upper boundary. This problem can be approximated by the normal distribution since $n = 30$, $p = 0.67$ and thus $np = 20.1$ and $n(1 - p) = 9.9$, which are both larger than 5. Since $(1 - p)$ is smaller than p, using the value of $(1 - p)$ we get the minimal number of students for which the approximation is valid as $> [5/(1 - 0.67)] = 15.15$, i.e. the minimal number of students for the approximation is 16.

Calculating μ and σ from Equations (3.28) leads to:

$$\mu = 30 \times 0.67 = 20.1, \qquad \sigma = \sqrt{[np(1 - p)]} = \sqrt{(30 \times 0.67 \times 0.33)} = 2.575$$

The standard variables z and the appropriate ϕ (from the table) are:

$$x_1 = 19.5 \quad \Rightarrow \quad z = \frac{19.5 - 20.1}{2.575} = -0.233 \quad \Rightarrow \quad \phi(-0.233) = 0.5 - 0.0910 = 0.4090$$

$$x_2 = 25.5 \quad \Rightarrow \quad z = \frac{25.5 - 20.1}{2.575} = 2.097 \quad \Rightarrow \quad \phi(2.10) = 0.9821$$

Thus the probability that from the chosen 30 students between 20 and 25 will be females is $0.9821 - 0.4090 = 0.5731$.

This should be compared to the value obtained by the rigorous calculation via the Bernoulli equation. However the Bernoulli equation requires several calculations and large powers which might also lead to large round-off errors. Thus, for example, the probability that exactly 20 of the students will be female is computed by a calculator as:

$$H_{30}^{20} = \frac{30!}{20!\ 10!}\ 0.67^{20}\ 0.33^{10} = 0.1529$$

Similarly for the other probabilities:

$$H_{30}^{21} = \frac{30!}{21!\ 9!}\ 0.67^{21}\ 0.33^{9} = 0.1478$$

$$H_{30}^{22} = \frac{30!}{22!\ 8!}\ 0.67^{22}\ 0.33^{8} = 0.1228$$

$$H_{30}^{23} = \frac{30!}{23!\ 7!}\ 0.67^{23}\ 0.33^{7} = 0.0867$$

$$H_{30}^{24} = \frac{30!}{24!\ 6!}\ 0.67^{24}\ 0.33^{6} = 0.0513$$

$$H_{30}^{25} = \frac{30!}{25!\ 5!}\ 0.67^{25}\ 0.33^{5} = 0.0250$$

Summing up the six values yields 0.587 compared to 0.5731 calculated by the approximation to the normal distribution. If we want to use the normal approximation for exactly 20 females we should calculate the probability of x being between 19.5 and 20.5:

$$x = 20.5 \quad \Rightarrow \quad z = \frac{20.5 - 20.1}{2.575} = 0.155 \quad \Rightarrow \quad \phi(z) = 0.5616$$

$$\phi(20.5) - \phi(19.5) = 0.5616 - 0.4090 = 0.1526$$

compared to 0.1529 by Bernoulli's equation.

3.5.3 Hypergeometric distribution

Bernoulli's equation for binomial distribution is based on n independent trials of an experiment, each having the same probability for success. However, if the experiment consists of selecting individuals from a finite population of trials, the probability is changing with each selection (in other words the trials are not independent). For example, in the last example the probability of selecting a female student from the school is the same for all the 30 selections only if the number of the students in the school is very large compared to 30. However if the school has only 100 students from which 67 are females, the probability of the first choice to be female is $67/100 = 0.67$. However if the first 20 selections were females, the probability of the 21st selection being a female student is only $47/80 = 0.5875$, since after the selection of 20 females there remains only 80 students from which 47 are females. For problems in which the population is small such that successive selections have different probabilities, the binomial distribution cannot be applied and it is necessary to apply the *hypergeometric disribution*.

Let N be the size of the population, in which the proportion of items who have a specific property is p, i.e. the number of items who have this property is Np. We randomly sample n items from the population. The question is what is the probability that exactly x of the n sampled items have this property? This probability is the ratio of the number of ways of obtaining our desired combination of x items with this property and $(n - x)$ items without this property, to the number of ways of obtaining any combination with n items. The number of ways of obtaining the desired combination is equal to the product of the number of ways to draw x items from the population which have the property, (Np), and $(n - x)$ items from the population which do not have this property, $[N(1 - p)]$. The number of ways for any combination of n items is the number of ways to sample n items from the total population, N. Generally the number of ways to draw a number of items from a population is the number of combinations:

$$\text{Hypergeometric distribution: } f(x) = \frac{C_{Np}^x \, C_{N(1-p)}^{n-x}}{C_N^n} \tag{3.29}$$

$$f(x) = \frac{(Np)! \, [N(1 - p)]! \, n! \, (N - n)!}{x! \, (Np - x)! \, (n - x)! \, [N(1 - p) - n + x]! \, N!} \tag{3.30}$$

If the sampling fraction (n/N) is small (≤ 0.10), then this distribution is well approximated by the binomial distribution. The mean $\mu = E(x)$ and variance

$\sigma^2 = V(x)$ of a hypergeometric random variable is given by (for a mathematical proof see Howell's book [1984]):

$$\mu = E(x) = np \qquad \sigma^2 = V(x) = \left(\frac{N-n}{N-1}\right) np\,(1-p) \tag{3.31}$$

Thus the mean is the same as for the binomial distribution, while the variance is smaller than in the binomial distribution by the factor $[(N-n)/(N-1)]$.

For example, a chemical laboratory must decide whether to buy chemicals from a new supplier. They ordered 40 chemicals and analyzed 5 of them. Since $n/N = 0.125$ a binomial distribution cannot be used. If it is known that 3 of the 40 chemicals are contaminated, what is the probability that one or two of the analyzed 5 are contaminated? Substituting $N = 40$, $n = 5$ and $p = 3/40 = 0.075$ in the formula for $f(x)$, we obtain for $x = 1$ (one contaminated):

$$f(1) = \frac{C_3^1 C_{37}^4}{C_{40}^5} = \frac{3!}{1!\,2!} \frac{37!}{33!\,4!} \frac{35!\,5!}{40!} = \frac{3 \times 66045}{658008} = 0.301$$

It should be mentioned that the combination number C_n^k should not be calculated by writing the factorial numbers since most modern scientific calculators calculate C_n^k directly.

For the probability that two samples are contaminated, Equation (3.30) yields:

$$f(2) = \frac{C_3^2 C_{37}^4}{C_{40}^5} = \frac{3 \times 7770}{658\,008} = 0.035$$

3.5.4 Poisson distribution function

The Poisson distribution function is a special case of the binomial distribution function when n is very large and p is very small, such that when $n \to \infty$ and $p \to 0$ their product approaches a positive constant, λ. The Poisson distribution is given by the equation:

$$\textit{Poisson distribution: } f(x) = \frac{e^{-\lambda}\lambda^x}{x!} \tag{3.32}$$

The Poisson distribution can be used to model the occurrence of rare events (small p) when the number of opportunities for the event is very small, e.g. the disintegration of a radioactive material when the number of disintegrations is small. For this distribution both the mean and the variance are equal to λ.

$$\mu = E(x) = \lambda \qquad \sigma^2 = V(x) = \lambda \tag{3.33}$$

3.6 References

Hogg, R. V. & Craig, A. T. 1971, *Introduction to Mathematical Statistics*, 3rd edn, Collier-Macmillan, London, pp. 87–88.

Howell, P. G. 1984, *Introduction to Mathematical Statistics*, John Wiley, New York, p. 127.

4 Confidence limits of the mean

4.1 Confidence limits

The mean of a sample of measurements, $\bar{x}$, provides an estimate of the true value, μ, of the quantity we are trying to measure. However, it quite unlikely that $\bar{x}$ is exactly equal to μ, and an important question is to find a range of values in which we are certain that the value lies. This range depends on the number of measurements done and on the question of *how certain we want to be*. The more certain we want to be, the larger the range we have to take. The larger the number of experiments done, the closer $\bar{x}$ is to μ, and a smaller range has to be taken for the same percentage of certainty. Some tables refer not to the number of repeated experiments but to the number of *degrees of freedom* (usually given by the symbols υ or df). In the calculation of the standard deviation, the number of degrees of freedom is $(n-1)$, where n is the number of repeated measurements. The number of degrees of freedom refers to the number of independent variables. Calculating the standard deviation we use n terms of $(x_i - \bar{x})^2$, but only $(n-1)$ terms are independent, since the nth term can be calculated from the equation

$$\sum_{i=1}^{n} (x_i - \bar{x}) = 0$$

In the normal distribution table we find that $\phi(1) = 0.841$ while $\phi(-1) = 0.159$, which means that for a normal distribution in the range $(\mu + \sigma) > x > (\mu - \sigma)$, $100 \times (0.841 - 0.159) = 68.2\%$ of the population are present. Thus if we take as the range $(\mu \pm \sigma)$ we are certain that 68.2% includes the real value of our quantity.

Similarly, for a normal distribution, $95.5\% = [100 \times (0.9773 - 0.0227)]$ of the values lie in the range $(\mu \pm 2\sigma)$ and 99.7% lie within $\pm 3\sigma$ of the mean. For 95% of the population the range should be $(\mu \pm 1.96\sigma)$, and for 99% of the population the range should be $(\mu \pm 2.58\sigma)$. The same is true also for the mean value of the number of repeating experiments, however their standard deviation has to be found from the central limit theorem.

4.2 The Central Limit Theorem – the distribution of means

This theorem deals with the sampling distribution of the sample mean. In other words it deals with the case where we have a large population from which we draw randomly several smaller samples, each of the same size. Each of these samples has its own mean and the Central Limit Theorem (CLT) is concerned with the distribution of the means.

Central Limit Theorem: If we have a population (which may or may not be a normal one) which has mean μ and variance σ^2, from which we draw random samples of n observations (y_1, y_2, ..., y_n), then if n is sufficiently large the distribution of the means of these samples can be approximated by the normal distribution function with $\mu_{\bar{y}} = \mu$ and standard deviation $\sigma_{\bar{y}} = \sigma/\sqrt{n}$. In a more mathematical style, this theorem states that if y_1, y_2, ..., y_n, is a random sample from an arbitrary distribution with a mean μ and variance σ^2, then if $n \to \infty$ the sampling distribution of $(\bar{x} - \mu)/(\sigma/\sqrt{n})$ converges to the $N(0,1)$ distribution.

The further the original distribution is from a normal distribution, the larger the n needed to obtain an approximation to a normal distribution, as can be seen in Fig. 4.1. For sample sizes larger than 30, the distribution of the sample means can be approximated reasonably well by a normal distribution for any original distribution. If the original population is itself normally distributed, then for any sample size n the samples are normally distributed.

It can be summarized in the following way. If random samples of size n are drawn from a normal population, the means of these samples will be distributed normally. The distribution from a non-normal population will not be normal but will tend to normality as n increases. The variance of the distribution of the means decreases as n increases according to the equation:

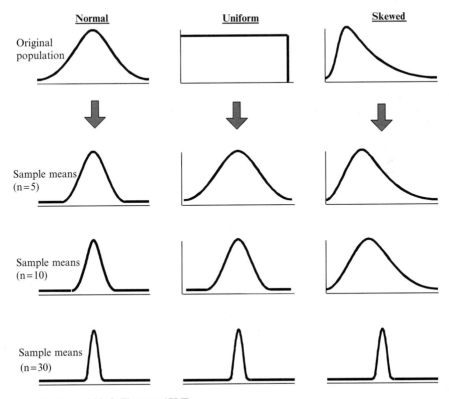

Fig. 4.1 The Central Limit Theorem (CLT).

$$\sigma_{\bar{x}}^2 = \frac{\sigma^2}{n} \quad \Rightarrow \quad \sigma_{\bar{x}} = \frac{\sigma}{\sqrt{n}}$$

where $\sigma_{\bar{x}}^2$ is called the variance of the means.

Just as $z = (x_i - \mu)/\sigma$ is a standard normal deviation from the normal distribution of x_i, then $z = (\bar{x} - \mu)/\sigma_{\bar{x}}$ is the standard normal deviation from the normal distribution of the means. The table of cumulative standard normal distribution (Table 3.1) can be used to answer a question like 'what is the probability of obtaining a random sample of nine concentration measurements with a mean larger than 30.0 mM from a population having a mean of 32.5 mM and a standard deviation of 9 mM?'

$$\sigma_{\bar{x}} = \frac{\sigma}{\sqrt{n}} = \frac{9}{\sqrt{9}} = 3\,\text{mM}$$

$$z = \frac{\bar{x} - \mu}{\sigma_{\bar{x}}} = \frac{30 - 32.5}{3} = -0.8133$$

From the table we get $\phi(-0.8133) = 1 - \phi(0.813) = 1 - 0.207 = 0.793$, i.e. there is probability of 0.793 of obtaining the required parameters.

4.3 Confidence limit of the mean

The important conclusion of the central limit theorem is that for sufficiently large n the mean is normally distributed with mean μ and standard deviation $\sigma/\sqrt{n}$. This means that the variable $z = (\bar{x} - \mu)/(\sigma/\sqrt{\mu})$ is distributed according to the standard normal distribution and its cumulative probability can be taken from the normal disribution table. If we want to have a certainty of 95%, we allow an error of 5%, which means that in each tail of the distribution function we allow an area of 0.025 (since the total area under the curve is unity) – see Fig. 4.2.

A 95% degree of confidence corresponds to uncertainty of $\alpha = 0.05$, half of it in each tail. The value of z which corresponds to this value of α for a standard normal distribution is $z_{0.025} = 1.96$. The value of z which fits the value of $\alpha/2$ is named the critical value. The range of confidence for this uncertainty is $-1.96 < z < 1.96$.

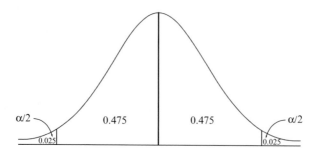

$\alpha/2$ 0.475   0.475 $\alpha/2$

0.025   0.025

Fig. 4.2

Substituting the definition of z from the central limit theorem yields :

$$-1.96 < \frac{\bar{x} - \mu}{\sigma/\sqrt{n}} < 1.96 \quad \Rightarrow \quad -1.96(\sigma/\sqrt{n}) < \bar{x} - \mu < -1.96(\sigma/\sqrt{n})$$

$$1.96(\sigma/\sqrt{n}) > \mu - \bar{x} > -1.96(\sigma/\sqrt{n})$$

and thus the confidence limit of the mean is given by :

$$\bar{x} - 1.96(\sigma/\sqrt{n}) < \mu < \bar{x} + 1.96(\sigma/\sqrt{n}) \tag{4.1}$$

which can be written as $\mu = \bar{x} \pm 1.96(\sigma/\sqrt{n})$.

We have to remember that the value of 1.96 is only for a 95% degree of confidence (95% certainty). For other values of degrees of confidence we have to look in the table for cumulative probability of the normal distribution. Thus for example for 90% confidence, we have to look for the value of z which will yield a cumulative probability of 0.95 since $[1 - (0.10/2) = 0.95]$, or a value of 0.45 in tables which give only the area to the right of the mean (i.e. tables in which $\phi(0) = 0$ and not $\phi(0) = 0.5$). The table yields $z = 1.645$. Similarly for 99% confidence we have to look for z for which $\phi(z) = 1 - (0.01/2) = 0.995$. The table shows it to be $z = 2.575$.

The problem with Equation (4.1) is that we don't know the value of σ which is needed in order to calculate the confidence limits. When the sample is large, mainly $n \geq 30$, the distribution is close enough to normal so we can substitute s for σ, and write :

$$(\bar{x} - z_{\alpha/2})(s/\sqrt{n}) < \mu < (\bar{x} + z_{\alpha/2})(s/\sqrt{n}) \tag{4.2}$$

where $z_{\alpha/2}$ is the critical value and s is the standard deviation of the values in our sample. However for small samples the distribution is broader than the normal distribution and the substitution of s instead of σ will lead to unrealistic narrow limits.

4.4 Confidence limits of the mean of small samples

Usually the number of repeating measurements is considerably below 30 and hence Equation (4.2) cannot be applied in order to calculate the confidence limits. As we wrote in Chapter 3 it was found that for small samples the real distribution is given by the student t-distribution. In this distribution the critical value is denoted by $t_{\alpha/2}$. The t-statistic is given by the equation for z normal statistics:

$$t = \frac{\bar{x} - \mu}{(s/\sqrt{n})}$$

which yields for the confidence limits :

$$(\bar{x} - t_{\alpha/2})(s/\sqrt{n}) < \mu < (\bar{x} + t_{\alpha/2})(s/\sqrt{n}) \tag{4.3}$$

or more concisely:

$$\mu = \bar{x} \pm t_{\alpha/2}(s/\sqrt{n}) \tag{4.4}$$

Table 4.1   Values for the two-sided student's t-distribution.

df	$t_{0.10}$	$t_{0.05}$	$t_{0.025}$	$t_{0.01}$	$t_{0.005}$
			Level of significance		
1	3.078	6.314	12.706	31.851	63.657
2	1.886	2.920	4.303	6.965	9.925
3	1.638	2.353	3.182	4.541	5.841
4	1.553	2.132	2.776	3.747	4.604
5	1.476	2.015	2.571	3.365	4.032
6	1.440	1.943	2.447	3.143	3.707
7	1.415	1.895	2.365	2.998	3.499
8	1.397	1.860	2.306	2.896	3.355
9	1.383	1.833	2.262	2.821	3.250
10	1.372	1.812	2.228	2.764	3.169
11	1.363	1.796	2.201	2.718	3.106
12	1.356	1.782	2.179	2.681	3.055
13	1.350	1.771	2.160	2.650	3.012
14	1.345	1.761	2.145	2.624	2.977
15	1.341	1.753	2.131	2.602	2.947
16	1.337	1.746	2.120	2.583	2.921
17	1.333	1.740	2.110	2.567	2.898
18	1.330	1.734	2.101	2.552	2.878
19	1.328	1.729	2.093	2.539	2.861
20	1.325	1.725	2.086	2.528	2.845
25	1.316	1.708	2.060	2.485	2.787
30	1.310	1.697	2.042	2.457	2.750
40	1.303	1.684	2.021	2.423	2.704
60	1.296	1.671	2.000	2.390	2.660
∞	1.282	1.645	1.960	2.326	2.576

As was seen in Chapter 3, $t_{\alpha/2}$ depends on both the number of degrees of freedom (df $= n - 1$) and the percentage of confidence. For convenience, the critical values of t are given again in Table 4.1.

No quantitative experimental result is of value if an estimate of the errors involved in its measurement is not given. The usual practice is to give $(\bar{x} \pm s)$ or $(\bar{x} \pm 2s)$. It should be stated clearly if the cited error is one or two standard deviations. Any confidence limit can then be calculated from Equation (4.3) if the number of experiments is known. If the number of experiments is not known, the standard deviation cannot give us the limit of confidence.

It should be emphasized that in many analytical papers results are reported as $(\bar{x} \pm 2s)$ and it is stated that they are with 95% confidence. Since many of these studies do not have more than 6–8 experiments, the factor for 95% confidence should be larger than 2. As was seen in Chapter 3, the t-distribution obeys the equation:

$$p = \frac{p_0}{[1 + t^2/(n - 1)]^{n/2}} \tag{4.5}$$

where p_0 is a constant normalizing factor whose value depends on n in such a way that the area under the curve is a unit area. For large n, p as a function of t looks very similar to the normal disribution. However, as n decreases, p decreases in the mean and increases in the tail.

The confidence limit can be used as a test for systematic errors in the measurement. If we know the confidence limit for the concentration of a compound but our observed value is not within the confidence limits, there is quite a large probability of a systematic error in the measurement. However we must remember that if our confidence limit is with a certainty of 95% there is still a 5% chance of being outside the confidence limits.

4.5 Choosing the sample size

If we want to know the mean with a required certainty, we can obtain this goal by choosing the right size of sample, i.e. the right number of repeated experiments, n. Usually we are not interested in the absolute value of the confidence limit but in the ratio between the confidence limit and the mean, usually given as a percentage. Thus, for example, if in analytical measurement of copper in drinking water we want to know the concentration within 1%, with level of confidence of 95%, we should write:

$$\frac{z_{0.025}\sigma}{\bar{x}\sqrt{n}} < 0.01 \quad \Rightarrow \quad \left(\frac{z_{0.025}\ \sigma\ 100}{\bar{x}}\right)^2 < n \tag{4.6}$$

Usually we do not know σ and will replace it by s. If we do not have estimates of s and $\bar{x}$ from previous knowledge, we can analyze a few samples in order to estimate from them s and $\bar{x}$ and use them to calculate the n required for our accuracy. Thus, for example, if we start with three measurements which give $\bar{x} = 0.752\,\mathrm{g}\ell^{-1}$ and $s = 0.022\,\mathrm{g}\ell^{-1}$. Using $z_{0.025} = 1.96$, the minimal number of repeating measurements for our accuracy is:

$$n > \left(\frac{1.96 \times 0.022 \times 100}{0.752}\right)^2 \quad \Rightarrow \quad n = 33$$

If the result for n is less than 25, then $z_{\alpha/2}$ should be replaced by $t_{\alpha/2}$. Attention should be paid to the different use of percentage. We want the limit of confidence to be 1%, but we know it to be true with uncertainty of 5%.

5   Significance test

5.1   Introduction

As we saw in the previous chapter, the measured mean value is not exactly the true value but a value close to it, within the confidence limits. So, how can we know if an analytical method is a good one? If the mean of our observations is not the same as the expected true value of the standard, used for quality control of the method, how do we know if it is due to systematic error of the method, or due to the spread of the results, i.e. a statistical error? For example, let us assume that we prepared a standard with a concentration of $0.5\,\mathrm{mg/m\ell}$, and measured it by a new analytical method repeatedly 10 times and observed $\bar{x} = 0.49$ with $s = 0.02$. Does this mean that this analytical method is inappropriate or is this difference acceptable with the required certainty due to the randomness of the sampling? A similar question can arise when comparing two materials. The first material was measured n_1 times and yielded a mean $\bar{x}_1$ with a standard deviation s_1; the respective values for the second material are n_2, $\bar{x}_2$ and s_2. Assuming that the means, $\bar{x}_1$ and $\bar{x}_2$, are different from each other, a question that remains is if this difference is significant or only due to the randomness of the associated errors. It should be remembered that always when we ask if a difference is significant it should be said at what level of certainty we want to know it.

The testing of whether differences can be accounted for by random errors is done by statistical tests known as *significance tests*. In making a significance test, we are testing the truth of a hypothesis, called the *null hypothesis*. This hypothesis assumes that there is no difference (null difference) between the two values we are comparing. Assuming that the null hypothesis is true, we can use statistical methods to calculate the probability that the observed difference arises solely as a result of random variations. The null hypothesis is denoted as H_0, If H_0 is rejected, then H_a (alternative hypothesis) is accepted. The evaluation of the hypothesis test involves an analysis of the chances of making an incorrect decision. There are two kinds of errors in the statistical decisions of rejecting or accepting the null hypothesis. If we reject a hypothesis while it should be accepted, this is called *an error of type I. An error of type II* is made when we accept a hypothesis when it should be rejected. In testing a hypothesis, the maximum probability with which we would be willing to risk a type I error is called the *level of significance* of the test. In practice, 0.05 is taken as the common level of significance, so there is about 5% probability (5 chances in 100) that we would reject the hypothesis when it should be accepted. In other words there is 95% confidence that we have made the right decision. In this case the common language is to say that the hypothesis has been rejected at the 0.05 level of significance. The statistical meaning of level of significance, α, with its critical value for the statistic (for example, the t-statistic) t_c, means that in the range

$-t_c < t < t_c$ there is $100(1 - \alpha)\%$ of the population. When comparing two values each of which can be larger than the other, so that the measured values can be on both sides of the mean, we call it a *two-tailed test*. However, sometimes we are interested only in extreme values to one side of the mean, i.e. only one 'tail' of the distribution. Such a test is called a *one-tailed test* or *one-sided test*. When using statistical tables for critical values, attention should be paid to whether the table is for a one- or two-tailed test. If the table is for a one-tailed test, a level of 2α (α level of significance) should be used for a two-tailed test; if a table for a two-tailed test is used for a one-tailed test, then the critical value for $\alpha/2$ should be taken.

The statistical test is to check if in the level of confidence required, the null hypothesis is a correct one. This is done by calculating from the data a statistical measure (usually called the statistic), e.g. t, F or χ^2, and comparing this to the critical values in the table.

5.2   Comparison of an experimental mean with an expected value (standard)

5.2.1   *Large sample*

One of the important steps in the development of a new analytical method (for example, a new method of wet chemistry for separation and enrichment or a new instrumental method) is the analysis of certified reference materials. There are certified reference materials of different kinds of matrices, e.g. geological, biological (either botanical or from animals), etc., prepared in large amounts, homogenized well to assure homogeneity and analyzed by several laboratories for their constituents. These standard materials can be purchased from several sources, e.g. The National Institute of Standards and Technology (NIST) in the USA. Assume repeating the measurement of the reference material by the new method (instrument) n times yields a mean $\bar{x}$ and a standard deviation s. The certified value, assumed to be true, is μ. How do we know if the difference $|\bar{x} - \mu|$ is significant, which means a systematic error in the new method, or insignificant and the new method is a reliable one?

In order to decide whether the difference between μ and $\bar{x}$ is significant, we have to calculate the statistics and compare it to the critical value of the statistic given in the table according to the level of significance we are interested in. If n is large enough, the test statistic can be z of the normal distribution:

$$z = \frac{\bar{x} - \mu}{\sigma_{\bar{x}}} = \frac{(\bar{x} - \mu)\sqrt{n}}{\sigma}$$

σ is a population parameter and since very seldom can we calculate population parameters, we must rely on estimating it from the random samples taken from the population. The best estimate for $\sigma_{\bar{x}}$ is $s_{\bar{x}}$, and the best estimate for σ is s.

$$s_{\bar{x}} = \frac{s}{\sqrt{n}}$$

where s is the standard deviation calculated for our sample by the equation:

$$s = \sqrt{\left(\frac{\sum (x_i - \bar{x})^2}{n-1}\right)} = \sqrt{\left(\frac{n\sum x_i^2 - (\sum x_i)^2}{n(n-1)}\right)}$$

For example, a new method to determine the concentration of Fe (iron) in rocks was studied. A certified geological reference material with certified concentration of $20.00\,mg/cm^3$ was used to check if the method is free of systematic error. Thirty samples were drawn from the certified reference material and in each sample the concentration of the Fe was measured by the new method. It was found that their mean was $19.41\,mg/cm^3$ while the standard deviation of the measurement was $1.08\,mg/cm^3$. Since the number of samples is sufficiently large, the distribution of the mean is approximately a normal distribution and hence we can use the z statistics (the statistics of standard normal distribution):

$$z = \frac{|\bar{x} - \mu|}{(s/\sqrt{n})} \quad \Rightarrow \quad z = \frac{|19.41 - 20.00|}{(1.08/\sqrt{30})} = 2.992$$

From the table of cumulative standard normal distribution (Table 3.1) we find $\phi(2.992) = 0.9986$. It means that there is $2 \times (1 - 0.9986) = 0.0032 = 0.32\%$ probability that for a true mean of $20\,mg/cm^3$ the observed mean will be either smaller than 19.41 or larger than 20.59 (Fig. 5.1). Usually we want to know our result with uncertainly of 5% and hence in this case we will say that our method is not a good one.

In most cases we are not interested in calculating exactly the risk of accepting a hypothesis (either accepting the null-hypothesis H_0 or rejecting it and thus accepting the alternative hypothesis, H_a), and it is sufficient for us to know that the risk is lower than a value we first decided. In this case instead of calculating $\phi(z)$ from the cumulative standard normal distribution table, it is simpler to use the table for critical values of z (see Table 5.1).

To explain the use of the numbers in this table, let us take the common risk of 5%, which means $\alpha = 0.05$. This is a two-tailed risk; we have a risk of 0.025 in each tail. The value of z_c in Table 5.1 is 1.96. Let us plot the normal distribution curve and mark the two vertical lines of $z = \pm 1.96$. The range of $-1.96 < z < 1.96$ is one in which we accept the null hypothesis, while the ranges of $z < -1.96$ or $z > 1.96$ are the ranges where we reject H_0. (see Fig. 5.2). The regions of rejection of H_0 are the areas of the alternative hypothesis, which assumes that $\bar{x} \neq$ true μ.

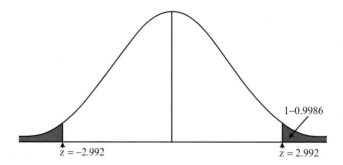

1–0.9986

$z = -2.992$ $z = 2.992$

Fig. 5.1

Table 5.1 Critical values of z (the standard normal distribution statistics).

Two-tail test	0.001	0.005	0.01	0.02	0.05	0.1	0.2	0.3
One-tail test	0.0005	0.0025	0.005	0.01	0.025	0.05	0.1	0.15
z_c	3.291	2.807	2.576	2.326	1.96	1.645	1.282	1.036

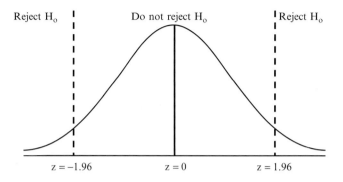

Fig. 5.2

An error of type I occurs if the test statistics (in our case z) fall in the rejection region (in our case $|z| > 1.96$), when in fact the null hypothesis is true. The probability of this happening is the *significance level*, which is the fraction of the distribution curve in the regions of rejecting H_0. This is the probability that $\bar{x} \neq \mu$ while in fact $\bar{x} = \mu$. As can be seen from Fig. 5.2, the condition for rejecting H_0 with the significance level of α is $|z| > z_c$, where z is taken from Table 5.1.

5.2.2 One-tail test

A random sample of 40 people in a village was studied for their daily intake of potassium, and it was found that the mean is 525 mg and the standard deviation is 112 mg. The recommended minimum daily intake (RMDI) is 550 mg. At 5% significance level, do the data provide sufficient evidence to conclude that all the village population receive not less than the RMDI?

Since the question only asks if the population receives not less than the RMDI, then we have only one side (or one-tail) of rejection for the null hypothesis which says that the true μ is not less than 550 mg.

H_0: $\mu \geq 550$ mg (mean intake is not less than 550 mg)
H_a: $\mu < 550$ mg (mean intake is less than 550 mg)

Since the alternative hypothesis has the less than sign ($<$), its region is only in the negative values of z, the left-tail. We want the significance level of 0.05 to be in the left-tail and hence we have to look at Table 5.1 in the row for one-tail test, $z_{0.05}$ (one-tail)=1.645. Calculating z from the observed mean $\bar{x}$, $\mu = 550$ and the standard deviation we obtain as:

$$z = \frac{525 - 550}{(112/\sqrt{40})} = -1.41$$

Since our $|z|$ is smaller than $z_{0.05}$ (one-tail), we do not reject the null hypothesis, which means that we cannot reject the hypothesis that the population receive not less than the RMDI (see Fig. 5.3).

If our observed results were $\bar{x} = 525$ and $s = 50$, then our z will be equal to -3.16 and hence our hypothesis must be rejected. The low value of s means that all our results are closer to $\bar{x} = 525$ and hence the observation $\bar{x} < \mu$ is a real one and cannot be due to statistical fluctuation (Fig. 5.4).

For example, the maximum allowed amount of element x in drinking water is 509 µg/mℓ. Forty samples of water were drawn and the concentration of x in each of them was measured. The results obtained were $\bar{x} = 52.1$ µg/mℓ and $s = 5.4$ µg/mℓ. Do the results indicate that the water can be used as drinking water at the 10% significance level? The null hypothesis is that the water can be used as drinking water and hence:

$$H_0: \mu \leq 50.9 \text{ µg/mℓ (mean } x \text{ concentration is not higher than 50.9 µg/mℓ)} \quad (5.1)$$
$$H_a: \mu > 50.9 \text{ µg/mℓ (mean } x \text{ concentration is higher than 50.9 µg/mℓ)} \quad (5.2)$$

Since the alternative hypothesis has the more than sign ($>$), its region is only in the positive values of z, the right-tail. The value of critical z for one-tail test with level of significance of 10% is 1.282, as given in Table 5.1 (see Fig. 5.5).

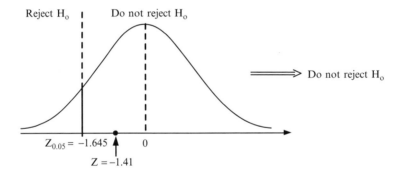

Fig. 5.3

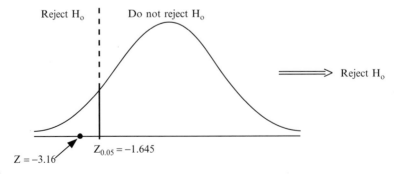

Fig. 5.4

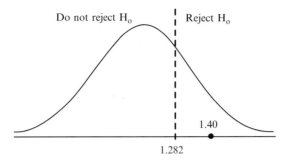

Fig. 5.5

Calculating the statistics for z yields:

$$z = \frac{\overline{x} - \mu}{(s/\sqrt{n})} = \frac{52.1 - 50.9}{(5.4/\sqrt{40})} = 1.40 \tag{5.3}$$

Since $z > z_c$ we are in the range of rejecting H_0. If we want a level of significance of 5% we will not reject H_0, since then $z_c = 1.649$.

5.2.3 Small samples

When a sample is small, usually below $n = 20$, there is no justification to assume that its distribution is normal and the student t-distribution (Table 5.2) is more justified. The procedure for a test is similar to the z-test since t is defined in the same way:

$$t = \frac{\overline{x} - \mu}{(s/\sqrt{n})} \tag{5.4}$$

However, instead of using the critical value from the z table (Table 5.1) it is taken from the table of critical values of t (Table 5.2). In this case the number of repeating experiments appears not only in the calculation of t but also in the critical value of t, i.e. t_c, as the number of experiments influence the number of the degrees of freedom through the relation:

$$v = \mathrm{df} = n - 1 \tag{5.5}$$

For example, a new method for measuring aluminum in biological samples was tested on a certified standard material of plant origin with a certified value of 1.37 ppm (μg/g). The five measurements gave a mean value of 1.25 ppm with a standard deviation of 0.12 ppm. Is the new method a good one, i.e. is there no systematic error in it? Calculating t according to Equation (5.4) we obtain:

$$t = |1.37 - 1.25|\sqrt{5}/0.12 = 2.24$$

Table 5.2   Values for the two-sided student's t-distribution.

	Level of significance				
df	$t_{0.10}$	$t_{0.05}$	$t_{0.025}$	$t_{0.01}$	$t_{0.005}$
1	3.078	6.314	12.706	31.851	63.657
2	1.886	2.920	4.303	6.965	9.925
3	1.638	2.353	3.182	4.541	5.841
4	1.553	2.132	2.776	3.747	4.604
5	1.476	2.015	2.571	3.365	4.032
6	1.440	1.943	2.447	3.143	3.707
7	1.415	1.895	2.365	2.998	3.499
8	1.397	1.860	2.306	2.896	3.355
9	1.383	1.833	2.262	2.821	3.250
10	1.372	1.812	2.228	2.764	3.169
11	1.363	1.796	2.201	2.718	3.106
12	1.356	1.782	2.179	2.681	3.055
13	1.350	1.771	2.160	2.650	3.012
14	1.345	1.761	2.145	2.624	2.977
15	1.341	1.753	2.131	2.602	2.947
16	1.337	1.746	2.120	2.583	2.921
17	1.333	1.740	2.110	2.567	2.898
18	1.330	1.734	2.101	2.552	2.878
19	1.328	1.729	2.093	2.539	2.861
20	1.325	1.725	2.086	2.528	2.845
25	1.316	1.708	2.060	2.485	2.787
30	1.310	1.697	2.042	2.457	2.750
40	1.303	1.684	2.021	2.423	2.704
60	1.296	1.671	2.000	2.390	2.660
∞	1.282	1.645	1.960	2.326	2.576

From Table 5.2 it is seen that for df $= 4$ (df $= n - 1$) the two close values of t are 2.132 for $t_{0.05}$ and 2.776 for $t_{0.025}$. If we look at the distribution curve (Fig. 5.6), for a level of significance of 0.025 we will not reject H_0 (H_0: the method is a good one and $\mu = 1.37\,\mu g/g$), but for a level of significance of 0.05 we will reject H_0, as $t > t_c$.

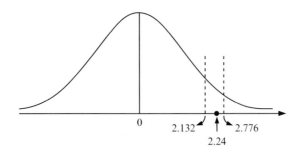

Fig. 5.6

5.3 Comparison of two samples

This test will be used to check if two different methods give the same results within our level of significance, by comparing their means and standard deviations. In the previous section the expected value was assumed to be accurate with zero standard deviation and it actually can be considered as a special case of the more general one treated in this section. This method is used also to compare if two samples measured by the same method are the same within our significance level.

If the first sample (or method) has n_1 repeated experiments with a mean $\bar{x}_1$ and standard deviation s_1 and the second sample has the same parameters with subscript 2, then for the case where s_1 and s_2 are of similar (close) values the joint variance and the statistics (test) t are defined by:

$$s^2 = \left[(n_1 - 1)s_1^2 + (n_2 - 1)s_2^2\right]/(n_1 + n_2 - 2) \tag{5.6}$$

$$t = (\bar{x}_1 - \bar{x}_2)/s\sqrt{(1/n_1 + 1/n_2)} \tag{5.7}$$

$$df = n_1 + n_2 - 2 \tag{5.8}$$

Equation (5.6) is obtained by adding the squares of the differences from the means of the two samples:

$$s_k^2 = \frac{\sum (x_i - \bar{x})^2}{n - 1} \quad \Rightarrow \quad ss_k = \sum (x_i - \bar{x})^2 = (n - 1)s_k^2$$

$$SSR = \text{sum of } SS_k = SS_1 + SS_2 = (n_1 - 1)\, s_1^2 + (n_2 - 1)s_2^2$$

$$s^2 = SSR/(n_1 + n_2 - 2) = \left[\sum_1 (x_i - \bar{x}_1)^2 + \sum_2 (x_i - \bar{x}_2)^2\right] \Big/ (n_1 + n_2 - 2)$$

which leads to Equation (5.6).

The degrees of freedom (number of independent variables) is the sum of the degrees of freedom of the two samples. This standard deviation is the pooled standard deviation and it is assumed to be the best estimate of σ^2:

$$s_{pooled}^2 = \frac{\sum_1 (x_i - \bar{x}_1)^2 + \sum_2 (x_i - \bar{x}_2)^2}{v_1 + v_2} = \frac{ss_1 + ss_2}{v_1 + v_2} \tag{5.9}$$

Here ss stands for sum of squares.

In some books Equation (5.7) is written as:

$$t = (\bar{x}_1 - \bar{x}_2)/\sqrt{\left(s_p^2/n_1 + s_p^2/n_2\right)} \tag{5.10}$$

where s_p^2 is the pooled standard deviation given in Equation (5.9). Equations (5.7) and (5.10) are identical but (5.7) is easier for calculation. However Equation (5.10) is close to the form of t in the general case where s_1 and s_2 are not close.

The sum $s_p^2/n_1 + s_p^2/n_2 = s_p^2(n_1 + n_2)/(n_1 n_2)$ is named the variance of the difference between the means and is denoted by $s_{\bar{x}_1 - \bar{x}_2}^2$:

$$s^2_{\bar{x}_1-\bar{x}_2} = \frac{s^2_p}{n_1} + \frac{s^2_p}{n_2} = \frac{s^2_p(n_1+n_2)}{n_1 \, n_2} \tag{5.11}$$

The testing of the assumption is done by calculation of t_c and comparison with the values in Table 5.2 in the same fashion as in the previous section.

5.3.1 *Example 1*

Two methods have been used to measure the concentration of fluoride in water collected in a well. The following results, all in ppm ($\mu g/\ell$), were obtained:

| Method I | 4.44 | 4.24 | 4.17 | 4.35 | 4.02 | 3.89 | 4.11 |
| Method II | 3.95 | 3.89 | 4.20 | 4.05 | 3.97 | 3.75 | 4.21 |

Do the two methods give results that differ for 0.05 level of significance?
Let us calculate for each method the mean and the standard deviation:

$$\bar{x}_1 = 4.17, \; s_1 = 0.19 \text{ and } \bar{x}_2 = 4.00, \; s_2 = 0.17$$

(n_1 and n_2 are equal in this case, but they can be different).
Using Equations (5.6)–(5.8) leads to:

$$s^2 = (6 \times 0.19^2 + 6 \times 0.17^2)/12 \quad \Rightarrow \quad s^2 = 0.0325$$

$$t = \frac{0.17}{0.182\sqrt{(2/7)}} \quad \Rightarrow \quad t = 1.75$$

$$\mathrm{df} = 12$$

From Table 5.2 it is found that for 0.05 level of significance $t_c = 2.179$. Since $t < t_c$ we cannot reject the null hypothesis which assumes that $\mu_1 = \mu_2$. Thus our results cannot prove that the two methods are different. Remember that we do not prove that the two analytical methods give the same results, but rather we only prove that we cannot say that they are leading to different results. The null assumption H_0 is $\mu_1 = \mu_2$ or $\mu_1 - \mu_2 = 0$.

The test for the three possible alternative assumptions are for level of significance = a:

For H_a:	$\mu_1 - \mu_2 \neq 0$	if	$	t	\geq t_{\alpha/2, \, v}$	then reject H_0
For H_a:	$\mu_1 - \mu_2 > 0$	if	$t \geq t_{\alpha, \, v}$	then reject H_0
For H_a:	$\mu_1 - \mu_2 < 0$	if	$t \leq -t_{\alpha, \, v}$	then reject H_0.

The derivation of the two-sample t-test assumes that both samples come at random from normal populations with equal variances. However, this must not be the case in our experiments. Nonetheless the t-test can be used for the two-sample inference since it was found that the t-test is robust enough to stand considerable departures from its theoretical assumptions, especially if the sample sizes are equal or nearly equal and especially when two-tailed hypotheses are considered (Box 1953; Boneau 1960; Posten *et al.* 1982; Posten 1992).

The larger the samples, the more robust is the test. The two-tailed t-test is affected very little by skewness in the sampled population, but there can be a serious effect on one-tailed tests. Equations (5.6)–(5.8) usually behave well for two samples with equal or similar variances. The comparison of the means from a normal population, without assuming equal variances, is known as the Behren–Fisher problem. Several solutions to this problem were suggested and the most common one is the procedure known as 'Welch's approximate t' (Smith 1936; Welch 1936, 1938; Davenport & Webster 1975):

$$t' = (\bar{x}_1 - \bar{x}_2) \Big/ \sqrt{\left(\frac{s_1^2}{n_1} + \frac{s_2^2}{n_2}\right)} \tag{5.12}$$

and the number of degrees of freedom is calculated by rounding off to the nearest whole number of the following term:

$$df = \frac{\left[s_1^2/n_1 + s_2^2/n_2\right]^2}{\dfrac{\left(s_1^2/n_1\right)^2}{(n_1 - 1)} + \dfrac{\left(s_2^2/n_2\right)^2}{(n_2 - 1)}} \tag{5.13}$$

5.3.2   Example 2

Two groups of patients suffering from disease A and B respectively were measured for their blood concentration of a special hormone. The results were that the seven people in group A have mean concentration of 3.10 units/mℓ with standard deviation of 0.12 units/mℓ, while the eight people in group B have mean concentration of 4.25 units/mℓ with standard deviation of 0.89 units/$\mu\ell$. The null hypothesis is that both diseases have no connection to the hormone concentration, i.e. H_0 is $\mu_1 = \mu_2$ or $\mu_1 - \mu_2 = 0$. The question is if our statistical data allow us to reject this hypothesis:

$$s_1^2/n_1 = 0.12^2/7 = 2.057 \times 10^{-3} \qquad s_2^2/n_2 = 0.98^2/8 = 120.05 \times 10^{-3}$$

$$s_1^2/n_1 + s_2^2/n_2 = 0.1221 \qquad t' = (4.25 - 3.10)/\sqrt{0.1221} \quad \Rightarrow \quad t' = 3.221$$

$$\nu' = df = \frac{0.1221^2}{\left[\dfrac{\left(2.057 \times 10^{-3}\right)^2}{6} + \dfrac{\left(120.05 \times 10^{-3}\right)^2}{7}\right]^{-2}} \cong \frac{0.1221^2 \times 7}{0.1221^2} = 7$$

Actually (s_1^2/n_1) is very small compared to (s_2^2/n_2) and can be neglected; thus the number of degrees of freedom is $n_2 - 1 = 7$.

When $n_1 = n_2$, Equation (5.13) can be written as:

$$df = \frac{\left(s_1^2 + s_2^2\right)^2 (n - 1)}{s_1^4 + s_2^4} \tag{5.14}$$

It is simple to see that when $s_1 \gg s_2$ then $df = n - 1$. When s_1 and s_2 have similar values, as in Example 1, then:

$$df = \frac{\left(0.19^2 + 0.17^2\right)^2 \times 6}{0.19^4 + 0.17^4} = 12.8 \quad \Rightarrow \quad df = 12$$

Thus, we get the same number of degrees of freedom as with the usual t-test. If $n_1 = n_2$ and $s_1^2 = s_2^2$, then $t' = t$ and $v' = v$.

Miller and Miller (1994) gave a slightly different equation for v', while having the same definition for t'. They gave for the number of degrees of freedom, v', the formula:

$$v' = \frac{\left(s_1^2/n_1 + s_2^2/n_2\right)^2}{\dfrac{\left(s_1^2/n_1\right)^2}{n_1 + 1} + \dfrac{\left(s_2^2/n_2\right)^2}{n_2 + 1}} - 2 \tag{5.15}$$

the result being rounded to the nearest whole number. However, for $s_1^2 = s_2^2$ and $n_1 = n_2$ while $t' = t$, $v' = 2(n_1 + 1) - 2 = 2n_1$ whereas $v = 2n_1 - 2$, indicating the advantage of Equation (5.13) over Equation (5.15). Similarly for Example 1, Equation (5.15) yields $v' = 14$ compared to $v' = 12$ by Equation (5.13) which is the same as was obtained from Equation (5.8) for v. For the case when $s_1^2 \gg s_2^2$, while n_1 and n_2 have close values, Equation (5.13) and (5.15) will yield the same value $(n_1 - 1)$.

Equation (5.13) can also be written in another form (Ott 1998) which makes it easier for calculation:

$$v' = \frac{(n_1 - 1)(n_2 - 1)}{(n_2 - 1)c^2 + (n_1 - 1)(1 - c)^2} \quad \text{where} \quad c = \frac{s_1^2/n_1}{s_1^2/n_1 + s_2^2/n_2} \tag{5.16}$$

If $n_1 \neq n_2$ and the population variances are very different, then t' will provide a better test than t. If the population variances are very similar, then t is the better test. If H_0 cannot be rejected we will use the results from both samples to calculate the mean and its confidence interval. The α level of significance in the rejection of t is parallel to the $(1 - \alpha)$ confidence limit. The $(1 - \alpha)$ confidence interval for the difference between the two population means in the case of equal variances is:

$$\bar{x}_1 - \bar{x}_2 \pm t_{\alpha/2,\ v}\ s_{\bar{x}_1 - \bar{x}_2} \tag{5.17}$$

The best estimate of the common mean of the samples is the 'weighted' (or 'pooled') mean, $\bar{x}$:

$$\bar{x} = \frac{n_1 \bar{x}_1 + n_2 \bar{x}_2}{n_1 + n_2} \tag{5.18}$$

which is the mean if we combine the two samples into one sample. The confidence limit for the mean is:

$$\bar{x} \pm t_{\alpha/2,\ v} \sqrt{\left(\frac{s_p^2}{n_1 + n_2}\right)} \tag{5.19}$$

In the case of very different variances the confidence interval for the difference of the means is:

$$\bar{x}_1 - \bar{x}_2 \pm t_{\alpha/2,\ v} \sqrt{\left(\frac{s_1^2}{n_1} + \frac{s_2^2}{n_2}\right)} \tag{5.20}$$

and the common mean is:

$$\bar{x}_2 \pm t_{\alpha/2,\ \nu} \sqrt{\left(\frac{s_1^2 + s_2^2}{2(n_1 + n_2)}\right)} \tag{5.21}$$

In Equations (5.20) and (5.21) ν' is calculated from Equation (5.13).

Sometimes two methods are tested not several times on the same sample (or aliquots of the same sample) but by measuring several different samples using each of the analytical methods. In this case the mean and the standard deviation have no meaning in each method since the samples (the repeated measurements) are different and each has its own value. For this case the paired t-test will be applied as described in the following section.

5.4 Paired t-test

In this test equal numbers of measurements must be done in both methods, since each sample is measured in both methods. We then calculate the mean of the differences between the two methods for each of the samples, $\bar{x}_d$, and the standard deviation of the differences, s_d. The statistic t is calculated by the equation:

$$t = \bar{x}_d \sqrt{n}/s \tag{5.22}$$

where n is the number of pairs (samples measured by both methods) and the degrees of freedom is df $= (n - 1)$.

5.4.1 Example

Let us look at the concentration of chlorides (in ppm) in rain water falling on different days in January, each day measured once by each method:

| Method I | 4.44 | 3.84 | 3.14 | 4.08 | 3.92 | 3.14 |
| Method II | 5.00 | 3.90 | 3.32 | 4.26 | 4.14 | 3.36 |

Do the results obtained by the two methods differ significantly?

If we use the methods of Section 5.3 we will find $\bar{x}_1 = 3.76$, $s_1 = 0.52$ and $\bar{x}_2 = 4.00$, $s_2 = 0.63$. Equations (5.6) – (5.8) will give $s^2 = 0.33$, $t = 0.72$, df $= 10$. Equation (5.14) will give also df $= 10$. Referring to Table 5.2, we can see that the results are the same even for a level of significance of 0.1, but there are no meanings to the values of $\bar{x}$ and s as the samples are different and the concentration can be different each day. The differences between the methods for the various days are: 0.56, 0.06, 0.18, 0.18, 0.22 and 0.22, which gives $\bar{x}_d = 0.24$ and $s_d = 0.17$, yielding, according to Equation (5.20), $t = 3.46$ with df $= 5$. According to Table 5.2, the difference will be not significant only at the level of 0.01. As the common requirement is a level of significance of 0.05, we must conclude that the two methods give different results.

It should be emphasized that this method of paired t-test assumes that any errors, either random or systematic, are independent of concentration. Over a wide range of concentration this assumption may no longer be true, and hence this method should not be used over too large a range of concentrations. When a new method is tested it should be tested over the whole range for which it is applicable. In order to perform this comparison, several samples with various concentrations should be prepared, and for each one several repeated measurements should be performed by each method, applying the usual t-test for means (Section 5.2) for each concentration.

Another method that is the preferred one in analyzing over a wide range of concentration is via linear regression, which will be treated later.

5.5   Comparing two variances – the F-test

The previous tests examined the significance of variation of two means using their standard deviations. These tests assumed that the two samples are distributed normally with the same σ (population standard deviation). The t-test was used rather than the normal distribution because of the small sizes of the samples that are usual in analytical chemistry. These tests for comparison of the means assume the same σ are used for detecting systematic errors. Yet, sometimes different methods have different precisions, i.e. different standard deviation $\left(\text{variance} = s^2\right)$. The comparison of two variances takes two forms depending on whether we ask if method A is better than method B (neglecting the possibility that B is better than A), i.e. a one-tailed test, or whether we allow also the probability that method B is better and ask only if the two methods differ in their precision, i.e. a two-tailed test. Thus, if we want to test whether a new analytical method is preferable to a standard one, we use a one-tailed test, because unless we prove that the new method is a better one we will stick to the old method. When we are comparing two new methods or two standard methods, a two-tailed test is the right one to use.

This test is based on the F-statistic, which obeys the F distribution (see Chapter 3). The F-statistic (test) is defined as the ratio of two variances:

$$F = s_1^2/s_2^2 \tag{5.23}$$

The subscripts 1 and 2 are chosen so that F is always equal to or greater than unity, $\left(\text{i.e. } s_1^2 \geq s_2^2\right)$. In other words, the larger variance is placed in the numerator and the smaller in the denominator. The calculated F are checked against the values in Table 5.3. If the calculated value is larger than that found in the table, the null hypothesis that the two methods are not different in their precision is rejected.

$$H_0: \sigma_1^2 = \sigma_2^2$$

For H_a: $\sigma_1^2 \neq \sigma_2^2$ if $\left|F_{\alpha/2,\ v_1,\ v_2}\right| > F_c$ then reject H_0

For H_a: $\sigma_1^2 < \sigma_2^2$ if $F_{\alpha,\ v_1,\ v_2} < -F_c$ then reject H_0

If H_0 is not rejected the variances of the samples can be used to calculate the common variance, which is s_{pooled}^2 as defined by Equation (5.9).

5.5.1   Example

A standard method for measuring copper ions in drinking water was repeated eight times and found to have a mean of $4.25\,\mu\text{g/m}\ell$ with standard deviation of $0.253\,\mu\text{g/m}\ell$. Seven measurements of a new method yield the same mean but with

Table 5.3 Critical values of the F-statistic for a significance level of $p = 0.05$

(a) One-tailed test

df_2	df_1												
	1	2	3	4	5	6	7	8	9	10	12	15	20
1	161.4	199.5	215.7	224.6	230.2	234.0	236.8	238.9	240.5	241.9	243.9	245.9	248.0
2	18.21	19.00	19.16	19.25	19.30	19.33	19.35	19.37	19.38	19.40	19.41	19.43	19.45
3	10.13	9.552	9.277	9.117	9.013	8.941	8.887	8.845	8.812	8.786	8.745	8.703	8.660
4	7.709	6.944	6.591	6.388	6.256	6.163	6.094	6.041	5.999	5.964	5.912	5.858	5.803
5	6.608	5.786	5.409	5.192	5.050	4.950	4.876	4.818	4.772	4.735	4.678	4.619	4.558
6	5.987	5.143	4.757	4.534	4.387	4.284	4.207	4.147	4.099	4.060	4.000	3.938	3.874
7	5.591	4.737	4.347	4.120	3.972	3.866	3.787	3.726	3.677	3.637	3.575	3.511	3.445
8	5.318	4.459	4.066	3.838	3.687	3.581	3.500	3.438	3.388	3.347	3.284	3.218	3.150
9	5.117	4.226	3.863	3.633	3.482	3.374	3.293	3.230	3.179	3.137	3.073	3.006	2.936
10	4.965	4.103	3.708	3.478	3.326	3.217	3.135	3.072	3.020	2.978	2.913	2.845	2.774
11	4.844	3.982	3.587	3.357	3.204	3.095	3.012	2.948	2.896	2.854	2.788	2.719	2.646
12	4.747	3.885	3.490	3.259	3.106	2.996	2.913	2.849	2.796	2.753	2.687	2.617	2.544
13	4.667	3.806	3.411	3.179	3.025	2.915	2.832	2.767	2.714	2.671	2.604	2.533	2.459
14	4.600	3.739	3.344	3.112	2.958	2.848	2.764	2.699	2.646	2.602	2.534	2.463	2.388
15	4.543	3.682	3.287	3.056	2.901	2.790	2.707	2.641	2.588	2.544	2.475	2.403	2.328
16	4.494	3.634	3.239	3.007	2.852	2.741	2.657	2.591	2.538	2.494	2.425	2.352	2.276
17	4.451	3.592	3.197	2.965	2.810	2.699	2.614	2.548	2.494	2.450	2.381	2.308	2.230
18	4.414	3.555	3.160	2.928	2.773	2.661	2.577	2.510	2.456	2.412	2.342	2.269	2.191
19	4.381	3.522	3.127	2.895	2.740	2.628	2.544	2.477	2.423	2.378	2.308	2.234	2.155
20	4.351	3.493	3.098	2.866	2.711	2.599	2.514	2.447	2.393	2.348	2.278	2.203	2.124

Table 5.3 (*Continued*)

(b) Two-tailed test

df$_2$	df_1												
	1	2	3	4	5	6	7	8	9	10	12	15	20
1	647.8	799.5	864.2	899.6	921.8	9371	948.2	956.7	963.3	968.6	976.7	984.9	993.1
2	38.51	39.00	39.17	39.25	39.30	39.33	39.36	39.37	39.39	39.40	39.41	39.43	39.45
3	17.44	16.04	15.44	15.10	14.88	14.73	14.62	14.54	14.47	14.42	14.34	14.25	14.17
4	12.22	10.65	9.979	9.605	9.364	9.197	9.074	8.980	8.905	8.844	8.751	8.657	8.560
5	10.01	8.434	7.764	7.388	7.146	6.978	6.853	6.757	6.681	6.619	6.525	6.428	6.329
6	8.813	7.260	6.599	6.227	5.988	5.820	5.695	5.600	5.523	5.461	5.366	5.269	5.168
7	8.073	6.542	5.890	5.523	5.285	5.119	4.995	4.899	4.823	4.761	4.666	4.568	4.467
8	7.571	6.059	5.416	5.053	4.817	4.652	4.529	4.433	4.357	4.295	4.200	4.101	3.999
9	7.209	5.715	5.078	4.718	4.484	4.320	4.197	4.102	4.026	3.964	3.868	3.769	3.667
10	6.937	5.456	4.826	4.468	4.236	4.072	3.950	3.855	3.779	3.717	3.621	3.522	3.419
11	6.724	5.256	4.630	4.275	4.044	3.881	3.759	3.664	3.588	3.526	3.430	3.330	3.226
12	6.554	5.096	4.474	4.121	3.891	3.728	3.607	3.512	3.436	3.374	3.277	3.177	3.073
13	6.414	4.965	4.347	3.996	3.767	3.604	3.483	3.388	3.312	3.250	3.153	3.053	2.948
14	6.298	4.857	4.242	3.892	3.663	3.501	3.380	3.285	3.209	3.147	3.050	2.949	2.844
15	6.200	4.765	4.153	3.804	3.576	3.415	3.293	3.199	3.123	3.060	2.963	2.862	2.756
16	6.115	4.687	4.077	3.729	3.502	3.341	3.219	3.125	3.049	2.986	2.889	2.788	2.681
17	6.042	4.619	4.011	3.665	3.438	3.277	3.156	3.061	2.985	2.922	2.825	2.723	2.616
18	5.978	4.560	3.954	3.608	3.382	3.221	3.100	3.005	2.929	2.866	2.769	2.667	2.559
19	5.922	4.508	3.903	3.559	3.333	3.172	3.051	2.956	2.880	2.817	2.720	2.617	2.509
20	5.871	4.461	3.859	3.515	3.289	3.128	3.007	2.913	2.837	2.774	2.646	2.573	2.464

standard deviation of $0.207\,\mu g/m\ell$. On first consideration the new method seems to be better and more precise, since the standard deviation is lower, but is this difference significant or it is just a random one?

For testing this we calculate the F-statistic. Since one method is a standard one, a one-tailed test should be used. The degrees of freedom of the larger s are $7 = (8 - 1)$ and of the smaller one 6, so we write this F as $F_{7,6}$. The first number is the df for the numerator and the second for the denominator:

$$F_{7,6} = \frac{0.253^2}{0.207^2} = 1.4938$$

From Table 5.3, $F_{7,6} = 4.207$ for 0.05 level of significance. Since the calculated value is lower that the critical value, the null hypothesis cannot be rejected and hence it is not proven that the new method is better that the standard one. At this level of certainty (that we might be wrong in 5% (1 from 20) cases), the lower standard deviation can be due completely to a random origin. However, if the new method has a standard deviation of 0.113, and we calculate $F_{7,6} = 5.01$, and since this value is larger that the critical value, we can say that the new method is more precise than the standard one.

5.6 Comparison of several means

The comparison of several means and the inference if they are equal or not is done by an F-test using two different variances, the *within-sample variance* and the *between-sample variance*. Comparing the mean variance of separate samples to the variance between the samples allows us to find if all the means are the same. Although we compare means we test them through their variance and consequently this method is called *analysis of variances* (acronymed ANOVA). If, for example, we want to measure the effect of various conditions of storage, e.g. effect of temperature, kind of material from which the containers are made, ratio of surface area to volume of the containers, then we will take from each container n samples and analyze them to obtain the mean $\bar{x}_i$ and standard deviation s_i, $i = 1, 2, \ldots, k$, where k is the number of different containers. In order to see if the containment conditions influence the chemical in question, we need to compare the data obtained. This is done by the following steps:

Step 1: Calculate the *within-sample* estimate of the variance, V_{ws}:

$$V_{ws} = \sum_{i=1}^{k} s_i^2/k \tag{5.24}$$

This is the arthmetic mean of the various variances for each sample. This is a combined estimate of the common variance and measures the variability of the observations within the k samples. The number of degrees of freedom is the sum for all samples. The general formula for the within-sample estimate of the variance, V_{ws} is:

$$V_{ws} = \sum_{i=1}^{k}\sum_{j=1}^{n}(x_{ij} - \bar{x}_i)^2/k(n-1) \tag{5.25}$$

V_{ws} is also known as the *mean square* since it involves the sum of squares divided by the total number of degrees of freedom of the k samples, i.e. $df = k(n-1)$.

Step 2: Calculate the *between-sample* estimate of the variance, V_{bs}. The total mean is calculated as the arithmetic mean of the individual means:

$$\bar{x} = \sum_{i=1}^{k} \bar{x}_i / k \tag{5.26}$$

The variance of the different individual means is calculated by the equation:

$$\text{variance of means} = \sum_{i=1}^{k} (\bar{x}_i - \bar{x})^2 / (k-1) \tag{5.27}$$

V_{bs} is calculated by multiplying the variance of the means by the number of measurements in each sample, n. The df is $(k-1)$. So:

$$V_{bs} = \frac{n}{k-1} \sum_{i=1}^{k} (\bar{x}_i - \bar{x})^2 \tag{5.28}$$

Equations (5.24) and (5.28) are for the special case when $n_1 = n_2 = \ldots n_k$. In the case of samples of different sizes, V_{ws} and V_{bs} will be calculated by the following equations, which is a similar calculation to that of s^2_{pooled} for two samples:

$$\left.\begin{aligned} V_{ws} &= \sum_{i=1}^{k} (n_i - 1)\, s_i^2 \Bigg/ \left[\left(\sum_{i=1}^{k} n_i\right) - k\right] \\[2ex] V_{ws} &= \frac{(n_1 - 1)\, s_1^2 + (n_2 - 1)\, s_2^2 + \ldots + (n_k - 1)\, s_k^2}{n_1 + n_2 + \ldots + n_k - k} \\[2ex] \text{and} & \\[1ex] V_{bs} &= \frac{1}{k-1} \sum_{i=1}^{n} n_i (\bar{x}_i - \bar{x})^2 \end{aligned}\right\} \tag{5.29}$$

ANOVA can also be done for samples of different sizes through Equations (5.29). The power of the test is strengthened by having sample sizes as nearly equal as possible. When all the samples differ by only one factor the analysis is termed a single factor ANOVA.

Step 3: Use these two variances to calculate the *F*-statistic:

$$F_{k-1,\, k(n-1)} = \frac{V_{ws}}{V_{bs}} \tag{5.30}$$

Step 4: If the calculated F is larger than the critical value (one-tailed) then the means are significantly different and the containment conditions are important. If the calculated value of F is lower than that from Table 5.3, the means are not different at that level of significance.

5.6.1 *Example 1*

Six containments are tested. From each, five samples were measured and the data are given below:

Containment	A	B	C	D	E	F
$\bar{x}$	100	104	96	106	98	103
s	2.5	4	3.5	5.1	3	5

Using Equations (5.24)–(5.29) we get:

$$V_{\text{ws}} = \left(2.5^2 + 4^2 + 3.5^2 + 5.1^2 + 3^2 + 5^2\right)/6 = 15.75$$

$$\text{total mean} = \frac{100 + 104 + 96 + 106 + 98 + 103}{6} = 101.1$$

$$\text{variance of means} = \left(1.1^2 + 2.9^2 + 5.1^2 + 4.9^2 + 2.1^2 + 1.9^2\right)/5 = 14.57$$

$$V_{\text{bs}} = 5 \times 14.57 = 72.83$$

$$F_{5,24} = \frac{72.83}{15.75} = 4.62$$

We do not have $F_{5,24}$ in Table 5.3 but it has to be lower than $F_{5,20}$, which is 2.711. Since the calculated F is larger than the critical one, we reject H_0 which hypothesizes that all the means are equal, i.e. we might conclude that the means are different from each other although we can only prove that the opposite is not true.

5.6.2 *Simplification of ANOVA calculation and separation*

The calculation of ANOVA can be simplified. Let us remember that we have k samples, each consisting of n measurements with a mean $\bar{x}_i$ and a standard deviation s. We calculated two kinds of variances, between-sample and within-sample. Let us call the total number of measurements N, i.e. $N = k \times n$. The sum of all measurements for sample i is:

$$T_i = \sum_{j=1}^{n} x_{ij} \tag{5.31}$$

and the sum of all measurements is:

$$T = \sum_{i=1}^{k} T_i = \sum_{i=1}^{k} \sum_{j=1}^{n} x_{ij} \tag{5.32}$$

It can be shown that:

$$V_{\text{bs}} = (1/n) \sum_{i=1}^{k} T_i^2 - T^2/N \quad (\text{df} = k - 1) \tag{5.33}$$

Be careful to distinguish between $\sum_{i=1}^{k} T_i^2$ and T^2, where $T^2 = \left(\sum_{i=1}^{k} T_i\right)^2$

$$V_{ws} = \sum_{i=1}^{k} \sum_{j=1}^{n} x_{ij}^2 - (1/n) \sum_{i=1}^{k} T_i^2 \quad (df = N - k) \tag{5.34}$$

Instead of calculating $\sum\sum x_{ij}$ and $\sum\sum x_{ij}^2$ directly from the observed data, it is easier to subtract a constant value which will bring x_{ij} to a smaller number. Of course if a computer program is used this is not necessary.

5.6.3 Example 2

Let us look at five containers with powdered coal ash. From each container five samples are taken and measured for their iron content (in %):

Container A	Container B	Container C	Container D	Container E
1.78	1.81	1.78	1.84	1.76
1.76	1.80	1.80	1.80	1.79
1.75	1.79	1.76	1.83	1.74
1.76	1.83	1.77	1.79	1.73
1.80	1.82	1.82	1.82	1.78

V_{bs} and V_{ws} can be calculated from Equations (5.33) and (5.34). It is easier to use the x_{ij} with a constant number of our choice subtracted for all them, rather than the original x_{ij}. Thus if we choose the constant number to be subtracted as 1.80 the values are:

System	x_{ij}					T_i	T_i^2	$\sum_j x_{ij}^2$
A	−0.02	−0.04	−0.05	−0.04	0	−0.15	0.0225	0.0061
B	0.01	0	−0.01	−0.03	0.02	0.05	0.0225	0.0015
C	−0.02	0	−0.04	−0.03	−0.01	−0.10	0.01	0.0030
D	0.04	0	0.03	−0.01	−0.02	0.08	0.0064	0.0030
E	−0.04	−0.01	−0.06	−0.07	−0.02	−0.20	0.0400	0.0106

$$T = \sum T_i = -0.32 \qquad \sum T_i^2 = 0.0814 \qquad \sum\sum x_{ij}^2 = 0.0242$$
$$n = 5 \qquad k = 5 \qquad N = 5 \times 5 = 25$$

Thus we obtain:

$$V_{bs} = \frac{1}{5} \times 0.0814 - \frac{1}{25} \times 0.1024 = 0.01218 \quad (df = 5 - 1 = 4)$$

$$V_{ws} = 0.0242 - \frac{1}{5} \times 0.0814 = 0.00792 \quad (df = 25 - 5 = 20)$$

$$F_{4, 20} = \frac{0.01218}{0.00792} = 1.538$$

At the beginning of the section we saw that *analysis of variances*, usually called by the acronym ANOVA, can show if several means differ significantly or not. The same analysis can also be used to separate the variance into its sources; into the part that is due to the random error of measurement and the part that is due to the difference between the samples.

It is assumed that the random error of measurement is normally distributed with a variance σ_0^2 and that the concentration of the material in the containers is also normally distributed with a variance σ_1^2. The means cannot tell us about σ_1^2 since they involve both variances. However, V_{ws} can yield σ_0^2 directly. It can be shown that $V_{bs} = \sigma_0^2 + \sigma_1^2$. If it can be shown that V_{ws} and V_{bs} do not differ significantly, then we can assume σ_1 to be zero. Otherwise we will get:

$$\sigma_0^2 = V_{ws} \qquad \sigma_1^2 = (1/n)(V_{bs} - V_{ws}) \qquad (5.35)$$

From Table 5.3 the critical value at the level of $p = 0.05$ is 2.866. Since the calculated F is less than the critical one, then $\sigma_1 = 0$. Calculating σ_0 and σ_1 from Equation (5.35) yields $\sigma_0^2 = 0.007\,92$ and $\sigma_1^2 = 0.001\,05$, considerably less than σ_0^2.

5.7   The chi-squared (χ^2) test

This test is used for testing if the distribution of an observed sample fits some theoretical distributions. Thus it can be used to test if the obtained data are distributed normally or not. The quantity χ^2 is calculated according to the equation:

$$\chi^2 = \sum_{i=1}^{n} \frac{(O_i - E_i)^2}{E_i} \qquad (5.36)$$

where O_i is the observed frequency of one of the results [number of times that this number (result) appears] and E_i is the expected frequency, as for example those expected from a normal distribution or other expectation, and n is the number of different values. The null hypothesis is that there is no difference between the expected and observed frequencies. If the calculated χ^2 is larger than the critical value, given in Table (5.4), then the null hypothesis is rejected. The critical value depends as in other cases on the significance level of the test and the number of degrees of freedom $= (n - 1)$.

Table 5.4   Critical values of χ^2 for level of significance of 0.05.

Number of degrees of freedom	Critical value
1	3.84
2	5.99
3	7.81
4	9.49
5	11.07
6	12.59
7	14.07
8	15.51
9	16.92
10	18.31

5.8 Testing for normal distribution – probability paper

In the previous section we saw that the χ^2 test can be used to test if the measurements are distributed normally by comparing the expected frequency from a normal distribution to the observed frequency for each value (or interval). Usually the χ^2 test should not be applied for samples smaller that 50. Another way to test for normal distribution is by plotting the distribution function on *probability paper*. This is, in principle, the same as semi-logarithmic paper. On semi-logarithmic paper the *x*-axis is linear while the *y*-axis is logarithmic, i.e. the distance between 1 and 2 (ratio 2) is equal to the distance between 3 and 6 or between 4 and 8, or the distance between 1 and 10 is equal to the distance between 10 and 100. On probability paper the distance is plotted according to the area under the normal curve.

The data are organized according to their values in increasing order starting from the lower value. Then their cumulative frequency is calculated by addition. If we plot the cumulative frequency vs. the value on ordinary graph paper, we will obtain a plot with a ⁄ shape. However, if we plot it on probability paper, a straight line should be obtained. The linearity of this line is a check for its normal distribution.

5.9 Non-parametric tests

All the previously mentioned statistical methods have one basic assumption – that the population (or at least the means of small samples) are normally distributed (Gaussian-distribution). However, for some cases it was proven than this assumption is not justified. In these cases a different kind of statistical test should be used. Statistical tests that make no assumption of the distribution of the population from which the data are taken are referred to as *non-parametric tests*.

Two other aspects that lead to the popular use of these methods in the social sciences are their simplicity and a very small amount of calculation, compared to those involved in the calculation of standard deviations in the parametric test. However, now that statistical programs are available on every spreadsheet program on almost all PCs and scientific calculators can easily be used to calculate the mean and the standard deviation of data, this point of speedier analysis has lost most of its significance. In order to see the simplicity of these methods we will demonstrate the use of two of these methods, the 'sign test' and the 'Wald–Wolfowitz runs test'.

While parametric analysis tests the mean of the samples, non-parametric analysis involves the median. While the value of the mean involves some calculation, i.e.

$$\bar{x} = \sum_i x_i/n$$

the finding of the median does not require any calculation, only ranking of the data points according to either decreasing or increasing order. The *median* is defined as the number for which half of the data points have larger (or smaller) or equal value. Thus if η is the median, then its definition is $p(x_i \le \eta) = 0.5$. For *n* data points ranked in a monotonic direction, the median is the $0.5(n + 1)$th observation if *n* is odd and the average of the $0.5n$th and $(0.5n + 1)$th observations for even *n*.

5.9.1   *Sign test*

Assume that there are eight observations which are (arranged by descending values) 6, 5, 4, 3, 3, 3, 3, 2. Since there are eight observations, the median is the average of the fourth and fifth observations and hence the median is 3. Now we subtract the median from all values and check the number of plus signs and minus signs (neglecting all differences that are equal to zero, i.e. all the values that are equal to the median). Thus we have three plus signs and one minus sign. To test the significance of this observation (three plus signs out of four – as we neglect those equal to the median in calculation of the signs but not in the calculation of the median) we use the binomial distribution. The binomial distribution uses Bernoulli's equation for the probability of k successes in n trials, where the probability of a success in a single experiment is p:

$$H_n^k = C_n^k p^k (1 - p)^n \tag{5.37}$$

where the binomial coefficient C_n^k,

$$C_n^k = \frac{n!}{k!(n - k)!} \tag{5.38}$$

There are equal chances of a plus sign or a minus sign and hence $p = 0.5$. Thus the probability of obtaining three plus signs out of four is:

$$H_4^3 = C_4^3 \times 0.5^3 \times 0.5^1 = 4 \times 0.0625 = 0.25$$

When we check the significance of the difference, we are checking the tail of the distribution, i.e. we check the probability of obtaining three or more equal signs. So we must calculate also the probability of obtaining four plus signs out of four:

$$H_4^4 = C_4^4 \times 0.5^4 \times 0.5^0 = 0.0625$$

Thus, the probability of obtaining three or more plus signs is $0.25 + 0.0625 = 0.3125$. However, the test should be two-tailed, i.e. what are the probabilities of obtaining three or more equal signs (either plus or minus) out of four? Since the probability of plus is the same as of minus, the total probability is $2 \times 0.3125 = 0.625$. This probability is much higher than our common level of significance, $p = 0.05$, and hence we cannot reject our null hypothesis, i.e. that the data come from a population with median $= 3$. The binomial probability does not have to be calculated, as was done here; it can be taken from Table 5.5.

Parametric analysis will treat this by the t-test. The eight data points will yield $\bar{x} = 3.625$ and $s = 1.302$. So:

$$t = \frac{|3.625 - 3|\sqrt{8}}{1.302} = 1.3572$$

Table 5.5 The one-tailed binomial probability for r successes in n trials (for a two-tailed test multiply by 2).

n	$r = 0$	1	2	3	4	5	5	7
4	0.063	0.313	0.688					
5	0.031	0.188	0.500					
6	0.016	0.109	0.344	0.656				
7	0.008	0.063	0.227	0.500				
8	0.004	0.035	0.144	0.363	0.637			
9	0.002	0.020	0.090	0.254	0.500			
10	0.001	0.011	0.055	0.172	0.377	0.623		
11	0.000	0.006	0.033	0.113	0.274	0.500		
12	0.000	0.003	0.019	0.073	0.194	0.387	0.613	
13	0.000	0.002	0.011	0.046	0.133	0.290	0.500	
14	0.000	0.001	0.006	0.029	0.090	0.212	0.395	0.605
15	0.000	0.000	0.004	0.018	0.059	0.151	0.304	0.500

For df $= 7$ the critical value of t for 95% confidence is 2.36. Since the calculated t is less than the critical one, we cannot reject the null hypothesis that the true mean is 3 (the median value).

5.9.2 Wald–Wolfowitz runs test

This test can be used to check if the operation of an instrument changes during the day, or if the performance of the analyst changes during the day. Running replicate analyses of the same sample, we write + any time the last measurement is larger than the previous one or − any time it is smaller. From n measurements we will have $n - 1$ signs. Let us denote the number of plus signs by P and the number of minus signs by M. We write the signs in the order of their appearance, i.e. the sign between the first and second experiment will be the first, while that between the second and third experimental will be the second, etc. Every time a sign is changed (from plus to minus or vice versa) we called it a *run*. Let us denote the number of runs by R. We want the number of runs to be random to show that the analysis is independent of the time of the measurement.

Table 5.6 shows the range of the number of runs for each P and M (note that they are interchangeable, so $P = 5$ and $M = 6$ is the same as $P = 6$ and $M = 5$ and hence the table gives only $M > P$), which can be regarded as due to a random distribution. If R is out of this range it is non-random and must be due to machine changes or human behavior change. For example, in one day of experiments the following results were obtained: +, +, +, −, −, −, −, −, −, +, +, +. We have six pluses so $P = 6$ and six minuses so $M = 6$. There are only two runs (changing of signs) and since the table shows that R must be larger than 4, we must conclude that the changes are not random, and the reason for it must be looked for.

Table 5.6   The Wald–Wolfowitz runs test.

P	M	At $p = 0.05$, the number of runs is significant if it is:	
		Less than	Greater than
2	10–20	3	n/a
3	6–14	3	n/a
3	15–20	4	n/a
4	5–6	3	8
4	7	3	n/a
4	8–15	4	n/a
4	16–20	5	n/a
5	5	3	9
5	6	4	9
5	7–8	4	10
5	9–10	4	n/a
5	11–17	5	n/a
6	6	4	10
6	7–8	4	11
6	9–12	5	12
6	13–18	6	n/a
7	7	4	12
7	8	5	12
7	9	5	13
7	10–12	6	13
8	8	5	13
8	9	6	13
8	10–11	6	14
8	12–15	7	15

n/a – the test cannot be applied in this case

5.10   References

Boneau, C. A. 1960, *Psychol. Bull.* vol. 57, p. 49.

Box, G. E. P. 1953, *Biometrika*, vol. 40, p. 318.

Davenport, J. M. & Webster, J. T. 1975, *Metrica*, vol. 22, p. 47.

Miller, J. C. & Miller, J. N. 1994, *Statistics for Analytical Chemistry*, 3rd edn, Ellis Harwood and Prentice Hall, New York, p. 57.

Ott, L. 1998, *An Introduction to Statistical Methods and Data Analysis*, 3rd edn, PWS – Kent publishing company, Boston, USA, p. 175.

Posten, H. O. 1992, *Commun. Statis. – Theor. Meth.*, vol. 21, p. 2169.

Posten, H. O., Yeh, H. C. & Owen, D.B. 1982, *Commun. Statist. – Theor. Meth.*, vol. 11, p. 109.

Smith, H. 1936, *J. Council SCI. Industr. Res.*, vol. 9, p. 211.

Welch, B. L. 1936, *J. Royal Statist. Soc.*, Suppl. 3, p. 29.

Welch, B. L. 1938, *Biometrica*, vol. 29, p. 350.

6 Outliers

6.1 Introduction

An important question every experimentalist meets quite often concerns one or two data points that deviate from the others and hence influence quite significantly the standard deviation, though not so much the mean. The tendency is usually to say that some exceptional error was involved in this measurement and to discard the data points concerned. However, is this justified? It may be that the deviation is just due to random error. We refer to measurements like these as outliers and this chapter discusses the question of whether or not they can be disregarded.

Outliers can result, for example, from blunders or malfunctions of the methodology or from unusual losses or contaminations. Assume that we measure aliquots from the same sample and obtain the values for manganese (Mn) content as 5.65, 5.85, 5.94, 6.52, 5.73, 5.80. At first sight we will tend to reject the fourth result and use only five values to calculate the mean and the standard deviation. Is this rejection justified? We will test this rejection assuming that the whole population is normally distributed. There are several tests for finding outliers (Barnet & Lewis, 1994), and only a few of them, the most common ones, are given here.

An outlier in a set of data is an observation (or subset of observations) which appears to be inconsistent with the remainder of that set of data. What is crucial is the phrase 'appears to be inconsistent' which implies a subjective judgment of the observer. To eliminate this subjectivity we need to have objective tests. Discordance is another word for the phenomenon of outliers. Some people use the term 'rejection of outliers', while others prefer the term 'test of discordance' as not all suspected outliers must be rejected and in some cases we are interested to find the causes for these outliers.

6.2 Dixon's Q-test

Dixon statistics (Dixon 1950, 1953; Likes 1996) are illustrated best if we arrange the various observations on a line (the number axis). This will arrange the values in a monotonous direction. Let us name the observations according to their ascending order.

In Fig. 6.1 we have 12 points. Both x_1 and x_{12} are extreme values, but while x_1 seems to belong to the set, x_{12} seems to be an outlier, since the distance between x_{12} and the adjacent value x_{11} is larger than the distance between any other adjacent

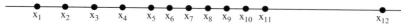

Fig. 6.1

two values. The possible outliers are the extreme values x_1 and x_n, where n is the number of observation in the set of data. Dixon suggested a test called the Q-test (Q stands for quotient) where the statistic Q is defined by the ratio:

$$Q = (x_n - x_{n-1})/(x_n - x_1) \quad \text{or} \quad Q = (x_2 - x_1)/(x_n - x_1) \tag{6.1}$$

where $x_1, x_2, \ldots, x_{n-1}, x_n$ are the observations arranged according to ascending value.

The calculated statistic Q is compared with a tabulated critical value of Q, Q_c. If the calculated value is larger than the tabulated value, rejection may be made with the tabulated risk.

Table 6.1 shows the values of the critical Q_c for different values of n and risk of false rejection. If either x_n or x_1 is rejected we should continue to check if x_{n-1} or x_2, respectively, is also an outlier. However, if x_n is not rejected there is no need to test for x_{n-1}.

For our example of six Mn determinations, $Q = (6.52 - 5.94)/(6.52 - 5.65)$ $= 0.666$. Next we look at Table 6.1 for the value of the critical Q_c for a sample size of 6. Thus for a risk of 1%, $Q_c = 0.698$, while for a 5% risk $Q_c = 0.560$. Thus for a 1% risk we should retain the value 6.52, but if we allow a 5% risk the value 6.52 is an outlier. If we did not perform the last two experiments, we would have the same Q, but since the sample size is 4, the critical value for a 5% risk is 0.765 and thus $Q < Q_c$ and hence the 6.52 measurement should be retained.

Actually the Q calculated by Equation (6.1) is correct only for sample size n in the range $3 < n < 7$. For larger sizes of sample, other ratios are calculated. If we arrange the x_i in either increasing or decreasing order, such that the outlier is the last one, x_n, the following ratios are calculated:

Table 6.1 Values for Dixon's Q-test for outliers.

Number of observations, n	Risk of false rejection (%)			
	0.5	1	5	10
3	0.994	0.988	0.941	0.886
4	0.926	0.889	0.765	0.679
5	0.821	0.780	0.642	0.557
6	0.740	0.698	0.560	0.482
7	0.680	0.637	0.507	0.434
8	0.725	0.683	0.554	0.479
9	0.677	0.635	0.512	0.441
10	0.639	0.597	0.477	0.409
11	0.713	0.679	0.576	0.517
12	0.675	0.642	0.546	0.490
13	0.649	0.615	0.521	0.467
14	0.674	0.641	0.546	0.492
15	0.647	0.616	0.525	0.472
16	0.624	0.595	0.507	0.454
17	0.625	0.577	0.490	0.438
18	0.589	0.561	0.475	0.424
19	0.575	0.547	0.462	0.412
20	0.562	0.535	0.450	0.401

$$3 \leq n \leq 7 \quad Q = |(x_n - x_{n-1})/(x_n - x_1)|$$
$$8 \leq n \leq 10 \quad Q = |(x_n - x_{n-1})/(x_n - x_2)|$$
$$11 \leq n \leq 13 \quad Q = |(x_n - x_{n-2})/(x_n - x_2)| \tag{6.2}$$
$$14 \leq n \leq 25 \quad Q = |(x_n - x_{n-2})/(x_n - x_3)|$$

However, n is rarely larger than 7, and Equation (6.1) is commonly used.

6.3   The rule of huge error

This test is still of a statistical nature but does not use tabulated values, and uses rather a kind of rule of thumb concerning the deviation of the suspected outlier from the mean. Calling the value of the suspected outlier x_0, the mean and the standard deviation of all the measurements by $\bar{x}$ and s, respectively, the size of the error is defined as:

$$E = \frac{|x_0 - \bar{x}|}{s}$$

Thus the error is the multiple of the standard deviation required in order to get the deviation of the suspected outlier. The size of the multiple required to decide whether x_0 is an outlier or not depends on the significance level (in other words the risk) required.

The size of E required for rejecting x_0 depends on the significance level and on the size of the sample, as the size of the sample influences how well we know the standard deviation. For a 2% level of significance, x_0 will be rejected for:

$$E > 6 \quad (5 < n \leq 8) \qquad E > 5 \quad (8 < n \leq 14) \qquad E > 4 \quad (n > 15)$$

6.4   Grubbs test for outliers

Grubbs (Grubbs 1950, 1969; Grubbs & Beck 1972) suggested a similar test to that of Dixon using a quotient for the statistics. In this test instead of the distance between the 'suspected' outlier and the adjacent value, Grubbs used the distance between the 'suspected' outlier and the mean value. Instead of using the range of the values in the denominator, he used the standard deviation. Taylor (1987) gave the following procedure for performing the Grubbs test for outliers.

Step 1: Rank the data in either ascending or descending order such that x_n is the outlier, i.e. if the outlier is the largest then $x_n > x_{n-1} > \ldots > x_1$, but if it is the smallest then $x_n < x_{n-1} < \ldots < x_1$.

Step 2: Calculate the mean and the standard deviation of all the data points.

Step 3: Calculate the statistic T from the equation:

$$T = |x_n - \bar{x}|/s \tag{6.3}$$

Step 4: Compare the calculated T with the critical values given in Table 6.2. For the example in the previous test, $\bar{x} = 5.915$ and $s = 0.13$. Hence:

Table 6.2 Values for Grubbs T-test for outliers.

Number of data points, n	Risk of false rejection (%)				
	0.1	0.5	1	5	10
3	1.155	1.155	1.155	1.153	1.148
4	1.496	1.496	1.492	1.463	1.425
5	1.780	1.764	1.749	1.672	1.602
6	2.011	1.973	1.944	1.822	1.729
7	2.201	2.139	2.097	1.938	1.828
8	2.358	2.274	2.221	2.032	1.909
9	2.492	2.387	2.323	2.110	1.977
10	2.606	2.482	2.410	2.176	2.036
15	2.997	2.806	2.705	2.409	2.247
20	3.230	3.001	2.884	2.557	2.385
25	3.389	3.135	3.009	2.663	2.486
50	3.789	3.483	3.336	2.956	2.768
100	4.084	3.754	3.600	3.207	3.017

$$T = \frac{6.52 - 5.915}{0.313} = 1.933$$

For $p = 0.05$, this value is larger than the critical value (1.822) in Table 6.2 and hence it should be rejected.

6.5 Youden test for outlying laboratories

Youden (1982) developed a statistical test applicable to a group of laboratories analyzing the same samples, as for example for the purpose of certifying the contents of standard reference materials, in order to reject the data of laboratories that are outliers. Each laboratory gets several different samples to analyze. For each sample the laboratories are ranked such that the one obtaining the highest value is ranked 1, and the second highest 2, etc. At the end, the different rankings of each laboratory are summed up. The expected ranges for the cumulative scores, for a random distribution at 0.05 level of significance (95% confidence), are given in Table 6.3. Any laboratory that is outside of this range is treated as an outlier.

6.5.1 Example

Let us look at the case of seven laboratories each analyzing five different samples of ores for their copper content (in %) with the results shown in Table 6.4.

Table 6.3 shows that with a 95% confidence limit the range is expected to be within 8–32. Accordingly laboratory C (Table 6.4) is considered to provide higher results and laboratory G lower results than the other laboratories. In calculating the certified value of reference standard materials the results of these laboratories will be rejected, at least for copper.

Table 6.3 Values for Youden's range to identify outlying laboratories (5% two-tailed limits).

Number of Participants	3	4	5	6	7	8	9	10	11	12	13	14	15
3		4	5	7	8	10	12	13	15	17	19	20	32
		12	15	17	20	22	24	27	29	31	33	36	38
4		4	6	8	10	12	14	16	18	20	22	24	26
		16	19	22	25	28	31	34	37	40	43	46	49
5		5	7	9	11	13	16	18	21	23	26	28	31
		19	23	27	31	35	38	42	45	49	52	56	59
6	3	5	7	10	12	15	18	21	23	26	29	32	35
	18	23	28	32	37	41	45	49	54	58	62	66	70
7	3	5	8	11	14	17	20	23	26	29	32	36	39
	21	27	32	37	42	47	52	57	62	67	72	76	81
8	3	6	9	12	15	18	22	25	29	32	36	39	43
	24	30	36	42	48	54	59	65	70	76	81	87	92
9	3	6	9	13	16	20	24	27	31	35	39	43	47
	27	34	41	47	54	60	66	73	79	85	91	97	103
10	4	7	10	14	17	21	26	30	34	38	43	47	51
	29	37	45	52	60	67	73	80	87	94	100	107	114
11	4	7	11	15	19	23	27	32	36	41	46	51	55
	32	41	49	57	65	73	81	88	96	103	110	117	125
12	4	7	11	15	20	24	29	34	39	44	49	54	59
	35	45	54	63	71	80	88	96	104	112	120	128	136
13	4	8	12	16	21	26	31	36	42	47	52	58	63
	38	48	58	68	77	86	95	104	112	121	130	138	147
14	4	8	12	17	22	27	33	38	44	50	56	61	67
	41	52	63	73	83	93	102	112	121	130	139	149	158
15	4	8	13	18	23	29	35	41	47	53	59	65	71
	44	56	67	78	89	99	109	119	129	139	149	159	169

Table 6.4 Application of Youden's test to seven laboratories analyzing the copper content of five different ore samples.

Material → Laboratory ↓	Measured results 1	2	3	4	5	Score 1	2	3	4	5	Sum of scores
A	9.8	15.8	21.3	18.6	24.9	4	6	3	4	5	22
B	10.0	16.1	20.5	18.7	25.3	2	4	7	3	1	17
C	10.2	17.2	29.7	18.8	25.2	1	1	1	2	2	7
D	9.7	15.9	21.5	18.9	25.0	5	5	2	1	4	17
E	9.6	16.8	20.7	18.3	25.1	6	2	5	5	3	21
F	9.9	16.3	20.8	18.1	24.8	3	3	4	7	6	23
G	9.3	15.4	20.6	18.2	24.0	7	7	6	6	7	33

6.6   References

Barnet, V. & Lewis, T. 1994, *Outliers in Statistical Data*, 3rd edn, John Wiley, Chichester, UK.
Dixon, W. J. 1950, *Ann. Math. Statist.*, vol. 21, p. 488.
Dixon, W. J. 1953, *Biometrics*, vol. 9, p. 74.

Grubbs, F. E. 1950, *Ann. Math. Statist.*, vol. 21, p. 27.

Grubbs, F. E. 1969, *Technometries*, vol. 11, p. 1.

Grubbs, F. E. & Beck, G. 1972, *Technometries*, vol. 14, p. 847.

Likes, J. 1996, *Metrica*, vol. 11, p. 46.

Taylor, J. K. 1987, *Quality Assurance of Chemical Measurements*, Lewis Publishers (CRC Press), Boca Raton, USA, p. 36.

Youden, W. J . 1982, *Journal Qual. Technol.*, vol. 4, p. 1.

7 Instrumental calibration – regression analysis

7.1 Errors in instrumental analysis vs. classical 'wet chemistry' methods

Many instrumental methods require also wet chemistry, since methods like atomic absorption spectrometry or plasma emission spectrometry require the measurement of solutions. Instrumental methods differ from the classical methods in calibration. The classical 'wet chemistry' methods like gravimetry and titration are absolute ones and completely linear, and hence do not require the use of standards and calibration curves. However, instrumental analysis has many advantages over the classical methods, such as sensitivity, versatility, price, large throughput, multi-element analysis and computer interfacing. Very few of the instrumental methods are absolute and usually the instrument signals, which are used for calculation of the amount of the analyte in the analyzed sample, depend on several of the instrument's parameters. The quantitative determination of the analyzed sample is achieved by comparison with standards. In most cases more than one concentration of standards is used and a calibration curve is plotted, a curve which gives the concentration vs. the electronic signal measured by the instrument. These calibration curves also provide the certainty (confidence limits) of our measurement and the range of concentrations for which the method can be applied accurately.

7.2 Standards for calibration curves

Standards can be purchased commercially or can be prepared by the analyst. There are commercially available standards for single elements or multi-element standards in various concentrations. However, an analyst will often need to prepare standards of his own or some part of them.

Great care should be taken in the preparation of standards used for calibration purposes. Only chemicals with exactly known stoichiometry should be used, and it is crucial that the amount of water for hydration is known accurately. It is preferable to use chemicals without any water of hydration. The standard should be dried, at sufficiently high temperature, until a constant weight is reached. A standard solution is prepared by weighing a known mass of this dried chemical into a known volume. Care should be taken that other standard solutions of lower concentration are not prepared by dilution of a principal standard solution, as an error in the preparation of the first solution will propagate to all other standard solutions. It is recommended that 3–4 different solutions are prepared by weighing and that these solutions are diluted. Since weighing is usually more accurate than pipetting, it is recommended that weighing is used whenever possible. If a standard solution is prepared by dilution of another standard solution it is recommended that a burette and at least 25 ml of the original solution are used. Another possible way is to weigh the transferred solution, however in this case the density of the original solution should be measured. If two chemicals with exactly known stoichiometry are available for the same element it is recommended that a standard of each one of them is used, thus verifying that both stoichiometries are actually known. *It should always be remembered that the accuracy of results cannot be better than the accuracy of the standards.* If using pure chemicals for standards, then we are preparing *primary standards*. However, for many analytical methods the signal depends also on the matrix. For this case we must use 'certified standards', which are mixtures that were analysed by several methods in several laboratories and for which the matrix is similar to our sample.

7.3 Derivation of an equation for calibration curves

Calibration curves and or functions are obtained by measuring several standard solutions or samples with varying elemental concentrations with an instrument and thus obtaining a set of data points each describing the concentration of the element (x_i) and the corresponding instrument signal (y_i). We can plot the data on millimetric paper and connect the points either by a broken line, or a line of best fit, or a curve by using a curved ruler, and then calculate the concentration of the unknown sample (the x-axis) from the instrument's signal (the y-axis), using interpolation from the plot. However, this method takes longer in routine work and may also lead to large and subjective errors depending on the analyst.

The best and most objective way to obtain a calibration curve is to use a set of coupled values to calculate a function $y(x)$ and obtain from it the concentrations of the analyzed samples according to their signal, y. Many analysts today are not bothered how this function is calculated, since in contrast to even 10–20 years ago, when the calibration function was calculated by the analyst, instruments today, all of them PC-based, are supplied with programs to calculate the calibration function. If not calculated by the instrument PC, the function is usually derived by a spreadsheet program (e.g. Quattro Pro or Excel). Several scientific calculators also have the ability to calculate this function, assuming it is linear. To understand how this is done on the computer, the principles are given here.

Finding the best functional equation which fits the experimental data is referred to as *regression*. Most calibration curves are linear and hence the search for the best linear function is called *linear regression*. How do we choose the best line? The basic assumption is that the values of x are exact (although actually this is not true, since there will be at least one random error in the volume in the standard solution taken for calibration), while the values of y are subject to errors usually assumed to have a normal distribution. In linear regression we are interested in finding a function $y = ax + b$ which best fits our data. What is the criterion for best fit? We cannot request that the sum of deviations between our observed values (y_{obs}) and our calculated values from the fitted equation $(y_{cal} = ax + b)$ will be a minimum [minimal value of $\sum_{i=1}^{n} y_i - (ax_i + b)$, where n is the number of observations (x_i, y_i) used to calculate the calibration equation], since in this case negative deviations will cancel positive deviations and minimal values of the sum do not warrant small deviations. In order to cancel out the effect of sign, instead of summing the deviations we can sum either the absolute values of the deviations or their squares. These deviations are referred to as residuals so we have two alternatives: either to look for the minimum sum of the absolute residuals (SAR) or for the sum of the squares of residuals (SSR). Which of the two alternatives is the better one? Without referring to an optimum quality the prevailing use is of the minimum SSR. The disadvantage of SSR, however, is that it has a larger influence from larger deviations, and many larger deviations occur at the limits of the instrument and hence will have larger associated errors. The main advantage of the SSR is that the parameters leading to the minimum values can be calculated by simple calculus methods. Simple equations for calculating a and b can be obtained in this way. This method of minimizing the SSR is called the *least squares method*:

$$\text{SSR} = \sum_{i=1}^{n}(ax_i + b - y_i)^2 \tag{7.1}$$

Differentiating SSR with respect to a and b and equating the derivatives to zero leads to the following equations where the summation is done for the n calibration points:

$$\sum x_i[y_i - (ax_i + b)] = 0 \quad \Rightarrow \quad a\sum x_i^2 + b\sum x_i = \sum x_i y_i$$
$$\sum y_i - (ax_i + b) = 0 \quad \Rightarrow \quad a\sum x_i + b\,n = \sum y_i \tag{7.2}$$

where n is the number of experimental pairs of values measured for the calibration. Solving these two equations for a and b yields:

$$\left. \begin{aligned} a &= \frac{n\sum x_i y_i - \sum x_i \sum y_i}{n\sum x_i^2 - (\sum x_i)^2} \\ b &= \frac{n\sum x_i^2 \sum y_i - \sum x_i \sum x_i y_i}{n\sum x_i^2 - (\sum x_i)^2} \end{aligned} \right\} \tag{7.3}$$

The equations for a and b can be written in another form by using $\bar{x}$ and $\bar{y}$:

$$a = \frac{\sum (x_i - \bar{x})(y_i - \bar{y})}{\sum (x_i - \bar{x})^2} \qquad b = \bar{y} - a\bar{x} \tag{7.4}$$

Using a calculator it is usually simpler to use Equation (7.3) since all scientific calculators give $\sum x_i$ and $\sum x_i^2$.

Assuming that the error for each y_i is the same, the variance of the calculated y is given by the minimal SSR divided by the number of degrees of freedom. Since n data points were used to calculate the two constants a and b, the number of degrees of freedom (independent variables) is $(n - 2)$, thus:

$$s_0^2 = \frac{\sum_{i=1}^{n}(y_{\text{obs}.i} - y_{\text{cal}.i})^2}{(n - 2)} \quad \Rightarrow \quad s_0^2 = \frac{\sum_{i=1}^{n}(y_i - ax_i - b)^2}{(n - 2)} \tag{7.5}$$

where a and b are those calculated by Equations (7.3) or (7.4). From the total variance s_0^2 given in equation (7.5), we can calculate the variance of the coefficients a and b.

Ignoring systematic errors and assuming only statistical errors in the determination of the signals in the data points, y_i, we can calculate the standard deviations of the least squares fitting parameters by referring to the general formula obtained in the section on the propagation of errors (Section 2.3). Thus if z is a function of y, then the standard deviation of z due to the statistical errors of the measurements y_i is:

$$\sigma_z^2 = \sum_j \sigma_{y_j}^2 \left(\frac{\partial z}{\partial y_j}\right)^2 \tag{7.6}$$

Assuming that all σ_{y_i} are equal, independent of i, $\sigma_{y_i} = \sigma_y = s_0$ and the last equation can be used to calculate σ_a^2 and σ_b^2 by calculating the derivatives of Equation (7.3):

$$\frac{\partial a}{\partial y_j} = \frac{n x_j - \sum x_i}{n \sum x_i^2 - \left(\sum x_i\right)^2} \qquad \frac{\partial b}{\partial y_j} = \frac{\sum x_i^2 - x_j \sum x_i^2}{n \sum x_i^2 - \left(\sum x_i\right)^2}$$

The sums are only taken over the subscripts i. The subscript j is another dummy subscript to be used in the substitution of the derivatives in Equation (7.6). Thus, for example, for the parameter a:

$$\sigma_a^2 = s_0^2 \sum_j \left[\frac{n x_j - \sum x_i^2}{n \sum x_i^2 - \left(\sum x_i\right)^2}\right]^2$$

$$\sigma_a^2 = \frac{s_0^2}{\left[n \sum x_i^2 - \left(\sum x_i\right)^2\right]^2} \left[n^2 \sum_j x_j^2 - 2n \sum_j x_j \sum_i x_i + \sum_j \left(\sum x_i\right)^2\right]$$

Summing the same constant value $\left[(\Sigma_i x_i)^2\right]$ n times, $\Sigma_{j=1}^n$, means multiplying this constant value by n. There is no difference if the dummy subscript is i or j and hence:

$$\sigma_a^2 = \frac{s_0^2}{\left[n \sum x_i^2 - \left(\sum x_i\right)^2\right]^2} \left[n^2 \sum_i x_i^2 - 2n \sum_i x_i \sum_i x_i + n \left(\sum_i x_i\right)^2\right]$$

$$\sigma_a^2 = s_0^2 \frac{n^2 \sum x_i^2 - n\left(\sum x_i\right)^2}{\left[n \sum x_i^2 - \left(\sum x_i\right)^2\right]^2} = \frac{n\, s_0^2}{n \sum x_i^2 - \left(\sum x_i\right)^2}$$

And similarly for σ_b^2:

$$\left.\begin{aligned}
s_a^2 &= \frac{n\, s_0^2}{n \sum x_i^2 - \left(\sum x_i\right)^2} = \frac{s_0^2}{\sum (x_i - \bar{x})^2} = \frac{s_0^2}{\sum x_i^2 - n(\bar{x})^2} \\
s_b^2 &= \frac{s_a^2}{n} \sum x_i^2
\end{aligned}\right\} \tag{7.7}$$

The *confidence limits of the regression parameters* are calculated as for the measurement of the mean by the student's t-statistic (Table 5.2) with the appropriate level of significance (probabilility level) using $(n-2)$ degrees of freedom. Thus the confidence limits of the parameters are:

$$a \pm ts_a \quad \text{and} \quad b \pm ts_b \tag{7.8}$$

In the calibration curve,

instrument's signal $= a \times$ concentration $+ b$

The parameter b is called the 'blank' of the method as this is the reading obtained for zero concentration. The parameter a is called the sensitivity of the method. As we will see later the parameter a influences the error of the determination, mainly for large concentrations. On the other hand, b influences mainly the limit of determination and the errors at lower concentrations. Ideally b should be zero, but in many applications there is a non-zero signal even for a zero concentration, which is due to impurities in the standards of supposedly pure matrix or owing to instrumental reasons.

7.4   Least squares as a maximum likelihood estimator

In the previous section we justified the use of the minimization of SSR rather than SAR due to the mathematical simplification of the procedure, since partial derivation with respect to the adjustable parameters and equating the derivatives to zero lead to explicit mathematical formulae for the parameters. However, if we assume that y_i has a measurement error that is independently random and is distributed as a normal distribution with the same standard deviation around the true $y(x)$ for all the points, we can show that minimization of the SSR is equal to the maximum likelihood estimation. The maximum likelihood estimation is a procedure in which we are looking for a set of parameters which will lead to a maximum of the probability that with these parameters our data (plus or minus some fixed Δy on the each point) could have occurred. The probability that our data could have occurred is termed 'likelihood'.

Since we assume that each data point has independent random error, the probability of our whole data set is the product of the probability of each point. Since a normal distribution is assumed the probability for the point y_i to be in the range $y(x_i) \pm \Delta y$ is given by:

$$p_i = \exp\left[-\frac{1}{2}\left(\frac{y_i - y(x_i)}{\sigma}\right)^2\right] \Delta y \tag{7.9}$$

The probability for the whole set is:

$$p = p_1\, p_2 \cdots p_n = \prod_{i=1}^{n} p_i = \prod_{i=1}^{n}\left\{\exp\left[-\frac{1}{2}\left(\frac{y_i - y(x_i)}{\sigma}\right)^2\right]\Delta y\right\}$$

$$p = \prod_{i=1}^{n} \exp\left\{-\frac{1}{2\sigma^2}[y_i - y(x_i)]^2\right\}\Delta y \tag{7.10}$$

Here $\prod_{i=1}^{n}$ is the common mathematical sign for the product of n terms (similar to the sigma sign $\sum_{i=1}^{n}$ which stands for a sum of n terms). Finding a maximum of a product is difficult because the derivative includes a sum of products. On the other hand finding the derivative of a sum is much simpler. Since the logarithm function for a base larger than unity is a monotonous increasing function, instead of maximizing p in Equation (7.10) we can maximize $\log p$ or minimize $(-\log p)$. Taking the logarithm of Equation (7.10) changes the product sign to a sum sign:

$$(-\log p) = \frac{1}{2\sigma^2}\sum_{i=1}^{n}[y_i - y(x_i)]^2 - n\log\Delta y \tag{7.11}$$

Since σ, n and Δy are constant, minimizing $(-\log p)$ is identical to minimizing the $\mathrm{SSR} = \sum_{i=1}^{n}[y_i - y(x_i)]^2$.

It should be emphasized that this treatment of the maximum likelihood estimator does not depend on the model (the function) we want to use for the calibration curve. For any function we will use for the calibration curve the minimization of SSR, i.e. least squares, which will lead to the maximization of the likelihood.

In this derivation we use the assumptions: (1) the measurement errors are independent; (2) they are normally distributed; (3) all the normal distributions have the same standard deviation. The cases where normal approximation is a bad approximation or cases where outliers are important are dealt with in a different statistics named robust statistics (not covered in this book). The case where assumption (3) above is not valid, i.e. when each data point (x_i, y_i) has its own standard deviation, σ_i is considered here. In this case the total probability is given by:

$$p = \prod_{i=1}^{n}\exp\left\{-\frac{1}{2\sigma_i^2}[y_i - y(x_i)]^2\right\}\Delta y \tag{7.12}$$

In Equation (7.11), σ^2 cannot be taken out of the sum sign:

$$(-\log p) = \frac{1}{2}\sum_{i=1}^{n}\left[\frac{y_i - y(x_i)}{\sigma_i}\right]^2 - n\log\Delta y \tag{7.13}$$

The maximum likelihood estimate of the model parameters will lead to minimization of the sum in Equation (7.13). The sum is called 'chi-square':

$$\chi^2 \equiv \sum_{i=1}^{n}\left[\frac{y_i - y(x_i)}{\sigma_i}\right]^2 \tag{7.14}$$

y_i is termed the *observed* y and $y(x_i)$ is the *calculated* y, and hence:

$$\chi^2 \equiv \sum_{i=1}^{n}\left(\frac{y_{\mathrm{obs},\,i} - y_{\mathrm{cal},\,i}}{\sigma_i}\right)^2 \tag{7.15}$$

The probability distribution for different values of χ^2 at its minimum can be shown to be the chi-square distribution for $(n - m)$ degrees of freedom, where m is the number of adjustable parameters in the model (function).

7.5   Tests for linearity

In the previous section we assumed that the calibration curve is linear and calculated its parameters, but is this assumption correct and how do we check it? The most common method for testing the linear correlation is by the statistic R^2, called the product-moment linear correlation coefficient. Often it is named just the *correlation coefficient* since this is the most used, although there are other types of correlation coefficients. The correlation coefficient is given by:

$$R^2 = \frac{[\sum(x_i - \bar{x})(y_i - \bar{y})]^2}{\sum(x_i - \bar{x})^2 \sum(y_i - \bar{y})^2} = \frac{[n\sum x_i y_i - \sum x_i \sum \bar{y}]^2}{[n\sum x_i^2 \, (\sum x_i)^2] \, [n\sum y_i^2 \, (\sum y_i)^2]} \qquad (7.16)$$

For absolute linearity, R^2 is equal to 1. The smaller R^2 is the larger the deviation from linearity. However, quite large R^2 are also obtained for non-linear curves, e.g. with one side of a parabola, and it is advisable to check the obtained linearity by observation of the calculated line versus the observed data points. Testing the significance of R can be done using the t-test, where t is calculated from Equation (7.17):

$$t = \left[\frac{R^2(n-2)}{\sqrt{(1-R^2)}}\right]^2 \qquad (7.17)$$

The calculated value of t is compared with the tabulated value (Table 5.2, a two-tailed test) with $(n-2)$ degrees of freedom. The null hypothesis in this case is that there is no correlation between x and y. If the calculated value of t is larger than the critical tabulated value then the null hypothesis is rejected and we conclude that within this level of confidence there is a correlation.

Another test for linearity is the F-test. In this test we compare the variance of the linear regression s_0^2 with the standard deviation for a single measurement, obtained by repeating measurements of the same standard solution and calculating the single measurement variance s_s^2. F is then calculated from the ratio $F = s_0^2/s_s^2$. If the calculated F value is larger than the tabulated critical value (Table 5.3) then the null hypothesis of a linear fit is rejected. If F is less than the tabulated value we retain the hypothesis of the linear correlation. But which of the single points of the calibration line do we repeat? It is best to repeat several of them (or even all of them) and pool together the standard deviations of all of them. The pooling together of the standard deviations is done by their weighted arithmetic average, when the weights are their degrees of freedom, i.e. the number of repeated measurements minus one:

$$s_{pooled}^2 = \frac{\sum_{i=1}^{n} v_i s_i^2}{\sum_{i=1}^{n} v_i} \qquad (7.18)$$

where v_i and s_i are for each calibration point that was repeated more than once and n is the number of calibration points (different concentrations). Equation (7.18) is the same as previously given in Chapter 5.

A further test of linearity is the comparison of the least squares linear regression parameter of dependence on x with its standard deviation. The smaller the ratio s_a/a, the better is the linearity fit.

All curve-fitting and spreadsheet programs calculate both the parameters, their standard deviations and R^2, and actually nowadays there is no need to calculate these values ourselves. The equations are given only so that we will understand the meaning of the program's output.

7.6   Calculation of the concentration

Once we have the calibration equation, $y = ax + b$, which actually stands for the equation:

$$\text{concentration} = (1/a) \times (\text{instrument's signal} - b) \tag{7.19}$$

then it is very simple to calculate the concentration from the instrument's signal. The calculation becomes more complicated when we want to state the confidence limit (interval) of the concentration. Remember that there are confidence limits for a and for b and those uncertainties in their values are reflected by uncertainty in the concentration. In many cases an approximate equation is used for the calculation of the standard deviation of the calculated concentration from Equation (7.19). If the experimental signal for the unknown solution is y_u and the linearization parameter a is calculated from n points (x_i, y_i), then the variance of the concentration, s_c^2 (the square of the standard deviation), is given by:

$$s_c^2 = \frac{s_0^2}{a^2}\left[1 + \frac{1}{n} + \frac{(y_u - \bar{y})^2}{a^2 \sum_i (x_i - \bar{x})^2}\right] \tag{7.20}$$

Instead of using $\sum (x_i - \bar{x})^2$ we can substitute it with $\sum x_i^2 - n(\bar{x})^2$.

If m readings were done on the unknown solution and y_u is their mean, then the first term in the brackets is changed from 1 to $(1/m)$:

$$s_c^2 = \frac{s_0^2}{a^2}\left[\frac{1}{m} + \frac{1}{n} + \frac{(y_u - \bar{y})^2}{a^2 \sum_i (x_i - \bar{x})^2}\right] \tag{7.21}$$

If the concentration obtained by substituting y_u in Equation (7.19) is c, then the confidence limits of concentration are $c \pm ts_c$.

Another way of calculating the confidence limits for the concentration is by taking the confidence limits for the parameters using Equation (7.8). Treating a and b independently we have four possibilities of extreme values of concentrations, i.e. using firstly $a + ts_a$ and $b + ts_b$ in Equation (7.19) and then changing to $a + ts_a$ and $b - ts_b$, and the same with $a - ts_a$. Calculation of these four extreme concentrations gives us the confidence limit around c, calculated from a and b. The disadvantage of the previous method is that it does not show how the confidence limit is changed if the signal for the unknown solution is read more than once. In this case for each reading the confidence limit is calculated and then the extreme values are averaged and the averaged confidence limit is divided by $\sqrt{m}$. However, this method requires many calculations, although they could be easily added to the least squares program or written separately with a, b, and s_b as the input values of the program.

The confidence limits can be narrowed (i.e. improved) by increasing either n, the number of points on the calibration line, or m, the number of repeating experiments on the unknown sample, or both.

7.7   Weighted least squares linear regression

We mentioned before that occasionally the points at the end of the calibration range are more erroneous than the others since we are tending to the limits of the instrument's capabilities. These points influence the fit more than the others since their deviations are larger. In order to overcome the disadvantage that the usual least squares linear regression gives similar importance to all points, both more accurate and less accurate, a weighted regression line can be calculated. In this regression method each point receives a weighting that depends on its accuracy. How do we assess the accuracy, or relative accuracy of each point? The common method to give a weight to a point is to measure each concentration several times, and then the weight is taken as inversely proportional to the variance of the various measurements of y for this x:

$$w_i = \frac{1}{s_i^2} \tag{7.22}$$

The weighted least squares linear regression minimizes the sum $\sum_i^n w_i(y_{obs,\,i} - y_{cal,\,i})^2$. This is the same equation we found before for the maximum likelihood estimator in the case of different standard deviations for each point (Equation (7.15)). This sum is usually named χ^2.

The parameters a and b are calculated from an equation similar to Equation (7.3) where each sum in Equation (7.3) has w_i also, and n is replaced by $\sum w_i$:

$$\left.\begin{aligned} a &= \frac{\sum w_i \sum w_i x_i y_i \sum w_i x_i \sum w_i y_i}{\sum w_i \sum w_i^2 - (\sum w_i x_i)} \\ b &= \frac{\sum w_i x_i^2 \sum w_i y_i - \sum w_i x_i \sum w_i x_i \sum w_i x_i y_i}{\sum w_i \sum w_i x_i^2 - (\sum w_i x_i)} \end{aligned}\right\} \tag{7.23}$$

Some analysts prefer to normalize the weights so that their sum is equal to n, such that in Equation (7.23), $\sum w_i$ is replaced by n, which is similar to Equation (7.3). In this instance w_i is defined by:

$$w_i = \frac{n}{s_i^2} \Big/ \sum_i \frac{1}{s_i^2} \tag{7.24}$$

7.8   Polynomial calibration equations

Several analytical methods give non-linear calibration plots, or they are linear only at limited range and at higher concentrations they become curved. Sometimes calibration curves seem to be linear over a large range with reasonable correlation coefficients, however different parameters of the calibration line are obtained for different range of concentrations (Alfassi et al. 2003), indicating that the calibration curve is not truly linear. In this case it is important to have a calibration curve in a narrow concentration range that extends only slightly below and above that of the concentration of the analyzed sample. Obtaining the same calibration parameters for low range and high range of concentration indicates a real linearity of the calibration curve. In some methods the calibration curves are close to linear but in other cases they are very far from linear as we will see in the next section. Usually the curves are curved downward, reducing the accuracy of the method in the higher concentration range. Thus, if possible, it is recommended to dilute the analyzed solution to the linear range.

However, there are cases where accuracy is sacrificed for shorter time of analysis, prohibiting any additional step of dilution. So, we would also like to fit our data to non-linear functions. There are computer programs that can fit the data to any function we choose. Most curved calibration lines are fitted to a quadratic equation $y = ax^2 + bx + c$ or to a third-power (cubic) equation $y = ax^3 + bx^2 + cx + d$. Fits to polynomial functions are present in most common spreadsheets and function plots such as Excel, Origin, Sigma Plot and Table Curve. The principle of least squares fitting to these polynomials is the same as for the linear function. Thus, for a quadratic regression, we want to minimize the SSR:

$$\text{SSR} = \sum_i [y_i - (ax_i^2 + bx_i + c_i)]^2 \tag{7.25}$$

By differentiating the SSR with respect to a, b and c and equating the partial derivatives to zero, three equations for the unknowns a, b and c are obtained. From these three equations we can derive formulae for the three parameters, similar to Equation (7.3). The general solution for polynomial regression will follow.

It is also possible to perform this polynomial regression with spreadsheets such as Quattro Pro, which does not have a polynomial regression option. Although they normally have only linear regression, such programs can perform multilinear regression where the regression assumes that y depends linearly not only on x but also on z, etc. For example, in bilinear regression the program minimizes the SSR of the function $y = az + bx + c$. However, if in the column of z_i values we insert values of x_i^2 we can perform a least squares analysis of the quadratic equation $y = ax^2 + bx + c$.

As explained in the last section the analysis of a calibration line as a polynomial or as a multi-variable, which means depending on more than one factor, is mathematically the same. The signal y is written as:

$$y = a_0 + a_1 x_1 + a_2 x_2 + \ldots + a_n x_n \tag{7.26}$$

where $x_1, x_2, \ldots, x_n$ can be different variables or different powers of the same variable. The least squares method (or most likely maximization) requires the minimum of the SSR or SWSR (sum of weighted square residuals), which is equal to χ^2 if the weights are re-chosen as $1/\sigma_i^2$:

$$\chi^2 = \sum \frac{1}{\sigma_i^2}(y_{obs,\,i} - y_{cal,\,i})^2 \tag{7.27}$$

$$\chi^2 = \sum \frac{1}{\sigma_i^2}(y_i - a_0 - a_1 x_{1,\,i} - a_2 x_{2,\,i} - \ldots - a_n x_{n,\,i})^2 \tag{7.28}$$

The parameters (coefficients) $a_0, a_1, \ldots, a_n$ which lead to the minimum value of χ^2, can be found by setting the partial derivatives of χ^2 with respect to each of the coefficients equal to zero. This leads to $(n+1)$ equations (remember n is the number of different dependent variables or the order of the polynomial; the number of the data points must be larger than $n+1$) similarly to the two equations of (7.2):

$$\frac{\partial \chi^2}{\partial a_0} = -2 \sum \frac{1}{\sigma_i^2}(y_i - a_0 - a_1 x_{1,\,i} - \ldots - a_n x_{2,\,i}) = 0$$

$$\frac{\partial \chi^2}{\partial a_1} = -2 \sum \frac{x_{1,\,i}}{\sigma_i^2}(y_i - a_0 - a_1 x_{1,\,i} - \ldots - a_n x_{n,\,i}) = 0 \tag{7.29}$$

$$\vdots$$

$$\frac{\partial \chi^2}{\partial a_n} = -2 \sum \frac{x_{n,\,i}}{\sigma_i^2}(y_i - a_0 - a_1 x_{1,\,i} - \ldots - a_n x_{n,\,i}) = 0$$

Rearranging the $(n+1)$ equations of (7.29) yields Equations (7.30). In order to simplify Equations (7.30) we omitted the factor σ_i^2 in each of them. Actually each term in the $\sum$ sign should have a denominator σ_i^2. If all the weighting factors are equal, i.e. σ is a constant independent of i, we do not have to insert it in Equations (7.30).

$$\sum y_i = a_0 \sum 1 + a_1 \sum x_{1,\,i} + a_2 \sum x_{2,\,i} + \ldots + a_n \sum x_{n,\,i}$$

$$\sum x_{1,\,i} y_i = a_0 \sum x_{1,\,i} + a_1 \sum (x_{1,\,i})^2 + a_2 \sum x_{1,\,i} x_{2,\,i} + \cdots + a_n \sum x_{1,\,i} x_{n,\,i}$$

$$\sum x_{2,\,i} y_i = a_0 \sum x_{2,\,i} + a_1 \sum x_{2,\,i} x_{1,\,i} + a_2 \sum (x_{2,\,i})^2 + \cdots + a_n \sum x_{2,\,i} x_{n,\,i} \tag{7.30}$$

$$\vdots$$

$$\sum x_{n,\,i} y_i = a_0 \sum x_{n,\,i} + a_1 \sum x_{n,\,i} x_{1,\,i} + a_2 \sum x_{n,\,i} x_{2,\,i} + \cdots + a_n \sum (x_{ni})^2$$

Notice that $\sum 1 = m$, where m is the number of calibration data points. The simultaneous $(n+1)$ equations for the $(n+1)$ adjustable coefficients (parameters) can be solved by the regular methods for solution of simultaneous linear equations, e.g. Kramer determinants method. The solutions are a ratio of two determinants for each coefficient. The denominator is the same for all coefficients.

$$\text{denominator} = \begin{vmatrix} \sum 1 & \sum x_{1,i} & \sum x_{2,i} & \cdots & \sum x_{n,i} \\ \sum x_{1,i} & \sum (x_{1,i})^2 & \sum x_{1,i}x_{2,i} & \cdots & \sum x_{1,i}x_{n,i} \\ \vdots & & & & \\ \sum x_{n,i} & \sum x_{n,i}\,x_{1,i} & \sum x_{n,i}x_{2,i} & \cdots & \sum (x_{n,1})^2 \end{vmatrix} \qquad (7.31)$$

The numerator is different for each coefficient. For the coefficient a_j the $(j+1)$ column in the $(n+1) \times (n+1)$ determinant of the denominator is replaced by the terms on the left side of Equation (7.30) in order to get the numerator. Thus for a_0 the numerator is:

$$\text{nominator} = \begin{vmatrix} \sum y_i & \sum x_{1,i} & \sum x_{2,i} & \cdots & \sum x_{n,i} \\ \sum x_{1,i}y_i & \sum (x_{1,i})^2 & \sum x_{1,i}x_{2,i} & \cdots & \sum x_{1,i}x_{n,i} \\ \sum x_{2,i}y_i & \sum x_{2,i}x_{1,i} & \sum (x_{2,i})^2 & \cdots & \sum x_{2,i}x_{n,i} \\ \vdots & & & & \\ \sum x_{n,i}y_i & \sum x_{n,i}x_{1,i} & \sum x_{n,i}x_{2,i} & \cdots & \sum (x_{n,i})^2 \end{vmatrix} \qquad (7.32)$$

$$a_0 = \text{numerator}/\text{denominator} \qquad (7.33)$$

We have $(n+2)$ determinants, each having $(n+1) \times (n+1)$ terms, however not all the terms are different. If we will look at Equation (7.31) we will see that $\sum x_{1,i}\,x_{2,i}$ and $\sum x_{1,i}\,x_{n,i}$ appear twice. In the case of polynomial linear regression we have only $(2n+1)$ different elements. In case of multi-parameter linear regression the number of different terms is:

$$\left[2n + 1 + \frac{n(n-1)}{2}\right]$$

Instead of finding the $(n+1)$ coefficients by solving the $(n+1)$ equations, it can be done by matrix inversion. Both procedures are not done manually but using computer programs. The matrix operation is done easily on various programs such as Matlab or Mathematica. Hence the set of Equations (7.30) can be written in matrix formalism:

$$\mathbf{y} = \mathbf{a}\,\mathbf{X} \qquad (7.34)$$

where $\mathbf{y}$ and $\mathbf{a}$ are column vectors $[(n+1) \times 1]$ and $\mathbf{X}$ is a matrix of dimensions $[(n+1) \times (n+1)]$. The elements of the vector $\mathbf{a}$ are the required coefficients. Multiplying Equation (7.34) by the inverse matrix of $\mathbf{X}^{-1}(\mathbf{X}\,\mathbf{X}^{-1} = \mathbf{I}$, the unit matrix) leads to $\mathbf{a} = \mathbf{y}\,\mathbf{X}^{-1}$.

Finding $\mathbf{X}^{-1}$ and the product of $\mathbf{y}\,\mathbf{X}^{-1}$ are simple operations in Matlab or Mathematica. The coefficients are:

$$a_k = \sum_{i=1}^{n+1} y_i\,\mathbf{X}_{k,i}^{-1} \qquad (7.35)$$

where $x_{k,i}^{-1}$ is the (k, i) element of the inverse matrix $\mathbf{X}^{-1}$.

If the uncertainties of the data points are not equal then we have to include the different uncertainties in the terms of Equation (7.30), i.e. each term will include $1/\sigma_i^2$ in the sums.

7.9 Linearization of calibration curves in nuclear measurements

In many cases the calibration curve is neither linear nor polynomial but rather a kind of exponential type. In this case the least squares fitting can be done iteratively as will be seen in the next section, or in some cases the equation can be transformed into a linear one, usually by taking the logarithm of the whole equation. We will see several examples of this type done in nuclear detection and measurement.

7.9.1 *Calibration of detector's efficiency*

The efficiency, ε, of an HPGe detector (i.e. the fraction of the γ-ray photons that produces a full-energy signal in the detector) as a function of the energy of the photons E can be approximated by the equation (Debertin & Helmer 1988; Knoll 1988):

$$\varepsilon = a \times E^b \tag{7.36}$$

where a and b are experimental parameters, characteristic for the specific setup.

The experimental data measured with calibrated standard radioactive sources, ε_i vs. E_i, can be fitted to the calibration curve either by a non-linear regression or by transforming the equation to a linear form in the parameters by using the logarithm of both sides of Equation (7.36).

$$\log \varepsilon = \gamma + b \log E \qquad (\gamma = \log a) \tag{7.37}$$

Here we have a linear equation in the parameters γ and b, and the equations of linear regression can be used. The variables in Equations (7.3) and (7.4) for the parameters are $\log \varepsilon_i$ and $\log E_i$.

In non-linear regression we have to find the values of a and b which will minimize the SSR:

$$\text{SSR} = \sum_i (\varepsilon_i - aE_i^b)^2 \tag{7.38}$$

The values of a and b, which minimize the SSR, cannot be found by substitution in equations as was done for linear regression so the appropriate values must be found by iteration. The most common method for non-linear regression is the one suggested by Marquardt (1963). This method is used in many of the new computer programs that deal with curve plotting and fitting like, for example, Sigma Plot, Origin, Table Curve, and several others. However, most people are used to the linear regression method, especially now that it is packaged with most of the spreadsheet programs such as Excel, Quattro Pro and Lotus. The concept of non-linear regression will be dealt in detail in a later section.

A slightly better approximating equation than Equation (7.38) is (Raman *et al.* 2000):

$$\log \varepsilon = a + b \ \log \ E + c(\log \ E)^2 \tag{7.39}$$

The data can be fitted to this equation by polynomial regression (or by multi-parameter linear regression, as discussed previously). In this instance, the minimization of the SSR is also done by substituting the data in the equations for the parameters and there is no need for iteration.

7.9.2 Calibration of decay curve

The decay of radioactive nuclides, similar to every first order process, proceeds according to the equation:

$$N(t) = N(0) \ e^{-\lambda t} \quad \text{or} \quad A(t) = A(0) \ e^{-\lambda t} \tag{7.40}$$

where N is the number of the radioactive nuclides and A is the activity of the radioactive nuclides [number of disintegration (or counts) per unit time]. The variable t is the time of the measurement and λ is a constant. The zero time can be chosen at any time which fits the measurer. In order to fit Equation (7.40) to a linear plot we have to take the logarithm of both sides:

$$\log \ A(t) = \log \ A(0) - \lambda t \tag{7.41}$$

The plot of $\log \ A(t)$ vs. t will yield a straight line with slope λ (the decay constant). Instead of calculating $\log \ A(t)$ we can plot the raw data (A and t) on semi-logarithmic paper. Semi-logarithmic paper is paper where the x-axis is linear, i.e. the distance between the lines is constant, while on the y-axis the distances between the lines are the differences between their logarithms. Thus on the y-axis the distance between 1 and 2 is equal to the distance between 2 and 4. In the equation in the previous section (7.36), the raw data will yield a straight line only if they are plotted on full-logarithmic paper, paper in which both axes are divided into lines according to their logarithmic values. Nowadays these special drawing papers are little used due to calculation and plotting by computer. Most programs calculate for Equations (7.36) and (7.40) not the non-linear least squares but the least squares after transformation to the linear form of Equations (7.37) and (7.41) respectively.

7.9.3 Fitting to a Gaussian function

Sometimes we need to fit data to a Gaussian function, e.g. the data of spectrum (i.e. counts or optical densities vs. energies or wavelength). This function has the mathematical form:

$$y(x) = y_0 \ \exp\left[-(x - x_0)^2/2\sigma^2\right] \tag{7.42}$$

where $y \ (x)$ is the count (or optical density) at channel (or wavelength) numbered x. There are three adjustable parameters, y_0, x_0 and σ. The last two are not in a linear form. They can be calculated iteratively or transformed into a linear form by the following procedure. Let us define a new function $z(x)$:

$$z(x) \equiv \frac{y(x - 1)}{y(x + 1)} \tag{7.43}$$

Substituting Equation (7.42) in Equation (7.43) yields:

$$z(x) = \exp\left\{\left[(x+1-x_0)^2-(x-1-x_0)^2\right]\bigg/2\sigma^2\right\}$$

and hence:

$$z(x) = \exp\left[2(x-x_0)/\sigma^2\right] \tag{7.44}$$

Taking the logarithm of both sides gives:

$$\ln\,z(x) = \frac{2}{\sigma^2}\,x - \frac{2x_0}{\sigma^2} \tag{7.45}$$

Thus $\ln\,z(x)$ is a linear function of x. The slope is $2/\sigma^2$ and the intercept is $-2x_0/\sigma^2$. Thus:

$$\sigma^2 = \frac{2}{\text{slope}} \quad\text{and}\quad x_0 = -\frac{\text{intercept}}{\text{slope}}$$

i.e. for line $y = ax + b$ then:

$$\sigma = \sqrt{(2/a)} \quad\text{and}\quad x_0 = -b/a \tag{7.46}$$

The parameter y_0 is obtained by substituting the obtained σ and x_0 in Equation (7.42).

7.9.4   Calibration curves for neutron capture radiography

Neutron capture radiography is used to measure instrumentally the concentration of elements which have high cross-section for the (n,α) reaction, e.g. $^{10}B(n,\,\alpha)^7Li$ and $^6Li(n,\alpha)T$. This method uses selectively the emitted α tracks induced by irradiation with thermal neutrons and their registration by an α-sensitive film. This method can measure boron concentration down to $10^{-9}\,g$ in samples of $10^{-3}\,g$ as, for example, in cell-culture experiments (Fairchild et al. 1986). The observed calibration curve of the number of tracks as a function of boron concentration is sub-linear (Alfassi & Probst 1997). The explanation for the sub-linearity is that existing tracks limit the area of the film that is available for new tracks.

For an infinitesimally small number of tracks the rate of formation of tracks dn/dt is proportional to the concentration c of the activable nuclide (^{10}B or ^6Li).

$$\frac{dn}{dt} = \alpha c$$

where α is a proportionality constant dependent on the flux of the neutrons and the cross-section for the formation of the α-particle. For final-size number of tracks the area of unavailable film is proportional to the number of already existing tracks and thus the fraction of available area is $(1 - \beta n)$, where β is a proportionality parameter. Taking these two factors together we obtain:

$$\frac{dn}{dt} = \alpha c(1 - \beta n) \xrightarrow{\text{logarithm}} \ln\,(1 - \beta n) = -\beta\alpha ct$$

$$c = -\frac{1}{\gamma} \ln (1 - \beta n) \qquad (7.47)$$

where $\gamma = \alpha\beta t$ and t is the irradiation time. Measuring n, the number of tracks, we can calculate the concentration, c.

The final calibration equation for c has two unknown parameters that cannot be fitted by a linear regression, but can be fitted by a non-linear regression analysis. One way to overcome the need for non-linear regression is by simple iterations of the linear regression. The linear equation chosen is $c = a_0 + a_1 \ln (1 - \beta n)$. The value of β is changed iteratively in order to obtain $a_0 = 0$. The initial value of $1/\beta$ was selected as slightly above the largest value of the number of tracks. The value of a_0 was found to be positive and $1/\beta$ was increased gradually until a_0 became zero. For larger $1/\beta$ values, a_0 will be negative (Alfassi & Probst 1997).

7.10 Non-linear curve fitting

In Section 7.8 we dealt with calibration functions which are linear in the adjustable parameters, although the functions themselves are not linear (e.g. polynomial). In Section 7.9 we dealt with calibration curves in which the adjustable (unknown) parameter is non-linear but can be transformed to a linear form by mathematical operation. In this section we will deal with the case of least squares of functions involving non-linear parameters (coefficients). In this case the equating of the partial derivatives to zero does not lead to explicit equations for the adjustable parameters, so finding the parameters which minimize the SSR (maximizing the likelihood) must proceed iteratively, i.e. by choosing various adjustable parameters searching for the ones with minimum SSR. Given trial values for the parameters, the different methods give a procedure that improves the trial solution. The procedure is then repeated until the SSR (or χ^2 if we include the uncertainties) decrease is less than a predefined value. The most common method today is the one called Marquardt or Levenberg–Marquardt method. It is based on a paper of Marquardt (1963) who put forth an elegant method related to an earlier suggestion by Levenberg. This method is based on two earlier methods (a) the gradient search method, which is ideally suited for approaching the minimum from trial values that are far away, but does not converge rapidly when in the immediate neighborhood; (b) analytical method of power series expansion. This method converges quite rapidly to the point of minimum SSR (χ^2) from points nearby, but it cannot be relied upon to approach the minimum with any accuracy from further points.

The Marquardt algorithm behaves like a gradient search for the first portion of the search and behaves like an analytical solution as the search converges. Marquhardt showed that path directions for gradient and analytical searches are nearly perpendicular to each other and that the optimum direction is somewhere between. In the following we will describe separately the two methods which are combined in the Levenberg–Marquardt method. We will start with another method which is easier to understand.

One of difficulties of searching for a minimum of SSR $(\text{or } \chi^2)$ is that this function may have more than one local minimum for a reasonable range of the various parameters, a_i, which together were considered as the vector **a**. In order to limit the range of the vector **a**, it is sometimes advantageous to divide the range of the various a_i into relatively large divisions and to map coarsely the value of χ^2 to the parameter space. If each a_i is divided into n_i divisions, it means that the number of computations of χ^2 is $n_1\, n_2\, n_3\, \ldots\, n_k = \prod_{i=1}^{k} n_i$, where k is the number of different parameters. This coarse mapping of the behavior of χ^2 as a function of all the parameters allows us to locate the main minima and identify the desired range of parameters over which to refine the search.

7.10.1   The grid-search method

In this method we minimize χ^2 with respect to each parameter separately. In the first step we use the initial values of all a_i (the vector **a**) and minimize χ^2 with respect to a_1. Using this value of a_i and all other initial values in the second step, we minimize χ^2 with respect to a_2, continuing for all k parameters. After one cycle, the minimization starts again with respect to a_i. The minimization of χ^2 with respect to any a_i is done by incrementing a_i by a step Δa_i and calculating χ^2. The sign of Δa_i is chosen such that χ^2 decreases. Then the parameter a_i is repeatedly incremented by the same amount Δa_i until χ^2 starts to increase. Taking the last three points (where in the last one χ^2 increases) the minimum is calculated assuming that there is a parabolic dependence of χ^2 on the last three values of a_i. The minimum of a_i is given by the equation:

$$a_i(\min) = a_i(3) - \Delta a_i \left[\frac{\chi^2(3) - \chi^2(2)}{\chi^2(3) - 2\chi^2(2) + \chi^2(1)} + \frac{1}{2} \right] \tag{7.48}$$

where $a_i(2) = a_i(1) + \Delta a_i$; $a_i(3) = a_i(2) + \Delta a_i$; $\chi^2(3) > \chi^2(2) \le \chi^2(1)$

This method behaves well if the variation of χ^2 with each a_i is fairly independent. When the variation of χ^2 with the various parameters is not independent, the convergence to the minimum may be very slow.

7.10.2   The gradient-search method

In this method all the parameters a_i are incremented simultaneously with relative magnitudes so that the motion in the parameter space is in the opposite direction of the gradient of χ^2. The gradient of χ^2, which is denoted by $\nabla\chi^2$, is the vector whose components are the partial derivatives of χ^2 with respect to the various a_i. The gradient is given by the equation:

$$\nabla\chi^2 = \sum_{i=1}^{k} \frac{\partial \chi^2}{\partial a_i} \mathbf{a}_i \tag{7.49}$$

where $\mathbf{a}_i$ is the unit vector in the direction of a_i. The gradient is the direction of maximum variation and taking its opposite direction leads to the maximum decrease of χ^2. This is the reason why this method is also named *the method of steepest descent*, since the vector of the parameters is moving along the direction of the steepest descent.

The partial derivatives in Equation (7.49) are approximated by a difference equation:

$$(\nabla \chi^2)_i = \frac{\partial \chi^2}{\partial a_i} \approx \frac{\chi^2(a_i + f\Delta a_i) - \chi^2(a_i)}{f\Delta a_i} \tag{7.50}$$

The variation of the parameter is given as a product $(f\Delta a_i)$, since for the calculation of the derivative we prefer to take a smaller value than the step size. The fraction f is usually not larger than 0.1. The step Δa_i is usually the step size for increments in the grid search. Since a_i must not all have the same dimension it is usual to replace the parameters of vector **a** by the vector **b**, whose components are defined by:

$$b_i = \frac{a_i}{\Delta a_i} \quad \Rightarrow \quad \frac{\partial \chi^2}{\partial b_i} = \frac{\partial \chi^2}{\partial a_i}\Delta a_i \tag{7.51}$$

The gradient vector derived from the vector **b** is normalized to give a dimensionless gradient vector **γ** of magnitude:

$$\gamma_i = \frac{\partial \chi^2 / \partial b_i}{\sqrt{\left[\sum_{j=1}^{n}(\partial \chi^2 / \partial b_j)^2\right]}}$$

Since **γ** is in the direction of increasing of χ^2, the direction which the parameter vector moves is the opposite direction of the vector **γ**. This direction is usually termed the direction of *steepest descent*:

$$\delta a_i = -\gamma_i \, \Delta a_i \tag{7.52}$$

All the parameters are incremented simultaneously. Δa_i in Equation (7.52) has the same magnitude as in Equation (7.51). In order to save calculation time, many programs do not calculate the direction vector **γ** after each step but rather keep the same **γ** while calculating only χ^2 in small steps until the value of χ^2 starts to rise. At this point the direction vector **γ** is recalculated and the search begins again with the new **γ**.

The gradient-search method suffers markedly as the search approaches the minimum. This is due to large errors in approximating the partial derivative by a difference in Equation (7.50).

7.10.3 *Analytic approximation method*

Instead of mapping the χ^2 hyper-surface for variation of the parameters, a simpler way might be to find an approximate analytical function for χ^2 and to use this function to locate the minimum directly. The main advantage of this method is that it produces fewer points for which we have to calculate χ^2. However, this method has the disadvantage that at each point the computations are considerably more complicated. The balance is still less computation time and in addition it has the advantage that the analytical method essentially chooses its own steps. Two different approaches were suggested for analytical approximation. In one method an analytical expression is found for χ^2, while in the other one an analytical approximation is sought for the fitting function itself.

7.10.4 Taylor's series expansion of χ^2

One way to fit an analytical expression for χ^2 is to take a Taylor's series of χ^2 and neglect the high powers (orders). If we take only the first order (neglecting all powers ≥ 2) we can write:

$$\chi^2 = \chi_0^2 + \sum_{i=1}^{k} \frac{\partial \chi_0^2}{\partial a_i} \, \delta a_i \tag{7.53}$$

where χ_0^2 is the value of χ^2 at some starting point and δa_i are the increments in the parameters required to move from the starting point (point zero) to the point where we want to compute χ^2. Differentiating Equation (7.53) with respect to a_j leads to:

$$\frac{\partial \chi^2}{\partial a_j} = \frac{\partial \chi_0^2}{\partial a_j} + \sum_{i=1}^{k} \frac{\partial^2 x_0^2}{\partial a_i \, \partial a_j} \, \delta a_i \qquad j = 1, \ 2, \ 3, \ \ldots, \ k \tag{7.54}$$

Equating the partial derivatives $\partial \chi^2 / \partial a_j$ to zero will result in a set of k linear equations in δa_i, which can be written in a matrix from:

$$\boldsymbol{\beta} = \boldsymbol{\alpha} \cdot \boldsymbol{\delta a} \quad \text{or} \quad \beta_j = \sum_{i=1}^{k} \delta \, a_i \, \alpha_{i,j} \qquad j = 1, 2, \ \ldots, \ n \tag{7.55}$$

where $\boldsymbol{\beta}$ and $\boldsymbol{\delta a}$ are column vectors and $\boldsymbol{\alpha}$ is a $k \times k$ matrix whose elements are equal to the coefficient of δa_i in Equation (7.54), i.e.

$$\alpha_{ij} \equiv \frac{\partial^2 \chi^2}{\partial a_i \, \partial a_j} \quad \text{and} \quad \beta_j \equiv -\frac{\partial^2 \chi_0}{\partial a_j} \tag{7.56}$$

In the treatment up to now we used Taylor's series of the first order. However, a similar solution to the matrix equation (7.55) (or set of simultaneous linear equations) can be obtained also if we use a Taylor's series of the second order for χ^2:

$$\chi^2 = \chi_0^2 + \sum_{i=1}^{k} \frac{\partial x_0^2}{\partial a_i} \, \delta a_i + \frac{1}{2} \sum_{i=i}^{k} \sum_{j=i}^{k} \frac{\partial^2 x_0^2}{\partial a_i \, \partial a_j} \, \delta a_i \, \delta a_j \tag{7.57}$$

The optimum values for the increments δa_i are those for which χ^2 is a minimum. We can obtain these values by equating the partial derivatives with respect to the increments $\partial \chi^2 / \partial \delta a_i$ to zero. Differentiating Equation (7.57) with respect to δa_i and setting the derivative equal to zero leads to:

$$\frac{\partial \chi^2}{\partial \delta a_j} = \frac{\partial \chi_0^2}{\partial a_j} + \sum_{i=1}^{k} \frac{\partial^2 x_0^2}{\partial a_i \, \partial a_j} \, \delta a_i = 0 \qquad j = 1, \ 2, \ \ldots, \ k$$

Equation (7.55) can be solved either by matrix inversion:

$$\boldsymbol{\delta a} = \boldsymbol{\alpha}^{-1} \, \boldsymbol{\beta}$$

or by Cramer's method for the solution of simultaneous linear equations. However, since most common computation tools such as Excel or Matlab can invert matrices it is simpler to solve Equation (7.55) by matrix inversion and multiplication of the inverse matrix by the vector $\boldsymbol{\beta}$. The elements of the matrix $\boldsymbol{\alpha}$ can be approximated by a difference equation in the neighborhood of the starting point:

$$\frac{\partial \chi^2}{\partial a_i} = \frac{\left(\chi_0^2(a_i + \Delta a_i)\right) - \left(\chi_0^2(a_i - \Delta a_i)\right)}{2\Delta a_i}$$

$$\frac{\partial^2 \chi_0^2}{\partial a_i\, \partial a_j} = \frac{\chi_0^2(a_i + \Delta a_i,\ a_j + \Delta a_j) - \chi_0^2(a_i + \Delta a_i,\ a_j) - \chi_0^2(a_i,\ a_j + \Delta a_j) + \chi_0^2(a_i,\ a_j)}{\Delta a_i \Delta a_j}$$

$$\frac{\partial^2 \chi_0^2}{\partial a_i^2} = \frac{\chi_0^2(a_i + \Delta a_i) - 2\chi_0^2(a_i) + \chi_0^2(a_i - \Delta a_i)}{\Delta a_i^2}$$

$$(7.58)$$

The Δa_i step sizes must be small enough to yield convergence near the maximum, but large enough to prevent round-off errors in the computation of the derivatives by Equation (7.58).

7.10.5 Taylor's series expansion of the calibration equation

In this method we treat analytically the calibration equation and minimize χ^2. We want to expand the calibration equation $y(x)$.

Expansion to first order yields:

$$y(x) = y_0(x) + \sum_{i=1}^{n} \frac{\partial y_0(x)}{\partial a_i} \delta a_i \qquad (7.59)$$

while expansion to second order yields:

$$y(x) = y_0(x) + \sum_{i=1}^{n} \frac{\partial y_0(x)}{\partial a_i} \delta a_i + \frac{1}{2}\sum_{i=1}^{n}\sum_{j=1}^{n} \frac{\partial^2 y_0(x)}{\partial a_i \partial a_j} \delta a_i\, \delta a_j$$

The derivatives are calculated as before by difference equations. Minimizing χ^2:

$$\chi^2 = \sum \frac{1}{\sigma_i^2}\left[y_i - y(x_i)^2\right]$$

yields similar equations for δa_i as obtained in the previous method.

7.11 Fitting straight-line data with errors in both coordinates

In all the previous sections we assumed that the x-coordinate values have no errors and only the y_i values have statistical errors. Assuming also errors in the x_i values leads to complicated mathematics in order to find the adjustable parameters. We will look only at the simple case of a straight line calibration curve, i.e:

$$y(x) = a + bx$$

We should remember that in the case of errors only in the y_i values we obtain simple equations for a and b which can be calculated easily. In the case of statistical errors in both x_i and y_i the situation is more complicated and various articles have appeared dealing with this problem mainly in the *American Journal of Physics*, which is the journal of the American Association of Physics Teachers. A summary of the results can be found in the paper of Reed (1992). Some investigators have calculated two lines, one that assumes that only y_i values have errors and another one in which it is assumed that only x_i values are errors and taking the arithmetic mean of the parameters. However, there is no scientific basis to this simplified procedure. An improved method was applied by Adcock (1878), who suggested finding a line such that the sum of the squares of the distances from the observed points to the line (the perpendicular distances as defined in geometry) will be minimized. This line was termed (Pearson 1901) the 'major axis'. This treatment yields explicit equations for the adjustable parameters (Kermack & Haldane 1950).

Assigning $u_i = x_i - \bar{x}$ and $v_i = y_i - \bar{y}$ yields:

$$b = \frac{\sum v_i^2 - \sum u_i^2 + \sqrt{\left[\left(\sum v_i^2 - \sum u_i^2\right)^2 + 4\left(\sum u_i v_i\right)^2\right]}}{2\sum u_i v_i} \tag{7.60}$$

$$a = \bar{y} - b\bar{x} \tag{7.61}$$

where the summation is done over all the observed n points. However, such an analysis yields a line which is not invariant under a change of scale. The invariance can be secured by using different types of 'standard coordinates'. One type of standardization is done by plotting y_i/σ_y and x_i/σ_x, where σ_y and σ_x are the respective standard deviations:

$$\sigma_y^2 = \sum_{i=1}^{n}(y_i - \bar{y})^2/(n-1) \quad \text{and} \quad \sigma_x^2 = \sum_{i=1}^{n}(x_i - \bar{x})^2/(n-1) \tag{7.62}$$

In these coordinates, b is given by the equation (Kermack & Haldane 1950):

$$b = \frac{\sigma_y}{\sigma_x}\left[1 \pm \sqrt{\left(\frac{1-r^2}{n}\right)}\right] \tag{7.63}$$

where r is the Pearson correlation coefficient defined by:

$$r = \sum (x_i - \bar{x})(y_i - \bar{y}) \bigg/ \left[\sum (x_i - \bar{x})^2 \sum (y_i - \bar{y})^2\right]^{0.5} \tag{7.64}$$

The parameter a is given by Equation (7.61). This line is usually termed 'the reduced major axis'. Its main advantage is the easy calculation of the slope from equation (7.63).

Another type of standardization of the coordinates (Teissier 1948) is obtained by plotting y_i/p_y vs. x_i/p_x, where p is the probable error, which is assumed to be constant, but p_y and p_x have different values.

Both modifications suffer from the disadvantage of using the same weights to each point, which will be the experimenter's estimates of the uncertainties in the various x_i and y_i (Worthing & Geffner 1946). Deming (1943) suggested that the *best* straight line is obtained by minimizing the following (weighted) Sum of Square Residuals (SSR):

$$\text{SSR} = \sum_{i=1}^{n} w(x_i)\left(x_{\text{cal},i} - x_{\text{obs},i}\right)^2 + w(y_i)\left(y_{\text{cal},i} - y_{\text{obs},i}\right)^2 \tag{7.65}$$

where the subscripts 'obs' and 'cal' denote the observed and adjusted (calculated) values of x_i and y_i. Deming expanded the SSR as a Taylor's series of the first order. However, neglecting the squared term and higher order terms can cause significant errors. An exact treatment was given by York (1966). Taking a differential of SSR and equating it to zero we get:

$$\delta(\text{SSR}) = \sum_{i=1}^{n} w(x_i)\left(x_{\text{cal},i} - x_{\text{obs},i}\right)\delta x_i + w(y_i)\left(y_{\text{cal},i} - y_{\text{obs},i}\right)\delta y_i = 0 \tag{7.66}$$

The calculated (adjusted) values lie exactly on the straight line and hence:

$$y_{\text{cal},i} = a + bx_{\text{cal},i} \tag{7.67}$$

Differentiating Equation (7.67) leads to:

$$\delta y_i = \delta a + b\,\delta x_i + x_i\,\delta b \quad \text{or} \quad b\,\delta x_i - \delta y_i + \delta a + x_i\,\delta b = 0 \tag{7.68}$$

There are n different equations (7.68), one for each of the n measured points. Multiplying each of them by its own undetermined multiplier λ_i, summing all the n equations and adding them to Equation (7.66) yields:

$$\sum\left\{\delta x_i\left[w(x_i)\left(x_{\text{obs},i} - x_{\text{cal},i}\right) + b\lambda_i\right]\right\} + \sum\left\{\delta y_i\left[w(y_i)\left(y_{\text{obs},i} - y_{\text{cal},i}\right) + \lambda_i\right]\right\}$$
$$+ \delta a \sum \lambda_i + \delta b \sum \lambda_i x_i = 0 \tag{7.69}$$

The coefficient for all the δ terms should be equal to zero and hence:

$$x_{\text{obs},i} - x_{\text{cal},i} = -\lambda_i b / w(x_i) \qquad y_{\text{obs},i} - y_{\text{cal},i} = -\lambda_i / w(y_i) \tag{7.70}$$
$$\sum \lambda_i = 0 \qquad \sum \lambda_i x_i = 0 \tag{7.71}$$

Substituting Equation (7.70) in Equation (7.67) yields:

$$\lambda_i = \left(a + bx_{\text{obs},i} - y_{\text{obs},i}\right) D_i \tag{7.72}$$

$$\text{where} \quad D_i = \frac{w(x_i)\,w(y_i)}{b^2 w(y_i) + w(x_i)} \tag{7.73}$$

There are n equations (7.72) and together with the two equations of (7.71) we have $(n+2)$ equations for a, b and the n various $\lambda_i(\lambda_1, \lambda_2, \ldots, \lambda_n)$. Solving them leads to the following cubic equation in b:

$$b^3 \sum \frac{D_i^2 u_i^2}{w(x_i)} - 2b^2 \sum \frac{D_i^2 u_i v_i}{w(x_i)} - b\left[\sum D_i u_i^2 - \sum \frac{D_i^2 v_i^2}{w(x_i)}\right] + \sum D_i u_i v_i = 0 \tag{7.74}$$

where $u_i = x_{\text{obs},i} - x_{\text{cal},i}$ and $v_i = y_{\text{obs},i} - y_{\text{cal},i}$.

If $w(x_i) = w(y_i) = c = $ a constant, then Equation (7.74) is transformed to a simpler equation:

$$b^2 + b\frac{c\sum D_i u_i^2 - \sum D_i v_i^2}{\sum D_i u_i^2 v_i} - c = 0 \tag{7.75}$$

which leads to the same formula for b found by Deming, for first order Taylor expansion (Deming 1943). If $c = 1$ then the equation for b is similar to Equation (7.60), i.e. the major axis solution, where we do not have different uncertainties.

In the general case, although Equation (7.74) seems to be a cubic equation and hence can be solved, it is not a real cubic equation, as D_i also involve b. However, using an approximate value of b we can solve Equation (7.74) and then substitute the new value of b in D_i and iterate the solution of Equation (7.74). You are reminded that the solution of the equation:

$$b^3 - 3\alpha b^2 + 3\beta b - \gamma = 0 \tag{7.76}$$

yields three real roots:

$$b_i = \alpha + 2\sqrt{(\alpha^2 - \beta)} \; \cos\frac{\phi + 2\pi(i-1)}{3} \qquad i = 1, 2, 3 \tag{7.77}$$

$$\text{where} \quad \phi = \cos^{-1}\frac{2\alpha^3 - 3\alpha\beta + \gamma}{2\sqrt{[(\alpha^2 - \beta)^3]}} \tag{7.78}$$

York (1966) claims that in all the examples considered by him the required root for b is b_3, i.e. $i = 3$ in Equation (7.77). After calculating b, the parameter a is calculated from the equation:

$$a = \frac{\sum D_i \, y_{\text{obs},i}}{\sum D_i} - b\frac{\sum D_i \, x_{\text{obs},i}}{\sum D_i} \tag{7.79}$$

Reed (1989) showed that York's method is exact and better than several attempts to handle this problem by modification of conventional least squares algorithms (Barker & Diana 1974; Orear 1982; Lybanon 1984). However, he does not agree with York's statement that there are three real roots to Equation (7.76). Reed found that in some instances the first few iterations may yield a complex third root. In this case he used only the real part for subsequent iterations. In any case he suggested that instead of solving the cubic equation for initial values of b and solving iteratively the cubic equation, it is simpler to scan $f(b) = b^3 - 3\alpha b^2 + 3^2 + 3\beta b - \gamma$ for various values of α, β and γ and look for the different roots (i.e. $f(b) = 0$). In a subsequent paper (Reed 1992), he showed that the 'cubic' equation for b can be transformed by algebraic manipulation to a 'quadratic' equation. Also in this quadratic equation the coefficients including b and an initial value should be guessed and a correct value is computed by iteration. The variances of the parameter were calculated approximately by York by the equation:

$$\sigma_b^2 = \frac{SSR}{N-2}\left(\sum D_i u_i^2\right)^{-1} \quad \text{and} \quad \sigma_a^2 = \left[\sum D_i x_i^2 \left(\sum D_i\right)^{-1}\right]\sigma_b^2 \tag{7.80}$$

However Reed (1992) gave more complex but also more accurate equations.

7.12 Limit of detection

Almost the first question asked by the analyst concerning an analytical method or an instrument is about its limit of detection, which means the minimum amount that can be detected by the instrument, within a given confidence level that this detection is not a false one. The limit of detection does not reveal what are the confidence limits of the concentration calculated, but just the confidence level indicating that the element is really present in the sample. The value of the limit of detection is not changed by repeated measurements, but the confidence limits of the measured concentration will be narrowed by repetition of the measurement. The limit of detection of the analyte is determined by the limit of detection of the signal. If the blank signal was zero with standard deviation of zero, any signal measured by the instrument would be significant, but actual measurements usually have blank signals larger that zero. The blank signals are not exact numbers and repeated experiments show dispersion. So how do we recognize if a small signal is really due to the presence of the analyte and is not just a blank signal?

Different people use different criteria for this limit of detection. The suggestion made by the National Institute of Standards and Technology (NIST) of the USA is to use the 95% level of confidence. So if the blank signal is distributed normally with standard deviation σ_0, then the limit of detection L_D of the signal is:

$$L_D = 3.29\,\sigma_0 \tag{7.81}$$

This means that only if our signal is $b + 3.29\,\sigma_0$ can we say that we have a true signal of the analyte with false positive and false negative risks, each of 5%. The 3.29 value is taken from $2z_{0.95}$, where z is the value of the probability of the normal distribution. Some claim that we have to use a one-tailed test since we are interested only in the probability that the signal is larger than the blank one. However, NIST's suggestion is to use the value for the two-tailed test, i.e. to include the factor 2. The factor 2 comes from the fact that not only does the blank signal have a normal distribution but so also does our measured signal. If we neglect the factor 2 we will obtain what is called a limit of decision. It will guarantee (with 95% confidence) that we do not claim the presence of the analyte when it is actually absent, but it will not assure us that we will not claim that it is absent when actually it is present. Many people round off the factor and use the factor 3 instead of 3.29. Using the normal distribution table shows that this factor ($z = 1.5$) means uncertainty of 7% of either false positive or false negative answer.

In most cases we do not know the real value of σ_0 and estimate it from the calculation of s, obtained from repeated measurements of the blank signal. In this case the Student's t-distribution should be used rather that the normal one giving:

$$L_D = 2 \times t \times s \tag{7.82}$$

where t is taken for degrees of freedom (df) $= (n - 1)$, where n is the number of repeated measurements. If the blank value is obtained from a least squares linear regression, then σ_0 is equal to $\sqrt{2}$ b, where b is the blank value.

The analyte detection limit, c_D, is related to the signal detection limit by the sensitivity of the measurement, and the parameter which transforms signals to concentration, i.e. the parameter a in the calibration linear equation:

$$c_D = L_D/a \tag{7.83}$$

Limit of determination and *limit of quantitation* (synonyms) are sometimes used for the lower limit of a precise quantitative measurement in contrast to limit of detection which deals with qualitative detection. A value of $10L_D$ was suggested as the limit of determination of the signal and $10L_D/a$ as the limit of determination of the concentration. It should be remembered that the accuracy of the measurement can be improved by repeated measurements of the sample. In any case very few reports include this term and usually only the limit of detection is given. It should always be borne in mind that it is actually almost impossible to measure quantitatively samples with a concentration close to c_D. We can only detect the presence of the analyte, but its real value will be considerably in error (large confidence intervals).

7.13   References

Adcock, A. J. 1878, *Analyst*, vol. 5, p. 53.
Alfassi, Z. B. & Probst, T. U. 1997, *Nucl. Instr. Meth. Phys. Res.*, vol. A428, p. 502.
Alfassi, Z. B., Huie, R. E., Milman B. L., & Neta, P. 2003, *Analytical and Bioanalytical Chemistry*, vol. 377, p. 159.
Barker, D. R., & Diana, L. M. 1974, *Am. J. Phys.*, vol. 42, p. 224.
Debertin, K. & Helmer, P. G. 1988, *Gamma and x-ray Spectrometry with Semiconductor Detectors*, North-Holland, Amsterdam.
Deming, W. E. 1943, *Statistical Adjustment of Data*, John Wiley, New York.
Fairchild, R. G., Gabel, D., Laster, B. H., Grenberg, D., Kiszenic, W., & Mica, P. L. 1986, *Med. Phys.*, vol. 13, p. 50.
Kermack, K. A. & Haldane, J. B. S. 1950, *Biometrika*, vol. 37, p. 30.
Knoll, G. F. 1988, *Radiation Detection and Measurement*, John Wiley, New York, 3rd edn, Chapter 4.
Lybanon, M. 1984, *Am. J. Phys.*, vol. 52, pp. 22 and 276.
Marquardt, D. W. 1963, *J. Soc. Ind. App. Math.*, vol. 11, p. 431.
Orear, J. 1982, *Am. J. Phys.*, vol. 50, p. 912.
Pearson, K. 1901, *Phil. Mag.* vol. 2, p. 559.
Raman, S., Yonezaka, C., Matsue, H., Iimura, H., & Shinohara N., 2000, *Nuc. Instr Meth. Phys. Res.*, vol. A45, p. 389.
Reed, B. C. 1989, *Am. J. Phys.*, vol. 57, p. 642.
Reed, B. C. 1992, *Am. J. Phys.*, vol. 60, p. 59.
Teissier, G. 1948, *Biometrics*, vol. 4, p. 14.
Worthing, A. G. & Geffner, J. 1946, *Treatment of Experimental Data*, John Wiley, New York.
York, D. 1966, *Can. J. Phys.*, vol. 44, p. 1079.

8   Identification of analyte by multi-measurement analysis

In several cases the nature of an analyte cannot be identified by one measurement alone and the identification of the analyte can be done only by the measurement of several values. There follows some examples of cases where the nature of the analyte or its amount (either qualitative or quantitative) can be determined only from several measurements.

(1)   Archeologists know that the source of antique ceramics is not always where it was found (discovered), as many ceramics in the ancient times were imported from other areas. Since the major elements are the same for ceramics from different sources, the origin of antique ceramics can be identified only from the composition of their trace elements. The concentration of one trace element might not be sufficient to determine the origin of the ceramics and only the concentrations of several trace elements or actually the ratio of the concentrations can lead to identification of the source of the ceramics. Let us say that we need to determine the concentrations of five trace elements. In order to identify the source of the ceramics we take a sample of earth or of known local old pottery from several, let us say ten, possible locations of source and determine in each of them the concentration of the same five trace elements. The question of identification is how do we determine to which set of these ten sets of five numbers of standard ceramics our five numbers of the unknown ceramics fit most closely.

(2)   A spill of oil was found in a port in which ten ships are anchored. How do we know which ship is to be blamed for the spill? Let us assume that we know how to measure four different physical or chemical properties of each oil. The question again is how do we determine to which from the ten sets of possible four values, for each ship, the set of four values of the spill fits the most.

(3)   The identification of many unknown chemicals in water or air samples are determined by their mass spectrum using electron impact (EI) ionization, followed by the determination of the intensities of the produced various ions characterized by their m/z values. The electron impact (usually electrons of 70 eV) forms not only the parent ions, and in many cases the parent ions are not formed at all, but also mainly fragmented ions. Actually, it is lucky that there are fragmented ions as well as parent ions because if there were only parent ions, then between $m/z = 16$ (for the simplest organic compound) and let us say $m/z = 500$, we could identify at most only 484 compounds. However, due to the pattern of the fragmented ions we can distinguish between more than 10000 compounds in this range since the pattern of fragmentation depends on the structural formula and isomers lead to different patterns of fragments. In many cases, the pattern of the

fragmented ions cannot indicate directly the analyzed compound and the identification of the compound is done by comparison to the product-ion spectra of known compounds. The catalog of ion spectra of EI-MS already contains more than 50 000 spectra (Stein *et al.* 1991; Lafferty 1991) and again the question is how to compare the EI-MS (electron impact mass spectra) of our unknown sample with all the data in the library catalog, searching for the one which fits best.

Better identification can be obtained in a system named the triple mass-spectrometer, which is actually two mass spectrometers separated by a collision chamber. Each of the ions separated in the first mass spectrometer (referenced as the parent ion) is accelerated separately toward an inert gas in the collision chamber. Due to these collisions the primary ion is fragmented and the secondary ions (daughter ions) are measured by the second mass spectrometer.

(4)   A similar situation exists in the case of measurement of radioactive contamination in the human chest (lungs). The contamination (U, Pu, ^{137}Cs, etc.) can be distributed homogenously in the lungs or can be concentrated in one point due to all the contamination caused by one particle carried by aerosols and stuck at one point in the lungs. Unless the location of the contamination is known, it is usually assumed that the distribution is homogenous (distributed uniformly). However, this assumption can lead to a considerable error in the calculation of the amount of contaminants, mainly for low-energy γ-emitting particles, due to the large absorption of them in the chest tissues. In the past, two large NaI-CsI detectors were used for lung contamination measurements. Recently, these detectors were replaced by HPGe detectors, due to their better resolution. However, HPGe detectors are smaller in size. The transformation of modern radiocontamination determination systems to HPGe detectors forces the use of four detectors instead of the two previously used, in order to cover the whole area of the lungs. The use of four detectors means that we can have more information which might allow us to learn more about the location of the radioactive particle. A phantom of the lungs with 56 holes which can be filled with either the phantom material or a radioactive source, can be used to find the count rate in each of the four detectors due to a known amount of radioactive material in each of these 56 points. We have 56 ordered four numbers for the standards and in order to find the location of the unknown active materials we have to search for the best fit of the four numbers of the unknown (the patient) with the four numbers of the 56 standard positions. It can very rarely fit exactly to one of the standards due to the statistical error of counting, and the question is to find the best fit (Alfassi 2003).

The mathematical treatment for all these problems can be done in the same way. Most of this work was done for example (3), i.e. the identification of an unknown molecule by comparing its mass-spectra to the large number of mass-spectra in the library catalog of known compounds, assuming that our compound is identical to one of the library spectra. Several tests (metrics) which were special to mass-spectrometry, such as the probability-based matching system (PBM) of McLaferty *et al.*

(1974(1), 1974(2), 1995) or that of Hertz *et al.* (1971) who defined a similarity index using both the masses that matched in the spectra of the unknown and the library molecule and their fraction relative to the total masses in the spectra (matched and unmatched), will not be discussed here and only the general methods which can be applied to all these problems will be treated in the following. The raw data, e.g. that of the ions' intensities of the various m/z peaks in the EI-MS, cannot be used directly, since the amounts of the unknown and the standard might be different leading to different total ion currents, and the ions' current should be normalized in most methods, except that of the angle between vectors, as will be explained in detail.

Rasmussen and Isenhour (1979) used several normalization methods. In the first method they divided the currents of all the ions by the current of the m/z with the highest current (base peak normalized). In the second normalization method, the sum of the currents of all the ions is taken as unity, i.e. every m/z current I_i is normalized by dividing by $\sum I_i$, which means taking the relative intensities. The third normalization factor is $\sum I_i^2$, which leads to each spectrum being a vector of unit length.

They studied several metrics (i.e. criteria) to find the best fit. The first metric test is the sum of the absolute intensity difference $\sum_{j=1}^{n} |I_{L, j} - I_{u, j}|$, where L and u stand for library and unknown, respectively. The summation is done over all the m/z components (the n various peaks) of the spectra. The metrics are calculated for all library spectra or spectra taken from the library after exclusion of spectra which from prior knowledge or gross mismatch cannot be the best fit. You are reminded that the various I_j are the normalized ones, according to one of the methods outlined previously.

The second metric test is the geometric (named the Euclidean) distance:

$$\sqrt{\left| \left[\sum_{j=1}^{n} (I_{L, j} - I_{u, j})^2 \right] \right|}$$

It should be mentioned that when the individual currents are normalized according to the first and second normalization method, the sum is not really the geometric distance, as will be explained later. The third metric test is that of Hertz *et al.* (1971) which is based on the averaged weighted ratio of matched and unmatched m/z in the unknown and library spectra. We will not deal with this method as we assume that the number of variables (components) is the same in the unknown and in the library. Rasmussen and Isenhour (1979) found that all the normalization methods and search methods gave similar results, although the total ion current normalization seems to be the best normalization method.

Stein and Scott (1994) gives a clearer presentation of the search methods. They suggested looking on the pieces of information (intensities of different m/z, count rates of the different detectors, the various numerical properties of the different oils, etc.) as the components of a vector and normalizing the vectors to unit length. Each individual normalized vector can be looked at as a single point on a sphere with unit radius in a hyperspace of n dimensions, where n is the number of components of the vector (e.g. the n different numerical parameters known for each oil, the number of detectors). If two vectors (points on the n-dimensional hypersphere) are identical for all the values of the components they will be a perfect 'match' and will be at the same point in the hyperspace. However, because of instrumental variability and instability and the statistical nature of the measurements, very rarely does the point of the unknown coincide with one of the points of the library of standards. They suggested looking at the problem of finding possible matches as the determination of which standard points will be within a specific volume element centered at the point of the unknown (i.e. at a specific distance from this point). The similarity of two normalized vectors can be seen as the inverse of their distance apart. In the same way as Rasmmussen and Isenhour (1979), they took both the geometrical distance (a square root of the sum of the squares of the differences between components of two vectors) and also an artificial distance (which is the sum of the absolute values of the differences). Thus, they suggested two criteria for the Matching Factor (MF):

geometric distance: $$\text{MF}_\text{g} = \left[1 + \frac{\sum (u_i - s_i)^2}{\sum u_i} \right]^{-1} \tag{8.1}$$

absolute difference: $$\text{MF}_\text{d} = \left[1 + \frac{\sum |u_i - s_i|}{\sum u_i} \right]^{-1} \tag{8.2}$$

where u_i and s_i are the ith component of the unknown sample and the standard one from the library, respectively. The addition of $1+$ in the brackets was done in order to prevent having a zero denominator in the case of identical vectors and to bring the range of the MF's to be from one to zero, when one means perfect identification, i.e. identical vectors. There is no explanation as to why the normalization was done only according to the unknown. A term which seems more reasonable for the geometric distance is:

$$\left[1 + \sum \left(\frac{u_i}{\sqrt{(\sum u_i^2)}} - \frac{s_i}{\sqrt{(\sum s_i^2)}} \right)^2 \right]^{-1} \tag{8.3}$$

or at least:

$$\left[1 + \frac{\sum (u_i - s_1)^2}{\sqrt{(\sum u_i^2 \sum s_i^2)}} \right]^{-1} \tag{8.4}$$

Stein and Scott (1994) following a technical report of the Finnigan Corporation (Sokolow 1978), suggested a completely different way of testing for similarity of the vectors. This method is based on the possibility of calculating the cosine of the angle between the two vectors from the components of the two vectors, through the use of their scalar product (also called the dot-product). The definition of the scalar product of two vectors is:

$$\mathbf{x} \cdot \mathbf{y} = |\mathbf{x}| \, |\mathbf{y}| \cos \theta \tag{8.5}$$

where $|\mathbf{x}|$ and $|\mathbf{y}|$ are the lengths of the vectors, respectively, and θ is the angle between the two vectors. Algebra of vectors shows that both the scalar product and the length can be calculated from the components of the vectors:

$$\mathbf{x} \cdot \mathbf{y} = \sum x_i y_i \qquad |\mathbf{x}| = \sqrt{\left(\sum x_i^2 \right)} \quad \text{and} \quad |\mathbf{y}| = \sqrt{\left(\sum y_i^2 \right)}$$

where the summation is done over all the components of the two vectors (which are of course of the same dimensions). Hence:

$$\cos\theta = \frac{\sum x_i y_i}{\sqrt{\left(\sum x_i^2 \ \sum y_i^2\right)}} \tag{8.6}$$

The third matching factor was defined by Stein and Scott (1994) as $MF_\theta = \cos^2\theta$.

Stein and Scott also indicated that $\cos\theta$ is the distance between the points (heads of the vectors) in the hyperspace, since if the vectors are normalized (of unit length), the cosine theorem gives for the distance between the two points, d:

$$d^2 = 1^2 + 1^2 - 2 \times 1 \times 1 \times \cos\theta \;\Rightarrow\; d^2 = 2(1 - \cos\theta) \;\Rightarrow\; \cos\theta = 1 - 0.5 d^2$$

Stein and Scott (1994) found that for identification of a compound through its mass spectra in comparison to a large library, MF_θ is better than both MF_g and MF_d (the latter was the worst of all three). It should be pointed out that the observation that MF_θ and MF_g did not yield the same rankings is due to the wrong normalization of the vectors in MF_g, leading to vectors of length different from unity and hence the distance is not the correct geometrical distance. If the correct geometrical distance, d, was calculated, then MF_g and MF_θ should give the same ordering, as MF_θ is given by $(1 - 0.5 d^2)^2$. The square should not change the ranking in the list, i.e. the order of matching of the library spectra to the unknown spectrum.

The matching factor of $\cos\theta$ does not require normalization at all and its meaning must not be the distance between the heads of the two vectors. Its real meaning is the cosine of the angle between the two vectors, called in later articles the spectral contact angle (Andrews & Richardson 1994; Gross et al. 2002). Thus for example in the case of the four detectors in chest counting, we can use directly the count rates, ignoring the specific activities of the unknown and the standards which can be different. Two vectors will be a good match even if their distance apart is large, as long as the angle between them is small which means that $\cos\theta$ should be close to unity, since in this case correct normalization will make the distance between the points small. In other words, instead of testing the distance between the heads of the normalized vectors (the points) in the hypersurface, we are testing the deviation of the two vectors from parallelism. Parallel vectors i.e. $\cos\theta = 1$ is an indication to a perfect match.

For normalized vectors $\sum x_i^2 = \sum y_i^2 = 1$, and Equation (8.6) becomes:

$$\cos^2\theta = \sum x_i y_i \tag{8.7}$$

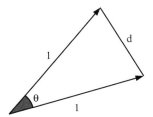

Fig. 8.1

As we noted previously for the $\cos^2\theta$ test, it is not mandatory to normalize the vectors. However, even for this test it is preferable to keep the data in the library as normalized vectors, since this means that there is no need to calculate the denominator of Equation (8.6) for every unknown (Alfassi 2003). In the case of normalized vectors, Equation (8.6) is replaced by Equation (8.7).

The requirement of parallel vectors can lead to another test, which has already been met in the discussion of linear regression to a linear function. The vectorial equation of two parallels vectors **x** and **y**:

$$\mathbf{y} = \alpha\,\mathbf{x} \tag{8.8}$$

where α is a scalar parameter, can be transformed into the functional form of $y = \alpha x$, where α is a constant. In this case x_i and y_i ($i = 1, 2, \ldots, n$) are the values of two variables which fulfil the dependence form $y = \alpha x$, instead of being components of two vectors. We have learnt that the fit of the experimental data to a linear function is given by Pearson's r (correlation coefficient) or by R^2, the determination coefficient. Although $-1 \le r \le 1$, we can use its absolute value which is equal to $\sqrt{R^2}$:

$$|r| = \sqrt{R^2} = \frac{\sum (x_i - \bar{x})(y_i - \bar{y})}{\sqrt{\left[\sum (x_i - \bar{x})^2 (y_i - \bar{y})^2\right]}} \tag{8.9}$$

Comparing this equation with the equation of $\cos\theta$ we see that it is the same equation, except that each point is translated to a system of axes for which the origin is in the mean of the two vectors. Calculating the two metrics, $\cos\theta$ and Pearson's r, for almost identical vectors, it can be seen that small deviation from identical vectors leads to larger deviation from unity in the case of Pearson's r, but it does not mean that this is a better matching factor. Thus, for example, for the two vectors (1, 2, 3, 4, 5) and (1, 2, 3, 4.1, 5), the Pearson's r is 0.999 675 while $\cos\theta$ is equal to 0.999 936, and for the vectors (1, 2, 3, 4, 5.2) and (1, 2, 3, 4, 5), $r = 0.998\,458$ while $\cos\theta = 0.999\,616$. The calculation of r (or rather $R^2 = r^2$) can be found in the library of common spreadsheets e.g. Excel, however the self-programming for calculation of $\cos\theta$ for Excel is very simple, almost trivial. (Actually there are two functions in Excel which give the value of r; these are Correl and Pearson.)

However, in the case of the lung contamination measurement it was found that $\cos\theta$ gave better ranking than r (Pelled *et al.* 2004). In this case each standard was measured four times. The mean of the four measurements (for each of the four detectors) was taken as the standard value (library value), while each of the unknown measurements was taken as the unknown value. Since we actually know to which point location each of the unknowns belongs, we can check if the tests (metrics) give the correct answer. It was found that the $\cos\theta$ test yielded the correct answer in 182 cases out of a total of 224 cases studied, while Pearson's r yielded the correct answer only in 159 cases. This is probably due to the fact that while the $\cos\theta$ test is measuring proportionality, R^2 is measuring only linearity but does not force the intercept to be zero.

It should be pointed out that the components of the vectors must not be the values measured directly; they can be weighted values according to various considerations. Thus for example, Stein (1999) for the mass spectrometric search chose for the components $x_i = m_i I_i^{1/2}$, where m_i and I_i are the mass and the intensity of the signal of ion i respectively. This was done in order to increase the weight of the ions with higher masses, which are closer to the parent ion, since these are the more specific peaks. The weighting of $I_i^{1/2}$ reduces the effect of the ions with the larger intensity which appear in the spectra of many compounds, and usually are not specific to any molecule.

Gross *et al.* (1983) suggested comparing spectra by means of a similarity index defined by:

$$SI = \sqrt{\left[\frac{\sum_{i=1}^{N} \left(\frac{x_i - y_i}{y_i} \times 100 \right)^2}{N} \right]} \qquad (8.10)$$

where x_i and y_i are the components of the two spectra, with y_i smaller than x_i, and N is the number of components. Later, Gross *et al.* modified this definition by dividing by $x_i + y_i$ instead of only by y_i. It should be mentioned that a better name would be a dissimilarity index since for complete similarity SI is equal to zero. Another way is to define the similarity index by $(100 - SI)$. In the case of lung decontamination it was found that the modified SI (however, for normalized vectors, while Gross *et al.* did not normalize their values) gave the best fit, slightly better than $\cos \theta$ (Alfassi 2003). The division by $(x_i + y_i)$ yields higher weighting for the smaller counts from the distant detectors:

$$SI = \sqrt{\left[\frac{\sum \left(\frac{x_i - y_i}{x_i + y_i} \times 100 \right)^2}{N} \right]} \qquad (8.11)$$

8.1   References

Alfassi, Z. B. 2003, *J. Am. Soc. Mass Spectrom.*, vol. 14, p. 262.

Alfassi, Z. B. 2003, *J. Am. Soc. Mass Spectrom.*, vol. 14, p. 261.

Andrews, R. W. & Richardson, H. 1994, *Journal of Chromatography A*, vol. 25, p. 628.

Gross, M. L., Jay, J. O. Jr, Zwinselman, J. J. & Nibbering, N. M. 1983, *Org. Mass. Spectrom.*, vol. 18, p. 16.

Gross, N. L., Wan, K. X. & Vidavsky, I. 2002, *J. Am. Soc. Mass Spectrom.*, vol. 13, p. 85.

Hertz, H. S., Hites, R. A. & Biemann, K. 1971, *Anal. Chem.*, vol. 43, p. 179.

McLaferty, F.W., Hertel, R.H. & Vilwock, R.D. 1974, *Org. Mass Spectrom.*, vol. 9, p. 690.

McLaferty, F. W., Stauffer, B., Ellis, R.D. & Peterson, D.W. 1974, *Anal. Chem.*, vol. 57, p. 771.

Mc Lafferty, F. W., Stauffer, D. B. & Loh, S.I. 1991, *J. Am. Soc. Mass Spectrom.*, vol. 2, p. 441.

McLafferty, F. W., Loh, Y. & Stauffer, D.B. 1995, *Computer-enhanced Analytical Spectroscopy*, Meuzelaar, H.C. (ed.), vol. 2, pp. 136–181, Plenum Press, New York.

Pelled, O., German, U., Tsroya, S. & Alfassi, Z. B. 2004, *Appl. Radiat. Isotopes.*

Rasmussen, G. T. & Isenhour, T. L. 1979, *J. Chem. Inf. Comput. Sci.*, vol. 19, p. 179.

Sokolow, S., Karnofsky, J. & Gustafson, P. 1978, *The Finnigan Library Search Program*, Finnigan Application Report 2, San Jose, California, USA.

Stein, S. E., Ausloos, P. & Lias, S. G. 1991, *J. Am. Soc. Mass Spectrom.*, vol. 2, p. 443.

Stein, S. E. & Scott, D. R. 1994, *J. Am. Soc. Mass Spectrom.*, vol. 5, p. 859.

Stein, S. 1999, *J. Am. Soc. Mass Spectrom.*, vol. 10, p. 770.

9   Smoothing of spectra signals

9.1   Introduction

In various fields of analytical chemistry we are measuring spectra, which means that we are not measuring a single value, as is the case in analysis by gravimeteric, titration, coloring measurements, etc., but rather we are measuring a function. This function is the dependence of the measured signal on a variable. It can be measurement of the intensity of the transmitted light vs. the wavelength in optical spectroscopy, or the transmitted microwave intensity vs. the magnetic field in continuous NMR, or the signal of various detectors vs. time in either chromatography (actually the dependence is on the volume of the effluent) or Fourier Transform Spectroscopy (for more details see Chapter 11), or the measurement of the ion current intensity vs. the mass (actually the magnetic or quadrupole field) in mass spectrometry.

Before the advent of mini- and micro-computers the functions were recorded on strip recorders or $x-y$ plotters. However, the digital revolution has made life easier now. Instead of having analogue signals from the various detectors they are fed into Analogue to Digital Converter (ADC) units which output digital values. The dependence on time [all other dependencies are actually transformed into time dependence, as, for example, the dependence on the wavelength is transformed to time dependence by the time of the turning of the dispersing grating or in the mass spectrometer the dependence on mass is changed to the time of changing of the magnetic (quadrupole) field] is digitized by the use of a multi-channel analyzer (MCA) in its mode as a multi-scaler. A multi-channel analyzer is an instrument (nowadays an electronic circuit on a card which is inserted into one of the slots of a PC) which has several memory cells (quite common today are several thousands of cells or even 32k cells). In its multi-scaler mode each cell is dedicated to some time during the measurement. For example, if we have 2k memory cells (usually named 'channels') and our total measurement is for 100 seconds, then each memory cell is for an interval of $100/2k = 0.05$ s. Thus, the first channel contains the measured result for the interval 0–0.05 s, while the second channel has the value measured during 0.05–0.10 s, and the k channel belongs to the result during the time interval $[(k-1) \times 0.05 - k \times 0.05]$ s. The MCA can be operated also as a pulse height analyzer (PHA), when each channel belongs to an interval of voltage rather than to an interval of time. The output of an MCA is an array which yields the discrete value of the function $f(t)$ where f is the signal measured by the detector and t is a successive number given by $k \, dt \ (k = 1, 2, \ldots, n)$, where dt is the time (or the voltage) interval of each channel and n is the number of channels.

9.2 Smoothing of spectrum signals

There are two reasons for spectra with peaks to have unsmoothed peaks. The main reason is that the observed signal is composed of the 'true' signal plus random noise. The amount and the structure of the noise depends on the experimental setup. However, some unsmoothed spectra can be formed also in the absence of noise due to the statistical nature of the number of the channel in which there is a peak. For example, in the measurement of gamma-ray spectrum, since the absorption of the whole original photon is not accomplished by one step (except for very low energy photons), but rather in several steps and the amount of energy absorbed in each step is determined statistically, the channel corresponding to the full energy is distributed statistically. Unless the number of counts is sufficiently large so that the statistics itself leads to a smooth peak, the observable peak is not smooth. In processing the spectrum we want to smooth the peaks before analyzing the spectrum with a computer to find the peaks and evaluate their areas. In manual analysis of digital spectra there is no necessity for prior smoothing, but for computer analysis it is almost always a must. In the case of noise, the smoothing algorithm is used to mathematically reduce the noise in order to increase the signal-to-noise ratio. A basic assumption in smoothing noise is that the noise is of higher frequency relative to the signal of interest.

Most modern spectrometers come with programs which already include the smoothing algorithms. However, many of them allow a choice between various algorithms. It is important to understand the algorithms in order to choose the best one for our case and to select judiciously the adjustable parameters.

Most smoothing methods use windows, number of adjacent channels (memory cells), for the smoothing algorithm. All the points in the window are used to determine the smoothed values and therefore the window width affects directly the results of the smoothing. We will present five methods of smoothing and all of them use a window. However there are methods like Fourier Transform smoothing that smooth the whole spectrum as one piece. We will not deal with these methods here.

9.2.1 *Method 1: Fixed window mean smoother*

The n points of the whole spectrum are divided into m windows ($m < n$) and each of the windows has $k = (n/m)$ points. The mean of the k points in each window is the value of the smoothed spectrum which has now only m points. This smoother was used considerably in the past in order to decrease the number of points in the spectrum for faster calculation. It is better than just taking each kth point in the spectrum, as it causes signal averaging. However, with the large capacity of modern computers this smoother is rarely used any more.

9.2.2   Method 2: Moving (sliding) window mean smoother

In this method the number of points in the smoothed spectrum is similar to that in the original spectrum. Actually in the smoothed spectrum there are $(m - 1)$ points less than in the original spectrum, where m is the width of the window. The algorithm moves the window each time across one point rather than across a full window as in the previous method. If the window width is m, then the algorithm takes in the first step the points $1, 2, \ldots, m$ and finds their mean. This is the first point of the smoothed spectrum. Then, it moves the window one channel and takes the mean of the points $2, 3, 4, \ldots, m + 1$. This procedure is repeated $[n - (m - 1)]$ times, where n is the total number of points in the original spectrum. This lead to a spectrum with $[n - (m - 1)]$ points. In cases where the end points on both sides of the original spectrum are zeros the loss of these $(m - 1)$ points does not change the spectrum considerably.

   A too-small window will not get rid of noise, while a too-large window will get rid of the noise but will also change the original peak, making the peak broader with lower maximum intensity. This is the reason why it is so important to choose the right width for the window.

9.2.3   Method 3: Moving window weighted mean smoother

This method differs from the previous one by calculating a different mean. Instead of calculating a sample mean, this method calculates a weighted mean, giving more weight to the central point, the one which is replaced by the mean. As will be seen in Method 5, there are analytical ways to choose the weights, but some cases use other weights, for example, the use of the Maestro emulation of the MCA of Ortec for smoothing a window of 5 channels and the use for weights the binomial coefficients for $n = 4$. Thus if x_i denotes the content of the unsmoothed channel and z_i the content of the smoothed channel then:

$$z_i = (x_{i-2} + 4x_{i-1} + 6x_i + 4x_{i+1} + x_{i+2})/16$$

The division by 16, which is equal to $2^4 = $ the sum of the binomial coefficients, leads to unchanged total area under the spectrum.

9.2.4   Method 4: Moving (sliding) window median smoother

This is similar to the moving mean smoother except that instead of calculating the mean, the median is used instead of the mean. The median is not as sensitive to extreme points as the mean (Hoaglin et al. 1983) and therefore the median smoother is very effective for removal of spikes from the data. However it is less efficient for filtering noise. Since the last three methods are complimentary, it is recommended both smoothing methods are tried and the three smoothed curves are compared.

9.2.5   Method 5: Moving (sliding) window polynomial smoother

This method uses also sliding windows but differs from the previous methods in that instead of finding the mean or the median of the points in the window, all the points in the window are fitted to a polynomial using least squares linear regression. The central channel element (jth) of the smoothed spectrum is set equal to the value calculated by the regression for element j. Khan (1987) performed this old method calculating for each window the least squares equation, while Savitzky and Golay (SG) (1964) had already showed that there is no actual need to perform the least squares minimization at each window in the case of data points which are evenly spaced along the abscissa axis, and in addition fulfilled the two following conditions: (1) the data curves are continuous, and (2) the uncertainty in the ordinate values is much larger than the uncertainty in the abscissa values – Savitzky and Golay showed that in this case the least squares minimization leads to the same result as summation of weighted values of the points in the window (Method 3). Thus, there is no need to do many least squares calculations but we need to know the weighting factors and that they were calculated by SG. The weighting factors depend on the width of the window and on the order (power) of the fitted polynomial. SG tabulated the coefficients; this enables us to calculate the value of the j component in the smoothed spectra using windows of $(2m + 1)$ points by writing:

$$y_j(\text{smoothed}) = \sum_{i=-m}^{m} C_i y_{j+i} \tag{9.1}$$

The coefficients C_i were calculated by Savitzky and Golay (SG) for extensive values of window widths and polynomial orders. However, as errors were found in the values of some of these coefficients, the coefficients should be taken from Steiner *et al.* (1972).

Savitzky and Golay's smoothing method is very common in chemistry, not only in analytical chemistry. Their papers were cited in the years 1965–1995 more than 2000 times. Ratzlaff and Johnson (1989) even mention the claim that Savitzky–Golay smoothing is the most common digital data processing technique used by analytical chemists with the exception of the fit of a straight line to data to generate a calibration curve (Madden 1978). For this reason we feel that it is not sufficient to just give the summarized equation, but rather discuss the mathematics of the method and how the coefficients are calculated.

9.3   Savitzky and Golay method (SG method)

The basis of this method involves fitting by least squares regression of a small subset of the data to a polynomial of the order n. Since in the polynomial there are $(n + 1)$ coefficients, the subset should include at least $(n + 1)$ points (actually for least squares we need at least $(n + 2)$ points). This method required originally that the points are equally spaced along the abscissa. The least squares coefficients are used to define the smoothed value at the center of the subset (window), and the process is repeated by moving the fitting interval (window) by one point, until the end of the original function (array) is reached. SG suggested that we take windows of $(2m + 1)$ points, i.e. m points at each side of the central point for which we want to calculate the smoothed value. Then to fit the data in the window to a polynomial of rank $n(n < 2m + 1)$ by referring to the central point as point number zero (the subset is going from $-m$ to m). Each point of the subset is fitted to an n-rank polynomial, i.e. fitted to the equation:

$$y_i = \sum_{k=0}^{n} b_{nk} i^k \qquad (9.2)$$

or more explicitly:

$$y_i = b_{n0} + b_{n1}\, i + b_{n2}\, i^2 + \ldots + b_{nn}\, i^n \quad [i = -m,\, -(m-1)\,,\, \ldots,\, 0\,,\, \ldots, m]$$

9.3.1   The original development

We have $(n + 1)$ coefficients which allows us to calculate the value of the function at the $(2m + 1)$ points of the window. However, it is used only to calculate the central point of the window, the value of y_0 in the smoothed spectrum. SG pointed out that since at the central point $i = 0$ then $y_0 = b_{n0}$, as this is the only coefficient which is not multiplied by zero. Actually for this to be correct we should define $0^0 = 1$, which is a regular definition. Since we are interested only in the value of the smoothed function at this point, it is sufficient to calculate b_{n0}; there is no need to calculate the other coefficients (however, if we are interested in calculating the derivatives of the smoothed function for the purpose of a peak search, as we will see in the next chapter, we have to calculate also b_{nk}, were k is the order of the derivative, so for the first derivative we need b_{n1}). The basis of the SG method is naming the window (subset) channels $-m, \ldots, 0, \ldots, m$ rather than $1, 2, \ldots, 2m + 1$, for which all the coefficients are required to calculate the value of the central point.

Requiring least sum of the squares of the deviations (residuals) between the calculated polynomial and the actual observed data means a minimum value of:

$$\text{SSR} = \sum_{i=-m}^{m} (y_{i,\text{cal}} - y_{i,\text{obs}})^2 = \sum_{i=-m}^{m} \left(\sum_{k=0}^{n} b_{nk}\, i^k - y_i \right)^2 = minimum$$

In order to get the coefficients b_{nk}, which will lead to the minimum value of SSR, we should differentiate SSR with respect to the various $b_{nk}(k = 0, 1, 2, \ldots, n)$ and equate the derivatives to zero. This will lead to $(n + 1)$ equations in $(n + 1)$ variables (the b_{nk}) as in usual least squares polynomial regression.

From the requirement $\partial(\text{SSR})/\partial b_{bn\ell} = 0$ we get the equation:

$$\sum_{i=-m}^{m} i^\ell \left(\sum_{k=0}^{n} b_{nk}\, i^k - y_i \right) = 0 \qquad (9.3)$$

As $\ell = 0, 1, \ldots, n$, we have $(n + 1)$ equations of the type of Equation (9.3). Rearranging the equation leads to:

$$\sum_{i=-m}^{m} \sum_{k=0}^{n} b_{nk}\, i^{k+\ell} = \sum_{i=-m}^{m} y_i\, i^\ell \qquad (\ell = 0, 1, \ldots, n) \qquad (9.4)$$

Changing the order of summation on the left side yields:

$$\sum_{k=0}^{n}\sum_{i=-m}^{m} b_{nk} i^{k+\ell} = \sum_{i=-m}^{m} y_i\, i^{\ell} \tag{9.5}$$

As b_{nk} is independent of i, it can be brought out of the sigma sign:

$$\sum_{k=0}^{n} b_{nk} \sum_{i=-m}^{m} i^{k+\ell} = \sum_{i=-m}^{m} y_i\, i^{\ell} \tag{9.6}$$

If $(k+\ell)$ is an odd number then $(-i)^{k+\ell} = -(i)^{k+\ell}$ and hence

$$\sum_{i=-m}^{m} i^{k+\ell} = 0$$

while for an even number $(k+\ell)$, then $(-i)^{k+\ell} = (i)^{k+\ell}$ and thus

$$\sum_{i=-m}^{m} i^{k+\ell} \neq 0$$

This difference between odd and even $(k+\ell)$ leads to each equation involving only some of the variables and enabling the partition of the $(n+1)$ equations into two parts; one set either equal in size [when $(n+1)$ is even] or having $(n/2)+1$ equations, and the other having only $n/2$ equations. Each set contains only part (about half) of the variables. The division of the set of $(n+1)$ equations into two parts allows the use of only about half of the equations in order to find b_{n0} and thus makes the solution much simpler. Savitzky and Golay (1964) suggested a short-hand notation for writing the equations more clearly. They define:

$$S_{\ell k} = \sum_{i=-m}^{m} i^{\ell+k} \quad \text{and} \quad F_{\ell} = \sum_{i=-m}^{m} y_i\, i^{\ell} \tag{9.7}$$

The set of equations becomes:

$$F_{\ell} = \sum_{k=0}^{n} b_{nk} S_{\ell k} \qquad \ell = 0, 1, \ldots, n \tag{9.8}$$

F_{ℓ} is calculated from the spectrum data for every ℓ, and $S_{\ell k}$ is calculated from its definition. When $(\ell + k)$ is an odd number then $S_{\ell k} = 0$, and thus for an even value of ℓ the appropriate equation will include only b_{nk} with even values of k [since then $(\ell + k)$ is not odd]. To make it more clear, let us look at the case of a polynomial of rank $n = 3$. We get four equations of the type in Equation (9.8) for the four coefficients:

$$F_0 = b_{30} S_{00} + b_{31} S_{01} + b_{32} S_{02} + b_{33} S_{03}$$
$$F_1 = b_{30} S_{10} + b_{31} S_{11} + b_{32} S_{12} + b_{33} S_{13}$$
$$F_2 = b_{30} S_{20} + b_{31} S_{21} + b_{32} S_{22} + b_{33} S_{23}$$
$$F_3 = b_{30} S_{30} + b_{31} S_{31} + b_{32} S_{32} + b_{33} S_{33}$$

Since $S_{01} = S_{03} = S_{12} = S_{21} = S_{23} = S_{30} = S_{32} = 0$, the four equations are divided into two sets of equations:

$$F_0 = b_{30} S_{00} + b_{32} S_{02} \quad (I) \quad \text{and} \quad F_1 = b_{31} S_{11} + b_{33} S_{13} \quad (II)$$
$$F_2 = b_{30} S_{20} + b_{32} S_{22} \qquad\qquad F_3 = b_{31} S_{31} + b_{33} S_{33}$$

Since we are interested only in the value of the polynomial at the center point (i.e. with $i = 0$), the value becomes:

$$y_i = \sum_{k=0}^{n} b_{nk} \, i^k \quad \Rightarrow \quad y_0 = b_{n0}$$

In order to find b_{30} (for $n = 3$), it is sufficient to solve the set (I) of the two equations:

$$b_{30} = \frac{\begin{vmatrix} F_0 & S_{02} \\ F_2 & S_{22} \end{vmatrix}}{\begin{vmatrix} S_{00} & S_{02} \\ S_{20} & S_{22} \end{vmatrix}} = \frac{F_0 S_{22} - F_2 S_{02}}{S_{00} S_{22} - S_{02} S_{22}}$$

Let us calculate the different terms which appear in this equation and let $m = 2$, i.e. window of $(2m + 1) = 5$ points (this is the minimum number of points for $n = 3$).

$$S_{\ell k} = \sum_{i=-m}^{m} i^{\ell+k} \quad \Rightarrow \quad S_{00} = (-2)^0 + (-1)^0 + 0^0 + 1^0 + 2^0 = 5$$

$$S_{20} = S_{02} = (-2)^2 + (-1)^2 + 0^2 + 1^2 + 2^2 = 10$$
$$S_{22} = (-2)^4 + (-1)^4 + 0^4 + 1^4 + 2^4 = 34$$

$$F_\ell = \sum_{i=-m}^{m} y_i \, i^\ell \quad \Rightarrow \quad F_0 = \sum_{i=-2}^{2} y_i = y_{-2} + y_{-1} + y_0 + y_1 + y_2$$

$$F_2 = \sum_{i=-2}^{2} y_i \, i^2 \quad \Rightarrow \quad F_2 = 4y_{-2} + y_{-1} + y_1 + 4y_2$$

Substituting these values in the equation for b_{30} yields the value of the smoothed polynomial:

$$y_{0,\,\text{smoothed}} = \frac{1}{35}(-3y_{-2} + 12y_{-1} + 17y_0 + 12y_1 - 3y_2)$$

As can be seen the SG method for polynomial smoothing is turning into a calculation of the weighted mean of the values in the window. SG calculated the coefficients for many values of n, the rank of the polynomial, and $(2m + 1)$, the number of points in the window. However since some of SG's numbers are wrong, it is better to use the tabulation of Steiner et al. (1972):

$$y_{j,\,\text{smoothed}} = \sum_{i=-m}^{m} \frac{C_i \, y_{j+i}}{N} \qquad\qquad (9.9)$$

where y_{j+i} is the original value of y in channel (point) numbered $(j+i)$, C_i are the coefficients and N is called the normalization factor. Thus for the case of $n = 3$ and $m = 2$, SG gives: $C_{-2} = -3, C_{-1} = 12, C_0 = 17, C_1 = 12, C_2 = 3$ and $N = 35$.

SG showed that $b_{ns} = b_{n+1,s}$ when n and s are either both odd or both even [i.e. $(n + s)$ is even], thus $b_{20} = b_{30}$ and $b_{40} = b_{50}$. If we choose $n = 2$, we will get three equations instead of four. The two equations of set (I) will remain the same, while set (II) will have only one equation. Thus:

$$b_{20} = b_{30}$$

If we choose $n = 4$, the five equations will be divided into three equations in set (I) and two equations in set (II), the same two as we obtained for $n = 3$, and hence:

$$b_{31} = b_{41}$$

In general, the SG coefficients *for smoothing* using even polynomial degree is equal to the next highest odd degree, e.g. 2/3 and 4/5, etc. For the *first derivative* we will soon learn that the opposite is true, 1/2, 3/4, etc.

The Savitzky–Golay method can, besides smoothing the function (spectrum), also find its derivatives. The first and second derivatives are instrumental in finding local maxima and minima of the function, i.e. it allows automatic searching of the peaks (see the next chapter). We approximate the window by polynomial

$$f_i = \sum_{\ell=0}^{n} b_{n\ell}\, i^{\ell}$$

Differentiating with respect to i (remember that our abscissa is i) leads to:

$$\frac{df_i}{di} = \sum_{k=1}^{n} \ell\, b_{n\ell}\, i^{\ell-1}$$

As we are interested only in the value of the derivatives at the center point $(i = 0)$, we get

$$\frac{df_i}{di} = b_{n1}$$

In the same way

$$\frac{d^2 f_i}{di^2} = \sum_{\ell=2}^{n} \ell(\ell - 1) b_{nk}\, i^{\ell-2}$$

and at the center point

$$\frac{d^2 f_i}{di^2} = 2b_{n2}$$

and the nth derivative is

$$\frac{d^n f_i}{di^n} = n! \, b_{nn}$$

9.3.2   Algebra of matrices method

Although the method up to now, as was performed by Savitzky and Golay, is sufficient, it is interesting to see the way Steiner *et al.* (1972) developed it by a matrix formalism which appears more clearly in the paper by Phillips and Harris (1990). This formalism leads to easier calculation of the coefficients. Let $\mathbf{y}$ be the $(2m + 1)$ dimensional column vector of our window of the spectrum (i.e. the window of the spectrum to be smoothed has $(2m + 1)$ points) and $\boldsymbol{\varepsilon}$ is the $(2m + 1)$ dimensional vector of the random 'error' variables. Since they are random it should fulfil the condition that the expectation value (the mean) is zero and that the dispersion matrix (the expectation value of the variance) is equal to the variance σ^2:

$$E(\boldsymbol{\varepsilon}) = 0 \tag{9.10}$$

$$V(\boldsymbol{\varepsilon}) = E(\boldsymbol{\varepsilon}\boldsymbol{\varepsilon}^T) = \sigma^2 I \tag{9.11}$$

where $\boldsymbol{\varepsilon}^T$ is the transposed vector of $\boldsymbol{\varepsilon}$ and I is the identity matrix. These requirements mean that the errors of the various points, i.e. ε_i, are uncorrelated, they all have a mean value of zero and the same variance σ^2. Let $\mathbf{D}$ be the vector of the exact values (smoothed values) of $\mathbf{y}$, then we can write:

$$\mathbf{y} = \mathbf{D} + \boldsymbol{\varepsilon} \tag{9.12}$$

Let us assume that the vector $\mathbf{D}$ can be expressed as the product of a specially designed matrix, $\mathbf{X}$ and the column vector of the coefficients of the polynomial $\boldsymbol{\theta}$. Since $\mathbf{D}$ is a vector of dimension $(2m + 1)$ and $\boldsymbol{\theta}$ is a column vector of dimension $(\ell + 1)$, where ℓ is the order of the smoothing polynomial, the matrix $\mathbf{X}$ should be of dimension $(2m + 1) \times (\ell + 1)$:

$$\mathbf{D} = \mathbf{X}\boldsymbol{\theta} \tag{9.13}$$

Substituting Equation (9.13) in Equation (9.12) yields:

$$\mathbf{y} = \mathbf{X}\boldsymbol{\theta} + \boldsymbol{\varepsilon} \tag{9.14}$$

The least squares method in order to estimate the best vector $\boldsymbol{\theta}$ (the vector of the parameters) is minimization of the matrix obtained by squaring the absolute value of the error vector. This is similar to the usual method, since the vector $\boldsymbol{\varepsilon}$ represents the diference between the calculated and the observed data. Squaring of the vector (scalar product) is done by multiplying it by its transpose. Thus the method requires minimization of the matrix obtained by the scalar product of the vector of the error $-\boldsymbol{\varepsilon}$ and its transpose:

$$SSR = \boldsymbol{\varepsilon}^T \cdot \boldsymbol{\varepsilon} \quad \Rightarrow \quad SSR = (\mathbf{y} - \mathbf{X}\boldsymbol{\theta})^T \cdot (\mathbf{y} - \mathbf{X}\boldsymbol{\theta}) \tag{9.15}$$

where the superscript T denotes the transpose (either vector or matrix).

The minimization is done by differentiating SSR with respect to $\boldsymbol{\theta}$ and equating the derivative to zero:

$$\frac{\partial \text{SSR}}{\partial \boldsymbol{\theta}} = 0 \Rightarrow \quad \mathbf{X}^T(\mathbf{y} - \mathbf{X}\boldsymbol{\theta}) = 0$$

$$\Rightarrow \quad \mathbf{X}^T \mathbf{y} = \mathbf{X}^T \mathbf{X} \boldsymbol{\theta}$$

(9.16)

By multiplying both sides by the inverse of $\mathbf{X}^T\mathbf{X}$, $\boldsymbol{\theta}$ will be left alone on the right side of the equation and thus the solution in Equation (9.16) is:

$$\boldsymbol{\theta} = (\mathbf{X}^T\mathbf{X})^{-1}\mathbf{X}^T\mathbf{y}$$

(9.17)

The estimated error in the coefficient vector $\boldsymbol{\theta}$ is contained in the variance – covariance matrix (Draper & Smith 1981; Deming & Morgan 1979):

$$\mathbf{V} = \sigma_0^2 (\mathbf{X}^T\mathbf{X})^{-1}$$

(9.18)

were σ_0^2 is the variance of the noise in the raw data.

The matrix $\mathbf{C} = (\mathbf{X}^T\mathbf{X})^{-1}\mathbf{X}^T$ contains the coefficients of the filter function. The convoluting coefficients for estimating the value of the smoothed function and the first derivative at the midpoint of the filtered window are the first and second rows of $\mathbf{T}$, respectively. Once these coefficients have been calculated and tabulated [depending on the order of the polynomial and the width of the window (filter length)], no further matrix operations are needed.

The specially designed matrix $\mathbf{X}$, should be similar to the equations of Savitzky and Golay. All the columns are of the form $(-m)^k \ldots (m)^k$, where $k = 0, 1, \ldots, \ell$ and $0^0 = 1$. Thus the matrix $\mathbf{X}$ has the form (general and a special case for $\ell = 3$ and $2m + 1 = 5$):

$$\mathbf{X}_{(2m+1,\,\ell+1)} = \begin{pmatrix} 1 & -m & (-m)^2 & \cdots & (-m)^\ell \\ \vdots & \vdots & \vdots & & \vdots \\ 1 & 0 & 0 & & 0 \\ \vdots & \vdots & \vdots & & \vdots \\ 1 & m & m^2 & & m^\ell \end{pmatrix} \Rightarrow \mathbf{X}_{(5,\,3)} = \begin{pmatrix} 1 & -2 & 4 & -8 \\ 1 & -1 & 1 & -1 \\ 1 & 0 & 0 & 0 \\ 1 & 1 & 1 & 1 \\ 1 & 2 & 4 & 8 \end{pmatrix}$$

The advantage of this method is the easy calculation of the matrix of the coefficients, which can be done by algebra of matrices. Nowadays this is easily done by personal computer using various software, e.g. Excel or Matlab.

Thus for case of $\ell = 3, (2m + 1) = 5$, we have to calculate $\mathbf{C} = (\mathbf{X}^T\mathbf{X})^{-1}\mathbf{X}^T$

$$\mathbf{C} = \left[\begin{pmatrix} 1 & 1 & 1 & 1 & 1 \\ -2 & -1 & 0 & 1 & 2 \\ 4 & 1 & 0 & 1 & 4 \\ -8 & -1 & 0 & 1 & 8 \end{pmatrix} \begin{pmatrix} 1 & -2 & 4 & -8 \\ 1 & -1 & 1 & -1 \\ 1 & 0 & 0 & 0 \\ 1 & 1 & 1 & 1 \\ 1 & 2 & 4 & 8 \end{pmatrix} \right]^{-1} \begin{pmatrix} 1 & 1 & 1 & 1 & 1 \\ -2 & -1 & 0 & 1 & 2 \\ 4 & 1 & 0 & 1 & 4 \\ -8 & -1 & 0 & 1 & 8 \end{pmatrix}$$

Using matrix calculation in MATLAB we get:

$$
C = \begin{pmatrix} -0.4857 & 0 & -0.1429 & 0 \\ 0 & 0.9208 & 0 & -0.2361 \\ -0.1429 & 0 & 0.0714 & 0 \\ 0 & -0.2361 & 0 & 0.0694 \end{pmatrix} \begin{pmatrix} 1 & 1 & 1 & 1 & 1 \\ -2 & -1 & 0 & 1 & 2 \\ 4 & 1 & 0 & 1 & 4 \\ -8 & -1 & 0 & 1 & 8 \end{pmatrix}
$$

$$
= \begin{pmatrix} -0.0857 & 0.3429 & 0.4857 & 0.3429 & -0.0857 \\ 0.0833 & -0.6667 & 0 & 0.6667 & -0.0833 \\ 0.1429 & -0.0714 & -0.1429 & -0.0714 & 0.1429 \\ -0.833 & 0.1667 & 0 & -0.1667 & 0.0833 \end{pmatrix}
$$

$$
= \frac{1}{35} \begin{pmatrix} -3 & 12 & 17 & 12 & -3 \\ 2.9167 & -23.3333 & 0 & 23.3333 & -2.9167 \\ 5 & -2.5 & -5 & -2.5 & 5 \\ -2.9167 & 5.8333 & 0 & -5.8333 & 2.9167 \end{pmatrix}
$$

Thus, the first line gives us the SG coefficients for cubic smoothing with a window of 5 points.

9.3.3 Formulae for SG coefficients

As will be discussed later with more details, Ernst (1966) suggested smoothing by least squares to orthogonal polynomials instead of smoothing by least squares to regular polynomials. In the case of continuous functions, Ernst suggested the use of Legendre polynomials, while for the case of a finite number of discrete points he suggested the use of Gram polynomials. We will discuss the Gram polynomial method later in connection with the calculation of the end points. Here we mention it only because it can lead to formulae for the calculation of SG coefficients, the convolution weighting factors, overcoming the need to solve the linear equations or doing algebra of matrices. Ernst gave only the formula for smoothing of $n = 2, 3$ and $n = 4, 5$, which appeared also in the book of Whittakar and Robinson (1964).

Madden (1978) extended these equations not only for smoothing but also to the calculation of the first to fifth derivatives. It is not certain whether the chances of error are greater in keying into the computer the table of coefficients or in keying the quite long formulae for the coefficients. The main explanation of Madden as to why these equations are needed is that Savitzky and Golay's tables are limited to windows of up to 25 points, and in some cases they found that the optimum width is larger than 25 points. They gave simple rules to check if the calculated coefficients are correct:

For smoothing: $$\sum_{i=-m}^{m} c_i = 0 \qquad (9.19)$$

For the qth order derivative: $$\sum_{i=-m}^{m} c_i^q \, i^q = q \qquad (9.20)$$

It is up to the reader to decide whether to key in to the computer the table of coefficients (Steiner *et al.* 1972) correction to SG (Savitzky & Golay 1964) or the Madden formulae (Madden 1978). In most modern spectroscopy instruments one of the methods has already been chosen by the manufacturer of the instrument; our purpose is to make the analyst aware of his choices.

9.3.4 *Smoothing of first points for kinetic studies*

While Savitzky and Golay developed their method to smooth the whole curve using a moving window, and calculate the coefficients of the midpoint, Leach *et al.* (1985) used the same method to calculate the smoothed first points in order to estimate the initial point and slope (first derivatives) as is usually needed in kinetic studies. They used the same method and tabulated the coefficients for the first point. However, Baedecker (1985) found errors in those tables for $(2m + 1) = 7$ or 9 and gave new correct tables.

9.4 Studies in noise reduction

Enke and Neiman (1976) and Leach *et al.* (1985) have studied noise reduction by using numerical simulation either separately for the noise and for the peak (Enke & Nieman) or for convoluted data consisting of a linear combination of normally-distributed noise with linear and exponential decays (Leach *et al.*). In a general way it can be said that as the window size increases the noise is continuously reduced but after certain window sizes the effect becomes quite small. On the other hand, when the window is too large sharp peaks may be removed and the remaining peaks distorted. Hence it is very important to choose the right window width (filter length). Most smoothing procedures used second or third order polynomials which were found quite acceptable for most noise removal. However, Barak (1995) suggested the use of a variable order polynomial as will be discussed later.

Enke and Nieman (1976) discussed the result of the enhancement of the ratio signal to noise by the SG method. The smoothing function will represent only approximately each local section of the data and therefore might lead to some distortion of the signal, due to band width reduction and loss of resolution along the abscissa. The compromise between noise reduction and signal distortion depends on the parameters of the smoothing process (the degree of the polynomial and the number of points) and the number of repetitions of the smoothing process. They limited their simulated experiments to quadratic–cubic smoothing ($b_{30} = b_{20}$), consisting of 5–23 points and from 1 to 128 repetitions. They measured the effect of smoothing on white noise, which is a mixture of signals of all frequencies with random amplitudes and phases. The probability distribution of white noise is a normal distribution. They measured the effect of smoothing by the reduction of the standard deviation (by definition the mean of the noise is zero). Enke and Nieman (1976) found (similar to Savitzky and Golay) that the ratio of the standard deviation after and before the smoothing is inversely proportional to the square root of the width of the window. In order to further reduce the standard deviation the smoothing can be repeated several times (multi-passes smoothing), although the most noticeable effect is of the first pass. Given enough passes, a narrow smooth can reduce the noise to the same extent as a wider smooth. In the case where multi-passes smoothing is done, care should be taken not to drop in each pass the $2m$ end points [m points on each end in the case where the window width is $(2m + 1)$], which usual SG smoothing omits. Enke and Nieman suggested, as will be discussed also later, in the case of multi-passes to use another narrow window smoothing on the ends. The smoothing not only reduces the noise but unfortunately also distorts the true signal, it becomes lower and broader. The degree of this distortion depends on the ratio of the width of the smoothing function to the FWHM (Full Width at Half Maximum height) of the peak. To keep the peak height distortion below 1% this ratio must be less than 0.9 for a Gaussian peak and less than 0.7 for a Lorentian peak. As long as the ratio is less than 1.0, the peak is not noticeably broadened. Enke and Nieman suggested for a single-pass smoothing to choose as window width for smoothing twice the FWHM of the peak. The signal-to-noise enhancement would be within 5% of the maximum signal-to-noise enhancement due to one-pass smoothing, and the area of the peak remains almost unchanged.

On the other hand Edwards and Wilson (1976) recommended that the smoothing window be chosen approximately as 0.7 FWHM of the narrowest single peak. Wider windows in their experience caused loss of resolution.

Leach *et al.* (1985) found that the relative improvement in the precision of the *initial data point* was independent of the magnitude of the noise added and the kinetic behavior of the function being fitted, but depended on the filter width. Better improvement was found for a wider window. They found proportional dependence on the square root of the window width, as was found also for the midpoints by Savitzky and Golay (1964) and by Enke and Nieman (1976).

Phillips and Harris (1990) used the matrix approach of Bialkowski (1989). They used symbolic programming software with the matrix formulation of polynomial filters to predict the precision of filtered results without the need for time-consuming numerical simulation. The diagonal elements of the variance-covariance matrix contained the variance of the filtered amplitude and derivative estimates, while the off-diagonal elements predicted the covariance between them. The analytical expression revealed the theoretical origin of the dependence of noise reduction on filter width and polynomial order.

For a polynomial of order two, the variance-covariance matrix will be:

$$\mathbf{V} = \sigma_0^2 (\mathbf{X}^2\mathbf{X})^{-1} = \begin{pmatrix} \dfrac{3(3m^2 + 3m - 1)}{(2m - 1)(2m + 1)(2m + 3)} & 0 & \dfrac{-15}{(2m - 1)(2m + 1)(2m + 3)} \\ 0 & \dfrac{3}{(2m - 1)(2m + 1)} & 0 \\ \dfrac{3(3m^2 + 3m - 1)}{(2m - 1)(2m + 1)(2m + 3)} & 0 & \dfrac{45}{(2m - 1)m(m + 1)(2m + 1)(2m + 3)} \end{pmatrix}$$

Symbolic manipulation produces exact results, in contrast to numerical methods which depend on the numerical stability of the algorithm. This can lead to the equation for the relative error, as for example quadratic smoothing with window width w ($w = 2m + 1$):

$$(\sigma/\sigma_0)^2 = \frac{3(3w^2 - 7)}{4(w - 2)w(w + 2)} \tag{9.21}$$

Khan (1987) studied molecular beam experiments with two kinds of analyzing systems, one which has some noise in the measurement and one which is free from noise. He tried to smooth the results of the first method to get similar data to that from the second instrument. He found that one pass of long window smoothing is much better than several iterations of short window (short filter length), mainly due to the observation that multi-pass smoothing (5 iterations of $(2m + 1) = 5$) leads to unreal peaks in the derivative plot, whereas one pass of $(2m + 1) = 21$ leads to the same smoothing without forming imaginary peaks. However, he was worried that with a 21-point fit one loses 10 points at each end of the data array and suggested that the edge point be smoothed in a similar method to Enke and Nieman (1976), although he detailed the procedure more. First he operated the 21-point window smoothing on all points of the data array and obtained $(n - 20)$ smoothed points. He then added the 10 first and 10 last points of the raw data and operated smoothing a 19-point window, starting from the first point only for one window. This way he got a value for the smoothed point numbered 10. He replaced the raw point by this value and operated a 17-point window from the first point, calculating the value of point numbered 9. He repeated this process down to a window of 7 points. The same procedure was done also for the other edge of the data array.

Whereas for smoothing of the whole curve a broad window is usually preferable, for the calculation of the initial slope it was found by Leach et al. (1985) and also by Khan (1988) that wide filters lead to large errors. Leach et al. suggested the use of a narrow window and high-order polynomials (keeping $(2m + 1) > \ell$). Khan found that this prescription although reducing the systematic error is inefficient in removing the noise. He found better results with smoothing of the midpoint.

9.5   Extension of SG method

Ratzlaff and Johnson (1989) extended the Savitzky – Golay method to functions of two variables using two-dimensional polynomial least squares. This kind of smoothing is important, for example in excitation-emission fluorescence matrices, in retention time/absorbancy surfaces in liquid chromatography and in spatial / spectral maps of emission from atomic spectroscopy sources, in two- and three-dimensional NMR spectroscopy and also other cases. They didn't give tables of coefficients but rather formulae for calculation of the coefficients, for some orders of polynomials.

Kuo et al. (1991) developed another method for multidimensional least squares smoothing. Instead of fitting to regular polynomials they used expansion of products of orthogonal polynomials (Gram polynomials), as will be described later. They showed that their results are equivalent to sequential one-dimensional fitting in each of the dimensions of the array. The sequential approach represents a significant saving in computation time over the multidimensional smoothing technique. Comparing this technique to tests developed by Ratzlaff and Johnson (1989), using the usual SG method of regular polynomial, it was found that the new method had additional terms in the formula for the calculation of the coefficients (weighting factors). These additional terms are due to the inclusion of the cross terms in the derivation, while Ratzlaff and Johnson treated each dimension separately. A comparison between the smoothing capabilities of the two techniques reveals that inclusion of these cross terms results in slightly better smoothing characteristics.

Gorry (1990) rejected the suggestion of Khan (1987) for calculation of the smoothing of the end points on the grounds that the calculation takes too long and he suggested a method similar to the Savitzky–Golay method but instead of using simple power series (polynomial) he used weighted expansions of discrete orthogonal polynomials known by the name Gram polynomials, as suggested initially by Ernst (1966). So instead of writing (as SG):

$$y_i = \sum_{k=0}^{\ell} b_{nk} \, i^k \tag{9.2}$$

Gorry wrote:

$$f_n(t) = \sum_{k=0}^{n} b_k p_k^m(t) \tag{9.22}$$

where $p_k^m(t)$ is the Gram polynomial of order k, over $(2m+1)$ points, evaluated at point t. The Gram polynomial is defined by (Ralston 1965):

$$p_k^m(t) = \sum_{j=0}^{k} \frac{(j)^{(j+k)}(j+k)^{(2j)}(m+t)^{(j)}}{(j!)^2 \, (2m)^{\theta}} \tag{9.23}$$

where $(a)^{(b)}$ is a generalized factorial function defined by $(a)^{(b)} = a\,(a-1)\,(a-2)$ ... $(a-b+1)$ and $(a)^{(0)} = 1$. Comparing this to the SG method in which the smoothed function and any derivatives of order s can be written as (ℓ is the order of the polynomial):

$$f_{\ell}^s(0) = \sum_{i=-m}^{m} h_i^s y_i \tag{9.1}$$

where h_i^s are the SG coefficients (convolution weights), Gorry's method used the same expression (9.1) but h_i^s are given by the following equation:

$$h_i^{t,s} = \sum_{k=0}^{\ell} \frac{(2k+1)(2m)^{(k)}}{(2m+k+1)^{(k+1)}} p_k^m(i) p^{m,s}(t) \tag{9.24}$$

where $\quad p_k^{m,s}(t) = \left[\frac{d^s}{dx^s} p_k^m(x)\right]_{x=t} \tag{9.25}$

The tables of SG are obtained from Equation (9.24) by setting $t = 0$. The tables of Leach et al. (1985) and the revised ones by Baedecker (1985) correspond to $t = -m$ (for $s = 0$ and $s = 1$). Equation (9.24) can allow us to smooth all the points in the data array. The Gram polynomial can be calculated from a recursive relationship.

Gorry gave tables for the convolution coefficients [where $(2m+1)$ is the number of the points in the window, from $-m$ to m]. Thus for example his tables for

Table 9.1

Smooth					
$i \longrightarrow$	-2	-1	0	1	2
-2	69	2	-3	2	-1
-1	4	27	12	-8	4
0	-6	12	17	12	-6
1	4	-8	12	27	4
2	-1	2	-3	2	69
norm	70	35	35	35	70

First derivative					
$i \longrightarrow$	-2	-1	0	1	2
-2	-125	-19	1	5	-29
-1	136	-1	-8	-13	88
0	48	12	0	-12	-48
1	-88	13	8	1	-136
2	29	-5	-1	19	125
norm	84	42	12	42	74

5 points for cubic polynomial smoothing and first derivatives are shown in Table 9.1. All the coefficients are in one column according to the value of i (the index of the calculated point going from $-m$ to m). The number in the leftmost column indicates the index of the original points.

For example, for smoothing the second point in the window ($i = -1$) we use the second column:

$$y_{-1} = \frac{2y_{-2} + 27y_{-1} + 12y_0 - 8y_1 + 2y_2}{35} \tag{9.26}$$

And for the slope (first derivative) of the first point ($i = -2$) we use the first column in the derivative table:

$$y'_{-2} = \frac{-125y_{-2} + 136y_{-1} + 48y_0 - 88y_1 + 29y_2}{84} \tag{9.27}$$

Gorry gave tables for quadratic fit ($\ell = 2$) for all points from $(2m + 1) = 5$ up to 21 for smoothing ($i = 0$) and for the initial point ($i = -m$). For other cases the coefficients have to be calculated from the recursive relationship between the Gram polynomials as given by Gorry.

In a second paper Gorry (1991) extended his method also to non-uniformly spaced data. In this case SG coefficients cannot be used and smoothing must be done with regular least squares of the polynomial to each moving window, which is a long process. Gorry claimed that in his method although non-uniformly spaced data required more computation time than the uniformly spaced data it is still considerably less than the least squares of regular polynomials.

Hui and Gratzl (1996) implemented Gorry's method for smoothing and derivative calculation for uniformly spaced data and tested it on titration curves by detecting the zero-crossing of the second derivative of the original data array. It is important to detect the end point (last point) in real time so that current titration can be terminated immediately and the next titration can be started. However it was found that Gorry's method yields some anomalies in the results produced. Hui and Gratzl found that in order to overcome this problem, one should choose a higher-order polynomial for the edge point in order to compensate for the lack of sliding window adaptivity. The increase of the order of the polynomial in the end sections increases the flexibility to match the real frequency content of the signal being processed. Thus, one must use separate sets of polynomials fitting for the center and end sections; the order of the polynomial used should be higher for the end sections than that used for the center section.

Barak (1995) suggested modifying the use of the SG method of fixed-order polynomials by using an adaptive-degree polynomial filter which will have different order polynomials in different sections of the data array. His reasoning is that for maximum smoothing of statistical noise, polynomials with low order are required, whereas high polynomial order is necessary for accurate reproduction of the real peaks. The choice of the order of the polynomial in each section is done by calculating the smoothing for two different fixed order degrees, n_1 and n_2 ($n_2 > n_1$). For each order he calculated the sum of the squares of the residuals:

$$SSR = \sum_{i=-m}^{m} (y_{i,\,obs} - y_{i,\,cal})^2 \tag{9.28}$$

where the subscripts obs and cal denote the observed raw data and the calculated smoothed data, respectively. Since Barak wanted the calculated value in all the $(2m + 1)$ points of the window he could not use SG coefficients (which are only for the midpoint), but had to use the convolution weights using Gram polynomials. From the SSR of two orders, n_1 and n_2, he calculated a test statistic which should follow the F-distribution with degrees of freedom $v_1 = n_2 - n_1$ and $v_2 = 2m - n_2$. The test statistic is given by:

$$F = \frac{SSR_{n_1} - SSR_{n_2}}{SSR_{n_2}/(2m - n_2)} \tag{9.29}$$

The resulting F-value is tested against the critical value of $F(v_1, v_2)$ with usual level of 5%. He usually started from the lowest polynomial degree and increased the degree until the order failed the significance test. The SSR can be calculated consecutively using the regression equation with Gram polynomials:

$$SSS_n = SSR_{n-1} - \sum_{i=-m}^{m} p_n^m(i)y_1 \Big/ \sum_{i=-m}^{m} \left[(p_n^m(i)\right]^2 \qquad n > 1 \tag{9.30}$$

Barak found that based on noise reduction for pure noise and signal reproduction for pure signal, the adaptive-degree polynomial filter (ADPF) performed nearly as well as an optimally chosen fixed-order SG filter and outperformed suboptimally chosen SG filters. For synthetic data consisting of signal and noise, ADPF outperformed even optimally chosen fixed-order SG filters.

9.6   References

Baedecker, P. A. 1985, *Anal. Chem.*, vol. 57, p. 1479.
Barak, P. 1995, *Anal. Chem.*, vol. 67, p. 2658.

Bialkowski, S. E. 1989, *Anal. Chem.*, vol. 61, p. 1308.

Deming, S. N. & Morgan, S. L. 1979, *Clin. Chem.*, vol. 25, p. 840.

Draper, N. R. & Smith, H. 1981, *Applied Regression Analysis*, 2nd edn, John Wiley & Sons, New York, Chapter 2.

Edwards, T. W. & Wilson, P. D. 1976, *Appl. Spec.*, vol. 28, p. 541.

Enke, C. G. & Nieman, T. A. 1976, *Anal. Chem.*, vol. 48, p. 705A.

Ernst, R. D. 1966, *Ad. Mag.*, vol. 2, p. 1.

Gorry, P. A. 1990, *Anal. Chem.*, vol. 62, p. 570.

Gorry, P. A. 1991, *Anal. Chem.*, vol. 63, p. 534.

Hoaglin, D., Mosteller, F. & Tukey, J. 1983, *Understanding Robust and Exploratory Analysis*, John Wiley & Sons, New York, p. 77.

Hui, K. I. & Gratzl, M. 1996, *Anal. Chem.*, vol. 68, p. 1054.

Khan, A. 1987, *Anal. Chem.*, vol. 59, p. 654.

Khan, A. 1988, *Anal. Chem.*, vol. 60, p. 369.

Kuo, J. E., Wang, H. & Pickup, S. 1991, *Anal. Chem.*, vol. 63, p. 630.

Leach, R. A., Carter, C. A. & Harris, J. M. 1985, *Anal. Chem.*, vol. 56, pp. 2304–307.

Madden, H. H. 1978, *Anal. Chem.*, vol. 50, p. 1383.

Phillips, G. R. & Harris, J.M. 1990, *Anal. Chem.*, vol. 62, p. 2749.

Ralston, A. 1965, *A First Course in Numerical Analysis*, McGraw-Hill Book Co., New York, pp. 235–54.

Ratzlaff, K. L. & Johnson, T. J. 1989, *Anal. Chem.*, vol. 61, p. 1303.

Savitzky, A. & Golay, M. J. E. 1964, *Anal. Chem.*, vol. 36, p. 1627.

Steiner, J., Termonia, Y. & Deltour, J. 1972, *Anal. Cem.*, vol. 44, p. 1906.

Whittakar, E. & Robinson, G., 1964, *The Calculus of Observation*, 4th edn, Dover Publications, New York, p. 235.

10 Peak search and peak integration

Spectra and/or chromatograms are functions which describe the intensity of the signal vs. time, where time can stand for time of separation (chromatogram) or other properties such as wavelength (which depends on the rate of the rotation of the monocromator) or mass (which depends on the rate of variation of the electric or the magnetic field) or magnetic field (in continuous NMR spectroscopy). In the past these functions were plotted on a chart recorder or x-y plotter and the areas or heights of the peaks were measured manually. Modern instruments use digital techniques. The analog signals from the detectors are digitized using ADC (analog to digital converters) and the time variation is given by the use of multi-channel analyzers (MCA) which are used as a multi-scaler. The MCA is usually a computer card inserted into a personal computer (PC) or external device connected to the computer via a UBS connector. Each channel is a memory cell. The sequential number of the channel is usually proportional to the time. Thus for example the signal intensity in the first second is registered in channel number 1 and the signal in the 57th second in channel number 57. The time width of each channel can be varied according to the length of the experiment and the total number of channels. In the previous chapter we discussed how this digital data can be smoothed. In this chapter we will discuss the automatic search for peaks and their integration.

Most instruments already have a computer program for automatic peak search and integration installed by the manufacturer. Yet it is important to learn the meaning of the various algorithms for peak search and integration in order to understand the meaning of the adjustable parameters. Wrong choice of parameters can lead to identification of statistical fluctuation as peaks or inversely to not identifying small peaks, such that only real peaks will be found while all the real peaks are found. In a recent experiment with a mass spectrometer a spectrum with 8 real peaks gave more than 100 peaks (four pages of results) when wrong adjustable parameters were inserted. Two main methods are used for automatic peak search (Najafi 1990). The first is a statistical one which compares the content of a channel in the window to the content of the whole window. The second one uses calculus methods, either first derivative or second derivative.

Most of the studies done on automatic peak search were done on analysis of γ-ray spectra as this kind of measurement was the first one to use multi-channel analyzers, although not as a multi-scaler but rather as a pulse height analyzer. Actually γ-ray spectra analysis is more difficult than other measurements due to the presence also of wider peaks than usual, due to Compton scattering. These peaks are rejected by a test imposed on the width of the peaks. In most studies it is assumed that the peaks have a Gaussian shape, although in many cases they are convoluted with polynomials in the tails.

10.1 A statistical method

This automatic method of peak search (Najafi 1990; Najafi & Federoff 1983; Federoff *et al.* 1973) is based on a statistical comparison of the contents of adjacent channels. A group of channels is considered together and in every step one channel at one end is deleted and one channel is added at the other end (the moving window method). A window of five channels was used. If n is the central channel number in the window with content $f(n)$, the spectrum is considered to have a peak in channel n if the following conditions are fulfilled:

Condition 1: Preventing false counts, which are very low:

$$f(n) > K \tag{10.1}$$

where K is the threshold value. In the case of measurement of γ-spectra, then K is usually an integer which generally is made equal to 4. However, this is an adjustable parameter which depends on the background.

Condition 2: $f(n)$ is statistically larger then $f(n-1)$:

$$f(n) - f(n-1) > \alpha[f(n-1)]^{1/2} \tag{10.2}$$

where α is generally equal to 3. The square root is taken since this is the standard deviation for Poisson distribution. If condition (10.2) is not satisfied we still do not reject the possibility that there is a peak in channel n but instead of requirement (10.2) we check the following two conditions:

$$f(n) \geq f(n-1) \tag{10.3}$$

$$f(n) - f(n-2) > \alpha[f(n-2)]^{1/2} \tag{10.4}$$

If either condition (10.2) or conditions (10.3) *and* (10.4) are satisfied [of course condition (10.1) must satisfied], the same conditions [either (10.5) or (10.6) *and* (10.7)] are tested for the other side of the window:

$$f(n) - f(n-1) > \alpha[f(n+1)]^{1/2} \tag{10.5}$$

$$f(n) \geq f(n+1) \tag{10.6}$$

$$f(n) - f(n+2) > \alpha[f(n+2)]^{1/2} \tag{10.7}$$

If these conditions are fulfilled on both sides of channel n, i.e. conditions (10.1), (10.2) or (10.3) *and* (10.4), (10.5) or (10.6) *and* (10.7), then channel n is considered as a maximum of the peak.

10.2 First derivative method

It is well known in calculus that the first derivative must be equal to zero in a maximum point of a function, changing sign from positive at the left side of the peak (an increasing function has positive first derivative) to negative at the right

side of the peak (a decreasing function has negative first derivative). Sasamoto *et al.*
(1975) gave three conditions which must be fulfilled in order to recognize the real
peaks:

Condition 1: The first derivative of the data changes from positive to negative,
crossing zero, with increasing channel number. The channel number where the
value of the derivative is zero is denoted x_p.
Condition 2: Assigning N to the distance, in channel units, between the local minimum and the local maximum of the first derivative of the data, N must fulfil the
requirement:

$$0.8\,\text{whm} \leq N \leq 3.0\,\text{fwhm} \tag{10.8}$$

where fwhm is the width at half maximum of a peak measured for a standard
containing only one peak. The fwhm of the standard is changing with x_p
so we must take a standard with a similar value of x_p. This condition is
effective especially to eliminate the Compton edge as its width is larger than
3.0 fwhm, but it is also useful for recognizing two overlapping peaks which appear
as one peak.
Condition 3: Determining that it is a real peak and not a statistical fluctuation. The
measured counts at each channel obey the Poisson distribution and hence the statistical error of f(x) is $\sqrt{[f(x)]}$, where f(x) is the contents of channel numbered x. In
the vicinity of a peak, f(x) can be approximated by:

$$f(x) = A\ G(x) + B(x) \tag{10.9}$$

where $G(x)$ is the Gaussian function, $B(x)$ is the background and A is a constant,
which is the amplitude of the Gaussian peak. For all common backgrounds,
$dG(x)/dx \gg dB(x)/dx$ and hence:

$$f'(x) = A\ G'(x) \tag{10.10}$$

for every x. Using the Gaussian function leads to:

$$|f'(x)|_{max} \cong A\ e^{-0.5x}/\sigma(x_p) \tag{10.11}$$

where σ is the standard deviation of the Gaussian function and x_p is the peak
location as determined by Condition 1. The first derivative is calculated in a similar
way to the smoothing by the SG method, as described in Chapter 9. The statistical
requirement for a real peak is:

$$A\ G(x_p) \geq n\sqrt{[f(x_p)]} \tag{10.12}$$

where n the sensitivity factor is an integer adjustable for obtaining only real
peaks. Using the Gaussian function together with Equations (10.11) and (10.12)
leads to:

$$|f'(x)|_{max} \geq n\ \sqrt{[f(x_p)]}\ e^{-0.5}/\sigma(x_p) \tag{10.13}$$

Equation (10.13) is the third condition.

10.3   Second derivative method

The assumption that the baseline in the window is linear means that the base line content of channel x is given by:

$$b(x) = a + bx \qquad (10.14)$$

Denoting the pure signal, usually a Gaussian or convolution of a Gaussian with polynomials in the tails, by $S(x)$ yields for the contents of channel x of the real signal $f(x)$:

$$f(x) = S(x) + a + bx \qquad (10.15)$$

which yields for the second derivative:

$$f''(x) = S''(x) \qquad (10.16)$$

Thus if we assume that $f(x)$ is a continuous function it is clear that the second derivative of $f(x)$ is independent of the background and it is equal to zero for any interval in which there is no peak. So an explicit requirement for a peak is $f''(x) \neq 0$. For the real case, our function is a discrete function $f(i)$, where i is an integer and $f(i)$ is the content of channel i. Since this is a discrete function we will denote the content by f_i. In this case the second derivative is replaced by a second difference, dd_i:

$$dd_i = -f_{i-1} + 2f_i - f_{i+1} \qquad (10.17)$$

This is the equation by Routti and Prussin (1969). Mariscotti (1967) used the negative value of dd_i. The second difference should be different from zero only in the vicinity of a peak. It can be shown that for a Gaussian signal, f, with peak centered at channel i_0, i.e. when f is given by the equation:

$$f = A \exp\left[-(i - i_0)^2/2\sigma^2\right]$$

then, unless for very large A, the second difference is comparable to its standard deviation sd_i:

$$sd_i = (f_{i+1} + 4f_i + f_{i-1})^{0.5} \qquad (10.18)$$

The term A, which is named the peak intensity, is the content of the central channel i_0 (i.e. the channel with the maximal content) and σ is related to the peak FWHM (FWHM $= 2.355\sigma$).

Let us denote by A_{min} the value of A (peak intensity) for which $dd_{i_0} = sd_{i_0}$. For a Gaussian peak, $A_{min} = 6.5\sigma^4$, thus for $\sigma = 4$, A_{min} is 1600.

In order to detect much weaker peaks it is necessary to reduce the standard deviation of $f''(i)$. This can be done by smoothing either the original function or the second difference function. Thus, for example in the analysis of gamma-ray spectra the programs AUTSPAN (Mills 1970) and GASAN (Barnes 1968) used polynomial data smoothing, as was shown in the previous chapter on smoothing. However, Routti and Prussin (1969) in their program SAMPO used smoothing of the second derivative as was suggested by Mariscotti (1967).

Mariscotti suggested smoothing the second difference by using the moving window with calculation of the mean. Thus the array dd_i is changed to the array $dd_i(w)$, which depends on the width of the window, w (equal to $2m + 1$).

$$dd_i(w) = \sum_{j=i-m}^{i+m} dd_j/(2m + 1) \qquad (10.19)$$

Again we can compare $dd_i(w)$ with its corresponding standard deviation and find a new value of A_{min}. In order to reduce A_{min} further we can repeat once more, or several times more, the operation for finding the mean of the moving window. If the number of times of repeating the operation of finding the mean is z, then the smoothed value at channel i is:

$$dd_i(z, w) = \underbrace{\sum_{j=i-m}^{i+m} \cdots \sum_{j=i-m}^{i+m} dd_h}_{Z \text{ sums}} \qquad (10.20)$$

This two-parameter smoothed function approaches the ideal function $f''(x)$ sufficiently if z and w are suitably chosen.

Equation (10.20) is not appropriate for calculation of the standard deviation and hence Mariscotti (1967) suggested another way of writing Equation (10.20) directly in terms of the original data f_i introducing the coefficients α_{ij} such that:

$$dd_i(z, w) = \sum_j \alpha_{ij}(z, w)f_j \qquad (10.21)$$

The α_{ij} weighting coefficients are given by the recursion formula for z:

$$\alpha_{ij}(z, w) = \sum_{k=i-m}^{i+m} \alpha_{kj}(z - 1, w) \qquad (z \geq 1) \qquad (10.22)$$

where for $z = 0$, α_{ij} are defined as:

$$-\alpha_{ij}(0, w) = \begin{cases} = 0 & \text{if } |j - 1| \geq 2 \\ = 1 & \text{if } |j - 1| = 1 \\ = -2 & \text{if } j = i \end{cases} \qquad (10.23)$$

This definition leads to, for each i, $\sum_j \alpha_{ij}(0, w) = 0$, as there are two terms equal to 1, one term equal to -2 and all the other terms are zero. Using the recursion formula (10.22) it can be shown that for any z for each i, the coefficients fulfil $\sum_j \alpha_{ij}(z, w) = 0$. Equations (10.22) and (10.23) lead also to $\alpha_{ij} = 0$ for every i and j which fulfil $|j - i| > z(m + 1)$. Mariscotti (1967) gives all the values of α_{ij} for $w = 5$ and $z \leq 5$.

The above definition of α_{ij} confirms that $dd_i(z, w)$ vanishes if f_i is linear and is similar to a second derivative of a Gaussian function if f_i is a Gaussian plus linear terms.

The standard deviation of $dd_i(z, w)$ is easily derived from Equation (10.21) since each term is independent of the others:

$$\mathrm{sd}_i(z,\ w) = \sqrt{\left| \left[\sum_j \alpha_{ij}^2(z,\ w)\, f_j \right] \right|} \tag{10.24}$$

In order to simplify computer calculation, Mariscotti suggested an approximate equation for Equation (10.24) using new coefficients β_{ij}, which are defined by:

$$\beta_{ij} = \sum_j \alpha_{ij}^2(z,\ w) \tag{10.25}$$

Mariscotti tabulated the values of β_{ij} for $z \le 5$ and $m \le 9$, where $w = 2m + 1$. The approximate equation is:

$$\mathrm{sd}_i(z,\ w) = \sqrt{\left[\beta_{ij}(z,\ w)\, f_j \right]} \tag{10.26}$$

However with the fast computers nowadays this approximation is not necessary. Mariscotti (1967) found that the optimum pair of $(z,\ w)$, which are denoted as $(z_0,\ w_0)$, is given by $z_0 = 5$ and $w_0 = 0.6$ FWHM, whereas Routti and Prussin (1969) gives $w_0 = 0.3$ FWHM $- 0.5$. They tabulated α_{ij} for this optimal case. If the background is strictly a succession of linear sections for the whole spectrum, then the presence of a peak can be determined by computing the significance of the second difference, i.e. the ratio between the second difference and its standard deviation $\mathrm{dd}_i(w_0,\ z_a)/\mathrm{sd}_i(w_0,\ z_b)$ and comparing it with a threshold value. Routti and Prussin chose two threshold values, 2 and 5. The lower value defines potential peaks and the higher one defines the acceptance level. The peaks between the two values have to pass several tests in order to be accepted. Block et al. (1975) suggested the value of 3 for the threshold for the Mariscotti method.

10.4   Computer – visual separation of peaks

Zlokazov (1982) summarized the shortcomings of the previous methods for analysis of gamma-ray spectra and suggested an algorithmization of the process of visual analysis of a gamma spectrum. The spectrum is partitioned into intervals, the bounds of which are local minima (background). Each interval is smoothed separately and then the background (the boundaries of the interval) is subtracted. The method is based on the concept of the curvature of δ the discrete function. The curvature is defined as the ratio of the second derivative (difference) to a term of the first one. The curvature is a point characteristic of geometric figures independent of their amplitudes. The curvature $c(x)$ of an integer-valued discrete function $f(x)$ is given by:

$$c(x) = f''(x)/\sqrt{\{1 + [f'(x)]^2\}} \tag{10.27}$$

or for channel i:

$$c_i = (f_{i+1} - 2f_i + f_{i-1})/\sqrt{[1 + (f_{i+1} - f_i)^2]} \tag{10.28}$$

where f_i is the content of channel i. The curvature as defined by Equation (10.28) has the important property that if $f(x)$ is replaced by $A\ f(x)$ then the curvature

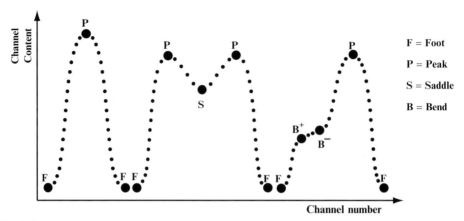

Fig. 10.1

remains almost constant. Zlokazlov wrote that this property exists if $|f_{i+1} - f_i| \gg 1$ but it should be added that even then it is not true for any A and the condition should be $A|f_{i+1} - f_i| \gg 1$. He distinguished between five kinds of pivotal points, as can be seen in Figure 10.1. The character of each point is found by the combination of three indicators:

(1) The amplitude, a_i: The amplitude is checked in order to remove peaks that are only statistical fluctuation. The amplitude is determined from the unsmoothed spectrum, for which s_i is the content of channel i (whereas f_i is the content of the smoothed spectrum). If s_i is smaller than a fixed number, err_i, chosen to be the statistical error (which can be dependent on the channel number), then $a_i = 0$. For all others cases, $a_i = 1$.

(2) The curvature indicator, c_i: The curvature c_i is compared to a constant (independent of i) that is computed from the results of pure peaks and is equal to the minimum of the absolute value of c_i at the peaks. By comparison to this constant we determine g_i according to the rules:

$$c_i > \text{constant} \quad \Rightarrow \quad g_i = 1$$
$$|c_i| \leq \text{constant} \quad \Rightarrow \quad g_i = 0$$
$$c_i < -\text{constant} \quad \Rightarrow \quad g_i = -1$$

(3) The amplitude's extreme, m_i: The amplitude's extreme is calculated from the unsmoothed spectrum by comparing the value of the content of channel i with the two adjacent channels (the one before and the one after). If s_i is the highest of the three then $m_i = 1$, and if it is the lowest then $m_i = -1$. If it is neither the highest nor lowest (i.e. it is between the other two values) then $m_i = 0$. It can be written in a more formal mathematical way:

$$s_i = \max(s_{i-1}, s_i, s_{i+1}) \quad \Rightarrow \quad m_i = 1$$
$$s_i = \min(s_{i-1}, s_i, s_{i+1}) \quad \Rightarrow \quad m_i = -1$$
$$s_{i+1} < s_i < s_{i-1} \text{ or } s_{i+1} > s_i > s_{i-1} \quad \Rightarrow \quad m_i = 0$$

We have two values of a_i, three values of g_i and three values of m_i and thus 18 possible combinations. Thirteen of these combinations yield one of the pivotal points while five combinations ($a_i = 0$ and $g_i = 0$ or $a_i = 0$ and $g_i = -m_i$) yield points that are not considered for peak search. The criteria for the pivotal points are:

$a_i = 0$	Independent of m_i and g_i		$\Rightarrow$	*Foot*
$a_i = 1$	$g_i = 1$	$m_i = -1$	$\Rightarrow$	*Saddle*
$a_i = 1$	$g_i = -1$	$m_i = 1$	$\Rightarrow$	*Peak*
$a_i = 1$	$m_i = 0$		$\Rightarrow$	*Bend:* $g_i = 1(B^+)$, $g_i = -1(B^-)$

A separate peak will be involved from a string of channels in which both ends are foot and one point is peak and all others are non-pivotal points. If there are FPSPF then we have two peaks which are not completely separated. The Bends indicate on multiplets, i.e. two peaks which are hardly separated.

These tests result in hypotheses about peaks and their boundaries. The supposed peaks are further checked by statistical and shape criteria that are different for different kinds of spectra and will not be discussed here.

10.5   Selection of the fitting interval and integration

The measured value that has the analytical content is the area under the peak. In principle, there are two computational methods to compute the area under the peak. The simple method is to sum the contents of all channels in the peak:

$$\text{total} = \sum_{i=L}^{u} f_i \tag{10.29}$$

where f_i is the contents of channel number i, and L and u are the lower and upper channels belonging to the peak. However this total count value also contains the background, which should be subtracted. The background is estimated by taking the average number of counts in 3–5 channels on both sides of the peak multiplied by the number of channels in the peak. Thus for 3 channels on each side:

$$\text{background} = \frac{u - L + 1}{2 \times 3} \left(\sum_{i=L-3}^{L-1} f_i + \sum_{i=u+1}^{u+3} f_i \right) \tag{10.30}$$

The value of 3 in this equation can be increased or decreased, but it is about the optimum value; a lower number will increase the statistical error in the evaluation of the background, and a larger number will increase the possibility that the following channels already belong to other peaks. The number of net content is given by:

$$N = \text{net content} = \text{total} - \text{background} = T - B \tag{10.31}$$

The standard deviation of the net content, σ_N, and the relative standard deviation, $\%\sigma_N$, are given by Equations (10.32) and (10.33):

$$\sigma_N = \sqrt{(T + B)} = \sqrt{(N + 2B)} \tag{10.32}$$

$$\%\sigma_N = \frac{100\sqrt{(T + B)}}{N} = \frac{100\sqrt{(N + 2B)}}{N} \tag{10.33}$$

Selection of the calculation intervals is made after the peak-search procedure. The simple calculation of the peak area by summation of the contents of the relevant channels can be done only for a peak which is sufficiently isolated from neighboring peaks, so that interference by neighboring peaks is negligible. However, the summation method cannot be used for overlapping peaks. For overlapping peaks they must be resolved (deconvoluted) using a suggested mathematical formula (usually a Gaussian or a Gaussian + polynomial) and fitting the smoothed measured data to this formula. In this case the integration is done on the mathematical expression. The mathematical fitting is done in many cases by computer automatic peak calculation also for isolated peaks, it being preferred to simple summation. Samamoto *et al* (1975) suggested dividing the whole spectrum into various fitting intervals, which may include one isolated peak or several peaks.

In order to resolve overlapping peaks accurately, it is necessary to fit them simultaneously. Samamoto *et al.* (1975) suggested that only two neighboring peaks must be fitted simultaneously if the separation between them is less than $1/2\sigma$, where σ is the width of the Gaussian representing the peaks.

10.6   References

Barnes, V. 1968, *IEEE Trans. Nuc. Sci.*, NS-15, no. 3.
Block, H. P., De Lange, J. C. & Schotman, J. W. 1975, *Nucl. Instr. Meth.*, vol. 128, p. 545.
Federoff, M., Blouri, J., & Ravel, G. 1973, *Nucl. Instr. Meth.*, vol. 113, p. 589.
Mariscotti, M. A. 1967, *Nucl. Instr. Meth.*, vol. 50, p. 309.
Mills, S. J. 1970, *Nucl. Instr. Meth.*, vol. 81, p. 217.
Najafi, S. I. & Federoff, M. 1983, *Radiochem. Radional. Lett*, vol. 56, p. 305.
Najafi, S. I. 1990 in *Activation Analysis*, Alfassi, Z. B. (ed.), vol. 1, CRC Press, Boca Raton, USA, p. 39.
Routti, J. T. & Prussin, S. G. 1969, *Nucl. Instr. Meth.*, vol. 72, p. 125.
Sasamoto, N., Koyama, K. & Tanaka, S. I. 1975, *Nucl. Instr. Meth.*, vol. 125, p. 507.
Zlokazov, V. B. 1982, *Nucl. Instr. Meth.*, vol. 199, p. 509.

11 Fourier Transform methods

Mathematical procedures known as transform methods are becoming more and more popular in chemistry. They can be used for different purposes, e.g. transforming from time-domain to frequency domain, smoothing or filtering to enhance signal-to-noise ratio, removal of any known irregularities in the excitation waveform, so that the corrected response reflects only the properties of the sample and not the effect of the measuring instrument. Three different transform methods are used in chemistry, namely Fourier, Hadamard and Hilbert Transforms (Marshall 1982, Griffith 1992). In this chapter we will limit ourselves only to Fourier Transform (FT) and its application in spectroscopy (IR and NMR).

11.1 Fourier Transform methods in spectroscopy

Fourier Transform (FT) methods are responsible for the vast use nowadays of infra-red and NMR in analytic organic chemistry. The instruments used before the development of the FT version have low intensities and require a relatively long time for measuring the spectrum. This drawback was changed in the FT instruments. In the case of IR, the FT instruments widen the range of wavelengths that can be measured in addition to the higher accuracy. In the case of NMR, they allow the measurement of other nuclides besides ^{1}H, e.g. ^{13}C. Actually all kinds of spectra can be improved by the use of FT techniques, but due to the higher cost, it is used only when without FT the accuracy or sensitivity is not sufficient.

A Fourier Transform is a mathematical recipe for transforming a function from the time-domain to the frequency-domain. For conventional spectra the dependence of the intensity amplitude on the frequency (or wavelength) is measured, whereas in the case of FT instruments we measure the variation of the intensity with time. Obtaining the variation with time is done differently for different kinds of spectra. Thus for example, in the case of NMR the variation with time is obtained by a short powerful RF pulse, and the spectrum is measured after the end of the pulse. In FT-IR instead of measuring a spectrum we are measuring an interferogram, using a Mikelson interferometer. FT then transforms the I vs. t plot to I vs. λ (or v) plot.

The advancement in the sixties in FT spectrometers was due mainly to the discovery of the fast FT which makes the computation time considerably shorter. The advancement of computers in the last decade and their lower price has made almost all NMR and IR instruments to be FT equipment.

11.2 Mathematics of Fourier Transforms

Let us first define the Fourier series. It is any periodic function in which the integral of $|f(x)|$ over one period converges and can be expanded into an infinite sum of cosines and sines.

$$f(x) = \frac{a_0}{2} + \sum_{k=1}^{\infty}\left[a_k \ \cos\left(\frac{2\pi kx}{T}\right) + b_k \sin\left(\frac{2\pi kx}{T}\right)\right] \tag{11.1}$$

where T is the period of the function. Actually there are additional conditions of continuity except for a finite number of points and there are only a finite number of maxima and minima in one period. The Fourier coefficients are:

$$a_k = \frac{2}{T}\int_{x_0}^{x_0+T} f(x)\ \cos\left(\frac{2\pi kx}{t}\right)dx \qquad b_k = \frac{2}{T}\int_{x_0}^{x_0+T} f(x)\cos\left(\frac{2\pi kx}{t}\right)dx \tag{11.2}$$

where x_0 can be any arbitrary value, but often it is taken as 0 or $-T/2$. In many cases in order to simplify the equation we use *angular frequency* instead of the period. The angular frequency is the defined by the equation:

$$w = \frac{2\pi}{T} \tag{11.3}$$

and hence:

$$\left.\begin{array}{l} f(x) = \dfrac{a_0}{2} + \displaystyle\sum_{k=1}^{\infty}[a_k \ \cos\ (kwx) + b_k \ \sin\ (kwx)] \\[4mm] a_k = \dfrac{w}{\pi}\displaystyle\int_{x_0}^{x_0+\frac{2\pi}{w}} f(x)\ \cos\ (kwx)dx \qquad b_k = \dfrac{w}{\pi}\displaystyle\int_{x_0}^{x_0+\frac{2\pi}{w}} f(x)\ \cos\ (kwx)dx \end{array}\right\} \tag{11.4}$$

As the Fourier series expansion contains both sine and cosine functions it can be written more compactly using a complex exponent notation. This compact form is defined as:

$$e^{ikx} = \cos\ kx + i\ \sin\ kx \tag{11.5}$$

where i is the unit for imaginary numbers, $i = \sqrt{-1}$. The complex Fourier series expansion is written as:

$$f(x) = \sum_{k=-\infty}^{\infty} C_k \ \exp\left(\frac{2\pi ikx}{T}\right) = \sum_{k=-\infty}^{\infty} C_k \ \exp\ (iwkx) \tag{11.6}$$

and the Fourier coefficients are given by:

$$\left.\begin{array}{l} C_k = \dfrac{1}{T}\displaystyle\int_{x_0}^{x_0+T} f(x)\ \exp\left(-\dfrac{2\pi ikx}{T}\right)dx \\[2mm] \text{or} \\[4mm] C_k = \dfrac{w}{2\pi}\displaystyle\int_{x_0}^{x_0+\frac{2\pi}{w}} f(x)\ \exp\ (-iwkx)dx \end{array}\right\} \tag{11.7}$$

The time domain and the frequency domain are interchangeable in both directions using the Fourier Transform given in Equation (11.8). The left equation (transforming from time to frequency) is called FT while the right one is named the inverse FT:

$$I(v) = \int_{-\infty}^{\infty} J(t)\, e^{2\pi ivt} dt \qquad J(t) = \int_{-\infty}^{\infty} I(v)\, e^{-2\pi ivt} dv \tag{11.8}$$

where v is the frequency (which in many books and articles is denoted by f), I and J are the intensity function in the time and frequency, respectively, and $i = \sqrt{-1}$ is the unit for imaginary numbers.

If the frequency v is replaced by the angular frequency w, which is defined as $w = 2\pi v$, then:

$$I(w) = \int_{-\infty}^{\infty} J(t)\, e^{iwt}\, dt \qquad J(t) = \frac{1}{2\pi} \int_{-\infty}^{\infty} I(w)\, e^{-iwt} dw \tag{11.9}$$

The FT is a linear operation, i.e. the transform of the sum of two functions is equal to the sum of the two transforms. $I(v)$ and $J(t)$ might have special symmetries. The symmetry of $I(v)$ causes special symmetry in $J(t)$ and vice versa as can be seen in the following table:

If:		then:		
	$J(t)$ is real		$I(-v) = [I(-v)]^*$	
	$J(t)$ is imaginary		$I(-v) = -[I(-v)]^*$	
	$J(t)$ is an even function		$I(v)$ is an even function	
	$J(t)$ is an odd function		$I(v)$ is an odd function	(11.10)
	$J(t)$ is real and even		$I(v)$ is real and even	
	$J(t)$ is real and odd		$I(v)$ is imaginary and odd	
	$J(t)$ is imaginary and even		$I(v)$ is imaginary and even	
	$J(t)$ is imaginary and odd		$I(v)$ is real and odd	

*The complex conjugate.

In the case of spectroscopy measurement in the time domain, $J(t)$ must be real, but it is neither an even nor an odd function. Since any function can be written as a sum of an odd function and an even function:

$$f(x) = \frac{f(x) + f(-x)}{2} + \frac{f(x) - f(-x)}{2} = f_{even} + f_{odd} \tag{11.11}$$

the transform of $J(t)$ can be written as the sum of real and even function $I_1(v)$ and imaginary and odd function $i\, I_2(v)$:

$$J(t) \quad \Rightarrow \quad I(v) = I_1(v) + i\, I_2(v) \tag{11.12}$$

where both I_1 and I_2 are real functions, I_1 is an even function while I_2 is an odd function.

11.2.1 Basic properties of FT and inverse FT

If the FT and its inverse are connected by the common sign $\Leftrightarrow$, and if $J(t) \Leftrightarrow I(v)$ then:

$$J(at) \quad \Leftrightarrow \quad \frac{1}{|a|} I\left(\frac{\nu}{a}\right) \quad \text{time and frequency scaling} \tag{11.13}$$

$$J(t - t_0) \quad \Leftrightarrow \quad I(\nu) e^{2\pi i \nu t_0} \quad \text{time shifting} \tag{11.14}$$

$$J(t) e^{2\pi i J_0 t} \quad \Leftrightarrow \quad I(\nu - \nu_0) \quad \text{frequency shifting} \tag{11.15}$$

11.2.2 The Convolution Theorem

The convolution of two functions is defined by:

$$f^*g = \int_{-\infty}^{\infty} f(t') \, g(t - t') \, dt' \tag{11.16}$$

where f^*g is a function in the time domain and consequently can be operated on by FT. The convolution theorem states that the FT of the convolution is equal to the product of the individual functions:

$$FT(f^*g) = FT(f) \, FT(g) \tag{11.17}$$

11.3 Discrete Fourier Transforms

In the measurement of a spectrum in the time domain, we do not have the value of the function $J(t)$ for every value of t, as we do not know a mathematical term for this function, but rather we have $J(t)$ in sampled data. The data is sampled (recorded) usually at evenly spaced intervals in time, i.e. $t_k = k \, \delta t$, $k = 0, 1, 2, \ldots$, where δt is the time interval between subsequent recordings. Consequently we have to estimate the FT from this finite number of recordings. Assuming that we have N recorded values and the sampling interval is δt, then $t_k = k \, \delta t$, where $k = 0, 1, 2 \ldots, N$ and $J_k = J(t_k)$. To simplify the treatment let us suppose that N is even. Since we have only N points of the original function we can calculate no more than N points of the FT. The points are arranged consecutively from $-N/2$ to $N/2$. The fact that this is $(N + 1)$ points is overcome by taking the two extreme points as equal, leaving only N independent points. The different frequencies are given by:

$$\nu_\ell = \frac{\ell}{N \, \delta t} \quad \ell = -\frac{N}{2}, \ldots, \frac{N}{2} \tag{11.18}$$

The FT for each frequency is approximated by the sum (which replaces the integral):

$$I(\nu_\ell) = \int_{-\infty}^{\infty} J(t) \, e^{2\pi i \nu_\ell t} \, dt \approx \sum_{k=0}^{N-1} J_k \, e^{2\pi i \nu_\ell t} \, \delta t \tag{11.19}$$

$$I(\nu_\ell) = \delta t \sum_{k=0}^{N-1} J_k \, e^{2\pi i k \ell / N} \tag{11.20}$$

The sum in Equation (11.20) is called the discrete FT of the N points of J_k, and we will denote it by FT_ℓ:

$$FT_\ell = \sum_{k=0}^{N-1} J_k \, e^{2\pi i k \ell / N} \tag{11.21}$$

FT_ℓ is periodic with respect to ℓ with a period N. Consequently $FT_{-\ell} = FT_{N-\ell}$. Generally using this convention, ℓ is taken as varying from 0 to $(N-1)$ instead of from $-N/2$ to $N/2$, so that ℓ and k have the same range.

The discrete FT, i.e. FT_ℓ, has almost the same symmetry properties as the continuous FT.

11.4   Fast Fourier Transforms (FFT)

As mentioned previously, the possibility to compute discrete FT is due to the discovery of FFT. While the usual FT computation of N points requires N^2 complex multiplications, the discrete FFT requires only $N \log_2 N$ operations. Thus for $N = 10^5$, regular FT requires 10^{10} operations while discrete FFT requires only 1.66×10^6 operations, i.e. reduction by a factor of 6×10^3 in the number of mathematical operations. For $N = 10^6$ the reduction is by a factor of $3.01 \times 10^5/6 = 5 \times 10^4$. The FFT is based on the observation that a discrete FT of length N can be written as the sum of two discrete FTs each of the length of $N/2$. One of the two FTs is formed from the even-numbered points of the original N, the other one is formed from the $N/2$ odd-numbered points, as can be seen in the following proof, concerning the ℓth component of the FT:

$$FT_\ell = \sum_{k=0}^{N-1} I_k \, e^{2\pi i \ell k / N} = \sum_{k=0}^{N/2-1} I_{2k} \, e^{2\pi i \ell (2k)/N} + \sum_{k=0}^{N/2-1} I_{2k+1} \, e^{2\pi i \ell (2k+1)/N}$$

$$FT_\ell = \sum_{k=0}^{(N/2)-1} I_{2k} \, e^{2\pi i \ell k /(N/2)} + \sum_{k=0}^{(N/2)-1} I_{2k+1} \, e^{2\pi i \ell k /(N/2)} \, e^{2\pi i \ell / N} \tag{11.22}$$

$$FT_\ell(N) = FT_\ell^{\text{even}}(N/2) + W^\ell \, FT_\ell^{\text{odd}}(N/2)$$

where $W = e^{2\pi i / N}$.

There are still N components of FT ($\ell = 0, 1, \ldots, n-1$), but each one of them is calculated from two $(N/2)$ terms, i.e. two cycles of $N/2$ terms are calculated. Using Equation (11.21) reduces the number of operations only by a factor of 2, as the number of operations is changed from N^2 to $2[(N/2)^2] = N^2/2$. However we can continue with Equation (11.21) to reduce further the number of operations by changing each FT to two FT with further smaller numbers of points. To make this explanation shorter let us use some shorter notation. Instead of FT we will write 'F', instead of even write 'e' and instead of odd write 'o'. Thus Equation (11.22) will have the form:

$$F_\ell(N) = F_\ell^e(N/2) + W^\ell \, F_\ell^o(N/2) \tag{11.23}$$

Then operating Equation (11.22) on both $F_\ell^e(N/2)$ and $F_\ell^o(N/2)$ gives for example:

$$F_\ell^e(N/2) = F_\ell^{ee}(N/4) + W^\ell \, F_\ell^{eo}(N/4) \tag{11.24}$$

where ee are the points that are divided by 2^2 and eo are the points that are divided by 2 but not by 2^2. Let us assume that N is a power of 2. If not we can add extreme points equal to zero to make it a power of 2. In the case where N is a power of 2 we can continue to apply Equation (11.23) until we have subdivided the data all the way down to one number (unit length or length $= 1$). How many subdivisions do we have to do in order to get to FT of unit length? Since $N = 2^n$ then $n = \log_2 N$ and hence we have to do $\log_2 N$ subdivisions. The FT of length one is just the identity operation that copies one input number into one output slot.

In spectroscopy we have to make real data of time series. The FT is composed of a real part and an imaginary part which have different shapes. Ideally the real spectrum corresponds to an absorption line shape (a Lorentian or Gaussian) and the imaginary spectrum to a dispersion line shape (a little like a derivative of an absorption line shape).

11.4.1 Sampling rates and Nyquist frequency

The time interval of sampling, δt, is connected to the interval in the frequency domain which can be resolved. If we measure N points and the sampling interval is δt in seconds, then the interval of each data point in the frequency domain is $\delta v = 1/(N \, \delta t) \, \mathrm{s}^{-1}$ (also called hertz). In order not to lose an absorption peak we must have at least two points in each cycle. Assume that we have a sine function, if we sample it at intervals equal to the period of the function we will have a constant value, as can be seen in the Fig. 11.1 if we measure only where points are marked or only at crosses. However, if we measure at both points and crosses, we can observe that there is a change with time and that it is periodic. Thus in order to measure the frequency, v, in hertz we must measure for a time interval which is not larger than $1/2v$.

$$\delta t \leq \frac{1}{2v} \quad \Rightarrow \quad v \leq \frac{1}{2\delta t} \tag{11.25}$$

The frequency obtained by taking the equal sign is called the Nyquist critical frequency.

$$v_c = \frac{1}{2\delta t} \tag{11.26}$$

Anything that oscillates faster than this frequency will appear to be at a lower frequency. Thus, the rate of sampling establishes the range of observable frequencies.

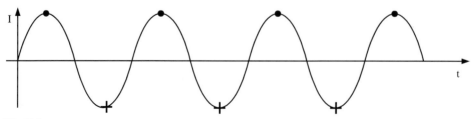

Fig. 11.1

The higher the sampling rate, the larger is the range of observable frequencies. Enlarging the sampling rate means that for equal measurement times more data points must be collected. Equation (11.25) can also be written in another form which correlates the number of data points collected, N, the acquisition time, T, and the range of observable frequencies, R:

$$N = 2RT \tag{11.27}$$

Thus for $N = 4000$ and $T = 4\,$s, R the range of observable frequencies is $500\,$Hz. A peak oscillating at $600\,$Hz will appear as $400\,$Hz in the transform.

11.4.2   Decaying peaks

In some forms of spectroscopy the signal decays with time, as for example in pulsed NMR. If this decay is not corrected it will lead to broader peaks. The faster the decay the broader the peak. In order to correct for the decay, the time series must have an additional term to describe the time decay, an exponential decay, $e^{-\lambda t}$:

$$J(t) = A \cos (wt)\, e^{-\lambda t} \tag{11.28}$$

In order to reduce the peak width the decaying term must be reduced. This is done by multiplying the measured time series by a correction factor of the form e^{+kt}. The corrected signal is:

$$J(t) = A \cos (wt)\, e^{(k-\lambda)t} \tag{11.29}$$

If $k < \lambda$ then we will still obtain a decaying time series but with slower decay, which results in a narrower line width in the transform than without the correction. However, although the multiplication by e^{kt} improves the resolution of the signal, it has the disadvantage that at the same time it amplifies the noise considerably. Without the problem of noise amplification we can try to optimize k to yield a peak of one data point width. A common method to filter the noise is to multiply the original time series by the factor $e^{kt - mt^2}$ and only then to calculate the transform. If k and m are chosen correctly the result will be an increased resolution without amplification of the noise.

11.5   References

Griffith, P. R. 1992, *Transform Techniques in Chemistry*, Plenum Press, New York.

Marshall, A. G. 1982, *Fourier, Hadamard and Hilbert Transforms in Chemistry*, Plenum Press, New York.

12 General and specific issues in uncertainty analysis

12.1 Introduction

In this chapter, general questions related to uncertainty and specific issues related to the calculation of uncertainties will be discussed. In Section 12.2, those general aspects that led to a situation in which we realized that there are fundamental limitations to uncertainties are presented. We also draw the reader's attention to the 'uncertainty era' in other human activities besides science. Sections 12.3 and 12.4 are devoted to the question – Do the laws of physics change? Section 12.5 is devoted to problems in statistical and systematic uncertainties. Section 12.6 deals with the 'bias factors' and Section 12.7 with the 'generalized bias operator method', both related to systematic uncertainties. Section 12.8 presents the calculation of uncertainties due to statistical uncertainties in the input parameters using sensitivities. In Section 12.9, the non-linear aspects of uncertainty analysis are considered. Section 12.10 deals with uncertainty analysis for several responses, and Section 12.11 gives data adjustments.

12.2 The uncertainty era

Our period can be characterized as the 'uncertainty era'. It is not that there were no uncertainties before; however, in our time, uncertainties have become both legitimate and unavoidable. In the past, we believed that uncertainties were temporary, and that we, as human beings, are moving in one direction – from an uncertain world to a certain one. In particular, this was considered to be true for the world of science.

As an example, consider the improved accuracy in determining the velocity of light. This accuracy was improved by a factor of more than 10^5 over about a hundred years, as demonstrated in Fig. 12.1 (today, the velocity of light is considered a constant value and the uncertainties are in the 'unit length' and the 'time unit').

However, the classical approach, as demonstrated by Fig. 12.1, is that uncertainties are reduced in time and, what is more important, that there is no limit to this process. To the contrary, at the beginning of the twentieth century we came to realize that movement from an uncertain to a certain world is utopian. We have learned that there are intrinsic limitations to our understanding and to our ability to comprehend our world.

I believe it all started in science, when Albert Einstein published his 'Relativity Theory' in 1905. Until then, it was believed that the basic concepts of science, such as *time* and *space*, were objective. *Time* was considered as part of objective reality, flowing regardless of anything. Einstein, however, demonstrated that *time* is different under different circumstances. Furthermore, *time* is meaningless without the

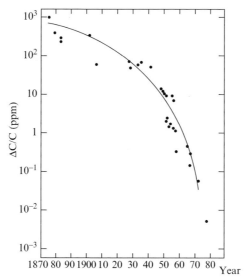

Fig. 12.1   The improvement of accuracy in determining the velocity of light.

means to measure it. Therefore, today we do not deal with *time per se*, but with *measured time*. The second blow to certainty in science came in 1927, when Heisenberg suggested the 'uncertainty principle'. According to this principle, it is impossible to specify precisely and simultaneously the values of both members of particular pairs of physical variables that describe the behavior of a physical system.

Two important pairs are the system (or the particle) momentum and its space coordinate or the system (particle) energy and the time at which it is measured. Quantitatively, the 'uncertainty principle' states that the product of the uncertainty regarding the knowledge of two such variables must be larger than Planck's constant, h, divided by 2π ($\hbar = h/2\pi = 1.054 \times 10^{-34} J\,s$). That is, for the above physical variables, we have:

$$\Delta x\, \Delta p \geq \hbar \tag{12.1}$$

and

$$\Delta E\, \Delta t \geq \hbar \tag{12.1a}$$

where x is the rectangular space coordinate of the system and p its corresponding component of momentum, E is the system energy and t is the time at which it is measured. The uncertainty of a variable A is defined as:

$$\Delta A = (\langle A^2 \rangle - \langle A \rangle^2)^{1/2} \tag{12.1b}$$

where $\langle A \rangle$ is the expectation value of A.

The meaning of Equation (12.1) is that the momentum of a system (particle) cannot be precisely specified without our loss of knowledge of the corresponding component and its position at that time and the location of the system (particle) in a particular direction cannot be located precisely without a loss of all knowledge of its momentum component in this particular direction. Equation (12.1a) means that to determine the energy of a system with an uncertainty ΔE requires at least a time interval of $\Delta t \approx \hbar/\Delta E$, in order to determine this accuracy in energy; on the other hand, for a system that is in a particular state for a time interval Δt, the uncertainty in its energy is at least $\hbar/\Delta t$.

The 'uncertainty principle' relates uncertainties in variables like energy and time to the Planck constant, $\hbar$. An interesting result is obtained if such uncertainties are considered in Planck's system of units, which are given by:

$$L_P = (\hbar G/c^3)^{1/2} = 1.62 \times 10^{-35}\,m$$
$$T_p = (\hbar G/c^5)^{1/2} = 5.39 \times 10^{-44}\,s \tag{12.2}$$
$$M_P = (\hbar c/G)^{1/2} = 2.18 \times 10^{-8}\,kg$$

Planck's system of units is the most basic system of units, since it is based on the three basic constants of nature, the speed of light c, the Planck constant, $\hbar$, and the gravitational constant, G. Using this system of units, for the 'uncertainty principle', we get:

$$\Delta E\, \Delta t = \Delta mc^2\, \Delta t \geq \hbar \tag{12.3}$$

and

$$\left(\frac{\Delta m}{M_P}\right)\left(\frac{\Delta t}{T_P}\right) \geq 1 \tag{12.3a}$$

In other words, the uncertainty in the mass in Planck's units multiplied by the uncertainty in time in Planck's units is of the order of one.

The 'uncertainty principle' eliminated one of the previous fundamental principles of science, namely, the 'causality principle', which claimed that if we know everything about a particular system we may know how this system will behave from now to eternity. Yet, the 'uncertainty principle' claims that we cannot know everything about a physical system, so we can never know accurately how it will behave in the future.

The third limitation to scientific 'objectivity' derives from mathematics. Mathematics is the language which science uses to describe reality. If the language is incomplete, our description of reality will be incomplete. In 1931, Kurt Gödel proved that mathematics is incomplete. Gödel proved that, in the framework of mathematics, you cannot have an absolute proof of consistency for any deductive system. He also demonstrated that there is an infinite number of true arithmetical statements which cannot be formally derived from any given set of axioms using a closed set of rules of inference. So the axiomatic approach, which is the basis of mathematics, cannot exhaust the domain of *mathematical truth*.

The 'trueness' of mathematical statements, that cannot be determined from a given set of axioms and a closed set of rules of inference, lies in the basic difference between 'true' and 'false', which are not just the opposite of each other. In order for a mathematical statement to be 'false', there must be at least one example in which it is false, even if we do not know of such an actual example. A similar argument is not valid for 'true' statements. Once there is an example in which the statement is false, then that statement can be determined to be 'false'. Following Gödel, such statements that cannot be determined to be either 'true' or 'false' must then be considered to be 'true'.

Uncertainty is not limited to science, but is equally valid in other aspects of human endeavor. Independently, and in a seemingly uncorrelated fashion, the role of uncertainties also began to play a role in the arts. The introduction of the abstract in art, and, in particular, in paintings and sculptures, marks the deviation from art which tries to be *certain* to art which introduces uncertainties and, ultimately, becomes completely *uncertain*. In the past, artists tried to convey the same message to their viewers; they tried to make their art as comprehensible as possible, to eliminate uncertainties, and to convey a specific message. This is not true for modern art, in which different viewers are expected to cull different meanings. Art is not *certain* any more. Artists do not necessarily have a specific intention or a predetermined message.

It is very surprising that the revolutions in science and art started at the same time. When Albert Einstein published his paper on relativity, Picasso in 1906 initiated modern art with his first cubist painting *The Mademoiselles from Avinion*. What began with Picasso in 1906/7, climaxed in the late twenties with the pure abstracts of Mondriaan, which exhibit complete uncertainty.

The wave of uncertainties in the visual arts did not spare other forms of art, such as music. Music is a less certain form of art. However, the structure, the rules and the melodies also play important roles in making music intelligible. Between 1908 and 1923, Arnold Schönberg introduced the twelve-tone system. Western music diverged from its classical basic harmonics and scales, from *certain* types of predictable melodic formulae. This tendency in music probably reached a peak of complete uncertainty in a piece called '4′ 3″ to Henri Flint' by the American composer John Cage in the year 1952. In this piece, the pianist sits before the piano in complete silence for precisely four minutes and three seconds.

Uncertainty was also introduced into other areas of life, like economics, at about the same time. The hyperinflation in Germany in 1923 and the collapse of the New York Stock Exchange in 1929 mark the loss of certainty in financial matters and introduced major uncertainties in economics, in general.

In our era, we have come to realize that we cannot avoid uncertainties. Uncertainties are legitimate aspects of our reality.

12.3 Uncertainties and the laws of nature

Uncertainties, as we understand them, are related to physical quantities and not to the laws of nature which determined their behavior. The laws of nature are usually considered to be *objective*, functioning regardless of any physical situation. However, we should ask ourselves if this picture a correct one? In order to try and answer this question, we should agree that the essence of any scientific relation or result lies in the ability (even if only in principle) to verify it experimentally.

The essence of scientific truth is embedded in experimental verification. If a scientific claim cannot be verified experimentally, it has no scientific meaning.

This restriction, which requires us to validate any scientific relation by experimentation, should also be applied to the laws of nature themselves. For example, consider one of the fundamental laws of nature, the 'law of electric charge conservation', which states that the *net electrical charge* should remain the same before and after any interaction in which charges are involved. However, an interesting question arises relating to conservation laws: Is the 'law of electric charge conservation' valid in a world with no electrical charge, as when the universe was in its very early stages? One can argue that the conservation of charge is valid even in a world with no charges at all, when the *net charge is zero*. However, such an argument cannot be validated by experiment and, as a result, has no scientific meaning. *In a world free of charges, there is no 'law of electric charge conservation'.*

Consider that there are some electric charges which are far away from any other electric charges. Nonetheless, the 'law of electric charge conservation' should still be applied to these distant electric charges, regardless of whether there are other interactive charges or not. In other words, the 'law of electric charge conservation' should be applied to the electric charges in question, even if there were no other charges in the universe.

Now consider a case in which the electric charges in question are a positron and an electron. The 'law of electric charge conservation' is also valid to these charges and their total sum is zero. If these particles were interacting in an *annihilation reaction*, this would result in no electron, no positron and a combined energy of at least $2m_e c^2$ (1.022 MeV), in the form of two or three gamma rays (two or three photons). You cannot apply the 'law of electric charge conservation' to gamma rays, so, we argue that, during an annihilation reaction, not only the electron and the positron disappear, but also the 'law of electric charge conservation' itself.

The same arguments are valid for the *pair production reaction*, in which the simultaneous creation of an electron and a positron is materialized by a photon (gamma ray), whose energy is greater than $2m_e c^2$.

Both the *annihilation* and *pair production reactions* cannot be explained in the framework of classical physics and are subject to the laws of quantum mechanics. So, as we will argue later, the creation and disappearance of a physical law are also subject to quantum mechanic considerations.

The theoretical basis for the explanation of the pair production and annihilation reactions was suggested by Dirac in 1928. Using special relativity, the total energy of a particle, like an electron, is given by:

$$E^2 = p^2 c^2 + m_e^2 c^4 \qquad (12.4)$$

where m_e is the electron rest mass and p its momentum. The energy, E, is given by:

$$E = \pm (p^2 c^2 + m_e^2 c^4)^{1/2} \qquad (12.5)$$

So, we have positive energies when $E > m_e c^2$ and negative energies when $E < -m_e c^2$. Dirac suggested that normally all negative energy states are filled with electrons. As a result, an electron in a positive energy state will be unable to go down to the negative energy state.

However, if an amount of energy $> 2m_e c^2$ were transferred to an electron in a negative energy state, it would be raised to a positive energy state and act like a normal electron. At the same time, a 'hole' would be created in the negative energy state. In the absence of a negative electron charge in the positive energy state, the 'hole' must have a positive charge; the 'hole' must also have an energy of $-E$. A negative energy in a negative energy state results in a 'hole' with a positive energy E and momentum $-p$. Thus, the 'hole' has the same energy as the electron, a positive charge and a negative momentum $-p$. This 'hole' is the positron. It should be mentioned that the energy required for this process is in the form of a gamma ray (photon), which also has momentum. So, in order for us to have conservation of energy and momentum simultaneously, the gamma photon must also interact with some other particles. Therefore, a *pair production reaction* cannot occur in a complete vacuum.

A similar, but opposite, reaction is the *annihilation reaction,* in which the electron and the positron annihilate each other, giving rise to two gamma photons, each with an energy $> m_e c^2$, or three gamma photons if the electron and the positron have parallel spins. Using the 'hole' description, the electron falls into the 'hole' and releases energy as a result.

The idea that we have negative energy electrons uniformly distributed throughout space is *meaningless*, since they do not give rise to any observable effect. Such a model acquires its meaning only during the process of the reaction and this is also in a *virtual* sense, as imposed by the 'uncertainty principle'.

Both the pair production and annihilation reactions are subject to the 'uncertainty principle', since the 'jump' of the electron from the negative energy state or the 'fall' of the electron into the 'hole' requires a jump over the forbidden energy interval obtained from Equation (12.5), namely:

$$-m_e c^2 < E < m_e c^2 \tag{12.6}$$

This reaction must be of the order of:

$$\Delta t \approx \frac{\hbar}{\Delta E} = \frac{\hbar}{2m_e c^2} = 6.44 \times 10^{-20}\, s \tag{12.7}$$

In conclusion, the 'law of electric charge conservation' is valid when we have electric charges and it is invalid when there are none. The transition period between the valid and invalid states is given by Equation (12.7). Let us define that when a law is 'valid', validity is defined as *one* and when a law is 'invalid', validity is defined as *zero*. We can plot the behavior or validity of the 'law of electric charge conservation' over time for a pair production reaction (Fig. 12.2) and an annihilation reaction (Fig. 12.3).

We have argued that the 'law of electric charge conservation' materializes and dematerializes during pair production and annihilation reactions. However, this is not the only conservation law which behaves in such a manner. The electron and the positron are leptons, whereas the photons are not. The 'law of lepton conservation'

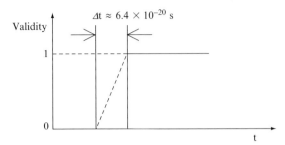

Fig. 12.2 Conservation of charge in a pair production reaction.

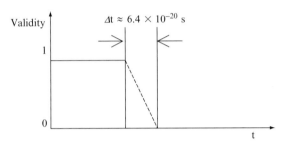

Fig. 12.3 Conversation of charge in an annihilation reaction

has meaning only when we have leptons. Thus, the 'law of lepton conservation' also appears and disappears during pair production and annihilation reactions.

Annihilation and pair production reactions are not limited to electrons and positrons, but also to bayrons. We can allocate a bayron number $B = 1$ and a $B = -1$ for anti-bayrons. Thus, both in pair production and annihilation reactions, we will have $B = 0$; the number of bayrons and anti-bayrons is the same. However, $B = 0$ has a meaning only if we do have bayrons, since $B = 0$ can also apply to the case in which we do not have any bayrons at all.

12.4   The creation of the universe and the law of energy and mass conservation

It is often argued that: 'You cannot get something from nothing', so the creation of the universe, in this respect, is most puzzling. If the universe was indeed 'created', and was actually created from 'nothing', then we do have an example of getting 'something' from 'nothing'. Therefore, *either the 'law of mass and energy conservation' is not valid or the universe was not created*, but rather underwent some kind of transformation.

However, this problem exists only if we believe that the laws of nature precede nature itself. The assumption is that the laws of nature were always there and that everything in the universe, and even the very creation of the universe, should be dictated by these laws. In some sense, this attitude towards the laws of nature is quite similar to the way 'time' was considered before the 'theory of relativity'. Time was considered to be a physical entity that flowed without change and regardless of anything. As a result, time could not be changed, stopped, interrupted, or made different for different observers. A similar approach is adopted when the laws of nature are considered. Such an approach, when applied to the laws of nature, results in this paradox in the matter of creation of the universe, which seems to disregard those very laws – being something from nothing. However, in the previous section, we argued that the laws of nature can be created or disappear, because *there is no law for the conservation of the laws of nature.*

Assume that the creation of the universe was from *nothing*. The 'law of mass and energy conservation' is inapplicable, senseless, when there is only *nothing*. We cannot measure anything in *nothing*, so it is meaningless to talk about the *conservation of mass and energy of nothing*! Such a law can only be applied when we have *something*, and not in the case of *nothing*.

The 'law of mass and energy conservation' is invalid within an uncertainty ΔE during a time interval Δt, where $\Delta E\,\Delta t \geq \hbar$. Let us consider that ΔE represents all the energy (mass) in the universe, namely a complete uncertainty of the total mass (energy) in the universe. This mass is estimated to be about 10^{53} kg, making the total energy of the universe about 10^{70} J. Given $\hbar = 1.054 \times 10^{-34}\,J\,s$, we have:

$$\Delta t \approx \hbar/\Delta E \cong 10^{-104}\,s \tag{12.8}$$

Thus, for an interval of 10^{-104} s, the 'law of mass and energy conservation' was inaccurate regarding the amount of total mass (energy) in the whole universe.

So, if we plot *validity* with respect to the 'law of mass and energy conservation', we will have Fig. 12.4.

Laws of nature can appear or disappear just like the materialization and dematerialization of physical quantities. The laws of nature play no independent role in relation to those things to which they apply. While there is a given relationship between a certain law and the subject to which it is applied, a law is not independent and does not precede its own subject.

In conclusion, we have argued that there is no meaning to a conservation law, if the physical entity which is supposed to be conserved does not exist. As a result, the *validity* of such a conservation law is zero. We have demonstrated this approach with respect to the 'law of electric charge conservation' during pair production and annihilation reactions, and to the 'law of bayron conservation'. We have also argued that such an approach can be adapted to the 'law of mass and energy

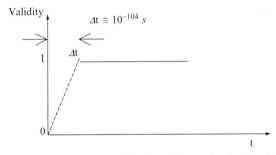

Fig. 12.4   The 'law of energy and mass conservation' from the creation ($t = 0$) of the universe.

conservation' when applied to the creation of the universe and to the specific question of creation from nothing – which became meaningless.

We have found that, in relation to three physical laws: the 'conservation of electric charge', the 'conservation of bayrons' and the 'conservation of mass and energy' during creation, the transition period between *zero validity* of a law to *full validity* of a law is governed by the 'uncertainty principle'. Is this always the case? Is it always true that the transition between the time when the law is valid to the time when the law is invalid is governed by the 'uncertainty principle'? I would speculate that the answer to this question is *Yes*.

12.5 Statistical and systematic uncertainties

We usually distinguish between statistical and systematic uncertainties or, as they are sometimes known, between statistical and systematic errors. The commonly used term 'error' might mean either *a mistake* or *an uncertainty*. When referring to statistical and systematic error, we mean an uncertainty and not a mistake. Thus, it is better to use 'uncertainty' rather than 'error'.

Systematic uncertainties arise due to negligence in the experiment or theory on which that experiment is based. Therefore, sometimes the systematic uncertainty is also called 'negligence uncertainty' or 'negligence error'. Statistical uncertainty is thought to arise from the finiteness of the input data ensemble.

However, in reality, the uncertainty assigned to the value of a parameter is related to the level of confidence we assign to this parameter, as compared to its true, unknown value. In this sense, the distinction between statistical and systematic uncertainties is an artificial one.

We would like to present in what follows a somewhat different approach to this problem. Let us measure a parameter α. This parameter is dependent on other parameters $(\beta_1, \ldots, \beta_N)$. We might divide these parameters into four groups. The effect of these parameters on α can be considered to be the effect of a change $\Delta\alpha$ in α, due to changes $\Delta\beta_i$ in the β_i parameters $(i = 1, \ldots, N)$. Thus:

$$\frac{\Delta\alpha}{\alpha} = \sum_{i=1}^{K} S_i \frac{\Delta\beta_i}{\beta_i}$$

$$+ \sum_{i=K+1}^{L} S_i \frac{\Delta\beta_i}{\beta_i} + \sum_{i=L+1}^{M} S_i \frac{\Delta\beta_i}{\beta_i} + \sum_{i=M+1}^{N} S_i \frac{\Delta\beta_i}{\beta_i} \qquad (12.9)$$

where

$$S_i = \frac{\partial\alpha/\alpha}{\partial\beta_i/\beta_i} \qquad (12.9a)$$

The first group, $i = 1, \ldots, K$, is the group of parameters for which we actually measure or calculate their effect on α. This group is characterized by large sensitivities, i.e. S_i $(i = 1, \ldots, K)$ are large. We usually like to have the value of K as small as possible, with $K = 1$ as the preferable situation. The second group, $i = K + 1, \ldots, L$, is also characterized by parameters with large sensitivities. However, any experiment is designed so that parameter α will be isolated from the parameters $\beta_{K+1}, \ldots, \beta_L$; i.e. by setting the experiment, we diffuse the effect of these parameters. Thus, by isolating the experiment from these parameters, we make S_i ($i = K + 1, \ldots, L$) equal to zero or reduce them to very small, negligible values. The determination of the value of an experiment in a system lies in our ability to reduce the number of parameters in the first group ($i = 1, \ldots, K$) and to isolate parameter α from the effect of the second group ($i = K + 1, \ldots, L$).

The third group, $i = L + 1, \ldots, M$, is also characterized by not large, but non-negligible sensitivities. Usually M is close to L. However, this third group is neither measured (does not belong to the first group) nor is it isolated from affecting α (does not belong to the second group). This third group of parameters is the source of the systematic uncertainty.

The fourth group of parameters, $i = M + 1, \ldots, N$, is characterized by two aspects: first, the sensitivities are very small, and second, N is large – approaching infinity. In this group, we find all the parameters which have very little effect on α and which we neglect. There is almost an infinite number of such parameters. This fourth group of parameters is the source of statistical uncertainty. According to the 'central limit theorem', variants, which are the sum of many independent effects, tend to be normally distributed as the number of those effects becomes large.

For example, let us consider an experiment designed to measure the acceleration constant, g. A steel ball is dropped from a given height. The time it takes the ball to reach the ground is measured. In order to reduce the effect of air friction, this experiment is performed in a vacuum. In this case, the two important parameters in the first group ($K = 2$) are the height and the time. In the second group of important parameters (from which the experiment is isolated), we have one parameter ($L = 3$), and it is the air friction. This factor is removed by performing the experiment in a vacuum. In the third group, which is the source for the systematic uncertainty, we may find signs of the fact that the vacuum is not complete, as we assumed. Another source of systematic uncertainty is the latitude of the experimental techniques used. However, there are many parameters which have very little effect on the outcome of an experiment, yet contribute to statistical uncertainty – the temperature, for instance, which might affect either the pressure of the incomplete vacuum or the determination of the height. Other examples of minor influences are the gravity of the moon and, to a lesser extent, the gravity of other planets. All of these make small, but significant, contributions. All of them contribute to statistical uncertainty.

12.6   Bias Factors (BF)

Let us calculate a response or a series of n responses, $R_{0,i}$ ($i = 1, \ldots, n$), of a given system. Assume that we have also measured these responses. These measured responses are denoted as R_i ($i = 1, \ldots, n$). If the difference between our measured responses and our calculated responses are within 'the uncertainties' of the calculated responses, then there is no problem. By 'the uncertainties' of the calculated responses, we mean their uncertainties due to the statistical uncertainties of the input parameters.

However, if there are differences between our calculated responses and our measured ones, which are not explained by the statistical uncertainties of the input parameters, this means that these are *systematic (errors) uncertainties.*

Another case which indicates that there are systematic errors is when we have measured the responses R_i ($i = 1, \ldots, n$) many times and the average measured responses $\overline{R}_i$ ($i = 1, \ldots, n$) are derived from the calculated responses $R_{0,i}$ ($i = 1, \ldots, n$). In such cases, we define the Bias Factor (BF) as:

$$\mathrm{BF}_i = \frac{R_i}{R_{0,i}} \quad or \quad \frac{\overline{R}_i}{R_{0,i}} \qquad i = 1, \ldots, n \tag{12.10}$$

BF_i are used in different cases that should be close to the original case which determined the initial BF. In a new system, BFs multiply the same responses in order to obtain a more accurate response. The underlying assumption in the application of the Bias Factors (BF) method is that the systematic errors, which cause the difference between the calculated and measured responses in a given system, will be about the same in a new system that is close to the original system.

The main advantage of the BF method is the ease with which it can be applied. On the other hand, the main disadvantage of this method is that it is arbitrarily applied. The only justification for applying this method lies in the assumption that systems which are close will have the same systematic errors. This leads to the question: How close is close enough? To such a question, there is no clear-cut answer, only our experience to guide us.

A different methodology, which is an alternative to the BF method, is the generalized Bias Operator (BO) method (Ronen *et al.* 1981; Navon & Ronen 1988). A single, generalized Bias Operator (BO) is obtained and used to improve the prediction of all performance parameters in new systems. The main advantages of this method are twofold: application of this method is simple and inexpensive with respect to computational requirements and the reasoning behind the method is less arbitrary than that of the BF method. As a result, the application of the BO method to a new system also necessitates closeness to the original system from which the BO was obtained. In several cases (Navon & Ronen 1988) in which both methods were compared, better results were obtained by the application of the generalized BO method.

12.7 The generalized Bias Operator (BO) method

Consider a system that is calculated by the numerical solution of the homogeneous equation:

$$\mathbf{A}_H \mathbf{f}_H^0 = \gamma_0 \mathbf{f}_H^0 \tag{12.11}$$

where $\mathbf{f}_H^0$ is the calculated eigenvector with N components, and the operator $\mathbf{A}_H$ is a $N \times N$ matrix. In practice, Equation (12.11) can take on a wide variety of representations. The calculated eigenvalue of this system is γ_0.

Now suppose now that 'reference' values for both the 'true' eigenvector $\mathbf{f}_H$ and the eigenvalue γ of the *same* system are also available. In general, the reference values $\mathbf{f}_H$ and γ do not coincide with the computed quantities $\mathbf{f}_H^0$ and γ_0, respectively. Consequently, $\mathbf{f}_H$ and γ may conceptually be considered to satisfy the following perturbed version of Equation (12.11):

$$(\mathbf{A}_H + \delta\mathbf{A}_H)\mathbf{f}_H = \gamma\mathbf{f}_H \tag{12.12}$$

The perturbation-like operator $\delta\mathbf{A}_H$, appearing in Equation (12.12), is defined as the generalized BO (Ronen *et al.* 1981; Navon & Ronen 1988) and is thought to account for all the sources of discrepancies (both known and unknown) in the solutions of Equation (12.11) and the reference quantities. If $\delta\mathbf{A}_H$ were known, the 'simple-to-solve' Equation (12.11) could be appropriately corrected to yield the reference values $\mathbf{f}_H$ and γ. It is, therefore, desirable to determine $\delta\mathbf{A}_H$. To this end, Equation (12.12) is rewritten as:

$$\delta\mathbf{A}_H \, \mathbf{f}_H = \mathbf{R}_H \tag{12.13}$$

where

$$\mathbf{R}_H = (\gamma - \mathbf{A}_H)\mathbf{f}_H \tag{12.13a}$$

or alternatively

$$\mathbf{R}_H = (\gamma_0 - \mathbf{A}_H)(\mathbf{f}_H - \mathbf{f}_H^0) + \mathbf{f}_H(\gamma - \gamma_0) \tag{12.13b}$$

is a known quantity.

The concepts presented so far are not restricted to homogeneous systems, but may also be employed to analyze source driven systems. In this vein, consider the equation:

$$\mathbf{A}_I\mathbf{f}_I^0 = \mathbf{S}_0 \tag{12.14}$$

to represent a calculated inhomogeneous system; the corresponding 'I' inhomogeneous system would then be described by:

$$(\mathbf{A}_I + \delta\mathbf{A}_I)\mathbf{f}_I = \mathbf{S} \tag{12.15}$$

Note that, in general, the operators $\delta\mathbf{A}_H$ and $\delta\mathbf{A}_I$ differ from one another, even though the operators $\mathbf{A}_H$ and $\mathbf{A}_I$ may be identical. The equation satisfied by $\delta\mathbf{A}_I$ is similarly obtained by subtracting Equation (12.14) from Equation (12.15). Thus:

$$\delta\mathbf{A}_I \, \mathbf{f}_I = \mathbf{R}_I \tag{12.16}$$

where

$$\mathbf{R}_I \equiv \mathbf{S} - \mathbf{A}_I\mathbf{f}_i = (\mathbf{S} - \mathbf{S}_0) - \mathbf{A}_I(\mathbf{f}_I - \mathbf{f}_I^0) \tag{12.17}$$

is a known quantity again.

At this stage, one can observe that the structures of both Equation (12.13) and (12.16) are formally identical. The respective subscripts can, therefore, be dropped, and either one of them can be represented in matrix form as:

$$\delta\mathbf{A} \, \mathbf{f} = \mathbf{R} \tag{12.18}$$

Note here that Equation (12.18) is exact. For subsequent manipulations, it is convenient to write Equation (12.18) in component form, as:

$$\sum_{j=1}^{N} a_{ij} f_j = R_i \quad i = 1, \ldots, N \tag{12.19}$$

where a_{ij}, f_j, and R_i are the components of $\delta \mathbf{A}, \mathbf{f}$, and $\mathbf{R}$, respectively. The objective of the procedure proposed here is to obtain the elements a_{ij} of $\delta \mathbf{A}$ for given $\mathbf{R}$ and $\mathbf{f}$. Since the number (N^2) of unknowns a_{ij} substantially exceeds the number (N) of equations available, the problem represented by Equation (12.19) admits a unique solution, only if additional constraints are imposed.

The elements a_{ij} of $\delta \mathbf{A}$ represent deviations from the corresponding elements A_{ij} of the original matrix $\mathbf{A}$. It is desirable to minimize these deviations. This can be achieved by considering the quantities a_{ij} as elements of an N^2 dimensional vector, so that a least squares procedure can be applied to:

$$Q' = \sum_{i,j=1}^{N} W_{ij} a_{ij}^2 \tag{12.20}$$

Since the quantities a_{ij} also have to satisfy Equation (12.19), they can be determined by minimizing the function:

$$Q = \sum_{i,j=1}^{N} W_{ij} a_{ij}^2 + \sum_{i=1}^{N} \lambda_i (R_i - \sum_{j=1}^{N} a_{ij} f_j) \tag{12.21}$$

with respect to a_{ij}. Here, the quantities λ_i are the conventional Lagrange multipliers, and W_{ij} are non-negative weighting factors. The assignment of numerical values to these weightings is determined by the importance we assign to the elements of the operator a_{ij}, if we can, otherwise $W_{ij} = 1$.

The minimum of the function Q is determined by requiring its first order derivatives with respect to a_{ij} to vanish. This yields:

$$\frac{\partial Q}{\partial a_{ij}} = 2 W_{ij} a_{ij} - \lambda_i f_j = 0 \quad i, j = 1, \ldots, N \tag{12.22}$$

The unique solutions λ_i and a_{ij} to the system of equations obtained by augmenting Equations (12.19) and (12.22) are:

$$\lambda_i = \frac{2R_i}{\sum_{j=1}^{N} f_j^2 / W_{ij}} \tag{12.23}$$

and

$$a_{ij} = \frac{R_i f_j}{W_{ij} \sum_{k=1}^{N} f_k^2 / W_{ik}} \tag{12.24}$$

Since the combination of terms arising from taking the second order derivatives of Q, with respect to a_{ij}, has a positive sign, Q is indeed minimized.

12.8   The statistical paradox

There is a lot of confusion between the notions of the *probability of an event* and the *a priori probability of an event*. Let us consider the example of throwing a die a 1000 times. We will get a series of 1000 numbers. The a priori probability of getting these numbers is $(1/6)^{1000}$ or 10^{-778}. Although this is an extremely low probability, it nonetheless happened and nobody is surprised. In the real world, events occur all the time that have extremely low a priori probabilities. However, there are events which occur, with extremely low a priori probabilities, yet people look for a different explanation, rather than accepting the 'random' one.

Perhaps one of the most typical examples is that of the creation of life. We do not know how life was created. However, if it was a result of some random processes, the probability for such a random process or processes is extremely low. As a result, people are looking for a different explanation, rather than accepting the possible occurrence of a random process. *It might be true that there is a non-random process which is responsible for the creation of life*, however, an event (life) which happened with an extremely low a priori probability does not exclude a random process, as in the case of dice. So, *if life indeed came to be due to a random process, its extremely low a priori probability makes it a unique event.*

Generally, low probability events do not happen. If so, then extremely low probability events certainly never occur. This leads us to the following observation, which I call the *statistical paradox*.

> *Extremely low probability events generally do not happen. Yet, all the time, events with extremely low a priori probabilities do happen!*

What is *extremely low probability*? In order to answer this question, let us consider what constitutes a *low probability* event. One of the important aspects of low probability is related to *risk*. *Risk*, for an individual, is defined as:

$$risk = \frac{\text{number of people killed in a year in a given accident}}{\text{number of people exposed to the danger of this given accident}}$$

In Table 12.1, we present the *risk* to an individual due to different types of accidents.

A *risk* with a probability of 10^{-5} or higher is considered to be a risk by people, and they take measures to decrease it. For instance, people will drive wearing safety belts. Cars have many safety features, and the safety level of a car is an important consideration when buying a car. People also consider drowning as a risk to be minimalized, so they generally swim in places with a lifeguard. However, traveling by train is usually not considered risky. So, *risks* with a probability of 10^{-7} or lower are not usually considered as such. In lotteries, however, people consider probabilities even lower than 10^{-8} as being worthwhile. However, events with a probability of 10^{-9} or lower are not considered to be tenable. So *we can define events with a probabilty of 10^{-10} or lower as being extremely low probabilities.*

Table 12.1   The *risk* to an individual as a result of different accidents.

Type of accident	The *risk*
Car accident	3×10^{-4}
Drowning	3×10^{-5}
Air accident	9×10^{-6}
Train accident	4×10^{-6}

12.9 The rejection test

Given an event that happens N times, we would like to know if it is random or if it is due to some possible non-random process. The rejection test (Ronen & Dubi 1983) can help us make such a determination.

Let us denote D_N for N different cases which have the event D. Let A_1 represent the system in a random state and A_0 in a non-random state. Consider the conditional probability $P(A_1|D_N)$. This conditional probability reflects the probability that, if the event D happened N times (D_N), the system is in a random state A_1.

On the other hand, $P(D_N|A_1)$ is the probability of obtaining the D event N times (D_N), if it is in a random state or the procedure is random. Therefore, $P(D_N|A_0)$ is the probability of obtaining the D event N times (D_N), if the system is in a non-random state A_0 and this state can yield these events. In such a case, $P(D_N|A_0) = 1$.

For example, look at dice. If a die is normal, and it is in A_1 state, the probability of getting an ace N times (D_N) will be $P(D_N|A_1) = (1/6)^N$. On the other hand, if the die is loaded, i.e. in a non-random state A_0, all the events will yield an ace and $P(D_N|A_0) = 1$ will be the case. Here the conditional probability is given by:

$$P(A_1|D_N) = \frac{P(D_N|A_1)}{P(D_N|A_0)P(A_0) + P(D_N|A_1)P(A_1)}$$

$$\leq \frac{P(D_N|A_1)P(A_1)}{P(D_N|A_0)P(A_0)} = P(D_N|A_1)\frac{P(A_1)}{P(A_0)} = P(D_N|A_1)\gamma \tag{12.25}$$

where $\gamma = P(A_1)/P(A_0)$ is the ratio of the probabilities of being in state A_1 or A_0. Consider a case in which only one die out of 1000 dice is loaded, then $\gamma = 10^3$.

We have discussed the case of low probabilities in our *statistical paradox* and claimed that *generally* very low probability events do not happen. Let us define probabilities of lower than β as such probabilities. In accordance with this, the rejection test claims that if $P(A_1|D_N)\gamma$ is lower than β for a given system, it cannot accept that that system is in a random state, A_1, and prefers non-random state, A_0, situations. Namely, it rejects state A_1 if:

$$P(D_N|A_1)\gamma \leq \beta \tag{12.26}$$

As an example, consider β to be 10^{-6} and set $\gamma = 1000$. Then the rejection test requires that:

$$\left(\frac{1}{6}\right)^N \times 1000 \leq 10^{-6} \tag{12.27}$$

and $N \geq 13$. So if an ace appears 13 times in a row, then 'draw your gun', because that die is loaded.

12.10   Uncertainty analysis based on sensitivity analysis

Consider a physical parameter, denoted as a *response* and symbolized by the letter R. This response is a calculated physical parameter and it is a real-valued function of the input parameters. These input parameters are real scalars $\alpha_1, \ldots, \alpha_k$. All the input parameters can be presented as a reactor $\boldsymbol{\alpha}$ namely:

$$\boldsymbol{\alpha} = (\alpha_1, \ldots, \alpha_k)^T \tag{12.28}$$

The nominal values of these input parameters, on the best values that we have for them, will be denoted as:

$$\boldsymbol{\alpha}^{(0)} = \left(\alpha_1^{(0)}, \ldots, \alpha_k^{(0)} \right)^T \tag{12.29}$$

The response R is a real function of the input parameters, namely:

$$R(\boldsymbol{\alpha}) = R(\alpha_1, \ldots, \alpha_k) \tag{12.30}$$

A change in any one of the input parameters, let us say the ith input parameter, is given by:

$$\delta\alpha_i = \alpha_i - \alpha_i^{(0)} \tag{12.31}$$

Changes in all (some of them can be zero) the input parameters will be denoted as:

$$\delta\boldsymbol{\alpha} = \boldsymbol{\alpha} - \boldsymbol{\alpha}^{(0)} \tag{12.32}$$

A change in one or more of the input parameters will cause a change in the response. An expansion of the new response $R(\boldsymbol{\alpha})$ in a Taylor series around the numeric values of the input parameters $\boldsymbol{\alpha}^{(0)}$, with the retention of terms up to the second order in the variation $\delta\boldsymbol{\alpha}$ around $\boldsymbol{\alpha}^{(0)}$, gives:

$$R(\boldsymbol{\alpha}) = R\left(\boldsymbol{\alpha}^{(0)}\right) + \sum_{i=1}^{k} \left.\frac{\partial R}{\partial \alpha_i}\right|_{\alpha_i^{(0)}} \delta\alpha_i \tag{12.33}$$

In other words, the new response is determined by the first order term of the Taylor series. Equation (12.33) can also be presented as:

$$\delta R(\boldsymbol{\alpha}) = R(\boldsymbol{\alpha}) - R\left(\boldsymbol{\alpha}^{(0)}\right) = \sum_{i=1}^{k} \left.\frac{\partial R}{\partial \alpha_i}\right|_{\alpha_i^{(0)}} \delta\alpha_i \tag{12.34}$$

Let us take the square value of both sides of Equation (12.34) and let us multiply it by the PDF, $f(\boldsymbol{\alpha})$, which is the Joint Continuous Probability Density Function. (We will elaborate on the PDF later.) In the second stage, we will perform an integration over all the space of the input parameters. The result is:

$$\int\limits_{-\infty}^{\infty} \cdots \int\limits_{-\infty}^{\infty} [\delta R(\boldsymbol{\alpha})]^2 \, f(\alpha) \, d\alpha_1, \ldots, d\alpha_k = \int\limits_{-\infty}^{\infty} \cdots \int\limits_{-\infty}^{\infty} \left[\sum_{i=1}^{k} \frac{\partial R}{\partial \alpha_i} \bigg|_{\alpha_i^{(0)}} \delta\alpha_i \right]^2 f(\boldsymbol{\alpha}) \, d\alpha_1, \ldots, d\alpha_k$$

(12.35)

The left-hand side (LHS) of Equation (12.35) is the definition of the variance of the response or the square of its standard deviation. Namely:

$$\text{var}(R) = \sigma^2(R) = \int\limits_{-\infty}^{\infty} \cdots \int\limits_{-\infty}^{\infty} [\delta R(\boldsymbol{\alpha})]^2 \, f(\boldsymbol{\alpha}) \, d\alpha_1, \ldots, d\alpha_k$$

(12.36)

The right-hand side (RHS) of Equation (12.35) gives:

$$\int\limits_{-\infty}^{\infty} \cdots \int\limits_{-\infty}^{\infty} \left[\frac{\partial R}{\partial \alpha_i} \delta\alpha_i \right]^2 f(\boldsymbol{\alpha}) \, d\alpha_1, \ldots, d\alpha_k$$

$$= \int\limits_{-\infty}^{\infty} \cdots \int\limits_{-\infty}^{\infty} \sum_{i=1}^{k} \sum_{j=1}^{k} \frac{\partial R}{\partial \alpha_i} \frac{\partial R}{\partial \alpha_j} \delta\alpha_i \, \delta\alpha_j \, f(\boldsymbol{\alpha}) \, d\alpha_1, \ldots, d\alpha_k$$

(12.37)

Since $\frac{\partial R}{\partial \alpha_i}$ and $\frac{\partial R}{\partial \alpha_j}$ are not statistical values and are not influenced by the statistical nature of the input parameters as determined by the PDF, $f(\boldsymbol{\alpha})$, we can take these terms out of the integration. Therefore, for Equation (12.37) we have:

$$\sum_{i=1}^{k} \sum_{j=1}^{k} \frac{\partial R}{\partial \alpha_i} \bigg|_{\alpha_i^0} \frac{\partial R}{\partial \alpha_j} \bigg|_{\alpha_j^{(0)}} \int\limits_{-\infty}^{\infty} \cdots \int\limits_{-\infty}^{\infty} \delta\alpha_i \, \delta\alpha_j \, f(\boldsymbol{\alpha}) \, d\alpha_1, \ldots, d\alpha_k$$

(12.38)

We denote $< \delta\alpha_i \, \delta\alpha_j >$ as:

$$< \delta\alpha_i \, \delta\alpha_j > = \int\limits_{-\infty}^{\infty} \cdots \int\limits_{-\infty}^{\infty} \delta\alpha_i \, \delta\alpha_j \, f(\boldsymbol{\alpha}) \, d\alpha_1, \ldots, d\alpha_k$$

(12.39)

which is the variance-covariance of the input parameters. Namely:

$$< \delta\alpha_i \, \delta\alpha_j > = \text{cov}(\alpha_i \, \alpha_j) \quad i \neq j$$

(12.40)

$$< \delta\alpha_i^2 > = \text{var}(\alpha_i) = \sigma_i^2 \quad i = j \quad (12.41)$$

The covariance of Equation (12.40) can also be represented by:

$$\text{cov}(\alpha_i \alpha_j) = \rho_{ij} \sigma_i \sigma_j$$

(12.42)

and ρ_{ij} is the corresponding off-diagonal element of the correlation matrix for the situation in which:

$$-1 \leq \rho_{ii} \leq 1$$

(12.43)

and the diagonal elements ρ_{ii} are equal to one.

Input parameters are *measured values*. If we have measured an input parameter N times, the nominal value of the input parameter $\alpha_i^{(0)}$ is, in fact, an average value. Namely:

$$\alpha_i^{(0)} = \frac{\sum_{j=1}^{N} \alpha_{i,j}^{(0)}}{N} \tag{12.44}$$

where $\alpha_{i,j}^{(0)}$ is the jth measurement of the input parameter $\alpha_i^{(0)}$. The standard deviation of the input parameter α_i is given by:

$$\sigma_i^2 = \frac{1}{N-1} \sum_{j=1}^{N} (\alpha_{i,j} - \alpha_i^{(0)})^2 \tag{12.45}$$

These concepts require further elaboration. When considering an input parameter α_i, its determined value is $\alpha_i^{(0)} \pm \sigma_i$, where $\alpha_i^{(0)}$ is the most probable value, and the uncertainty of this value is characterized by its standard derivation σ_i. Usually, the most probable value and its uncertainty are determined by experiments, as we have shown. In other cases, the input parameter is calculated, but the input parameters for this calculation are validated experimentally. Thus, we can even consider calculated input parameters, such as these, to be experimental ones. The implication of saying that the value of an input parameter α_i is $\alpha_i^{(0)} \pm \sigma_i$ is that when repeating the measurements of α_i with the same experimental device, the *a priori* probability of obtaining α_i within the range of $d\alpha_i$ is usually given by the normal distribution PDF:

$$f(\alpha_i)d\alpha_i = (2\pi\sigma_i^2)^{-1/2} \exp\left[-1/2\left(\frac{\alpha_i - \alpha_i^{(0)}}{\sigma_i}\right)^2\right] d\alpha_i \tag{12.46}$$

In this sense, a given experimental device, created to measure the input parameter α_i, may also be considered as a generator of random numbers from the normal distribution Equation (12.46), with a mean value of $\alpha_i^{(0)}$ and a standard deviation of σ_i. In such a situation, it is justifiable to treat the possible result of any measurement as a random variable and the actual experiment data as a sample drawn from a well-defined distribution. Furthermore, we have:

$$\langle \delta\alpha_i^2 \rangle = \left\langle \left(\alpha_i - \alpha_i^{(0)}\right)^2 \right\rangle = \int_{-\infty}^{\infty} \left(\alpha_i - \alpha_i^{(0)}\right)^2 f(\alpha_i)\, d\alpha_i$$

$$= \int_{-\infty}^{\infty} (\alpha_i - \alpha_i^{(0)})^2 (2\pi\sigma_i^2)^{-1/2} \exp\left[-1/2\left(\frac{\alpha_i - \alpha_i^{(0)}}{\sigma_i}\right)^2\right] d\alpha_i = \sigma_i^2 \tag{12.47}.$$

Equation (12.47) is valid for a single input parameter α_i. In a case where we have k input parameters, and each input parameter is measured independently, the PDF will be a generalization of the normal distribution given in Equation (12.47), namely:

$$f(\boldsymbol{\alpha})\, d\alpha_1, \ldots, d\alpha_k = [(2\pi)^k \sigma_1^2 \sigma_2^2 \ldots \sigma_k^2]^{-1/2}$$

$$\exp\left[-1/2\sum_{i=1}^{k}\left(\frac{\alpha_i - \alpha_i^{(0)}}{\sigma_i}\right)^2\right] d\alpha_1, \ldots, d\alpha_k \tag{12.48}$$

Combining all the terms derived from Equation (12.35), we end up with the following substitute for Equation (12.38):

$$
\begin{aligned}
\operatorname{var} R(\boldsymbol{\alpha}) &= \sum_{i=1}^{k}\sum_{j=1}^{k} \left.\frac{\partial R}{\partial \alpha_i}\right|_{\alpha_i^{(0)}} \left.\frac{\partial R}{\partial \alpha_j}\right|_{\alpha_j^{(0)}} \langle \delta\alpha_i\, \delta\alpha_j \rangle \\
&= \sum_{i=1}^{k}\sum_{j=1}^{k} \left.\frac{\partial R}{\partial \alpha_i}\right|_{\alpha_i^{(0)}} \left.\frac{\partial R}{\partial \alpha_j}\right|_{\alpha_j^{(0)}} \operatorname{cov}(\alpha_i\, \alpha_j) \\
&= \sum_{i=1}^{k}\sum_{j=1}^{k} \left.\frac{\partial R}{\partial \alpha_i}\right|_{\alpha_i^{(0)}} \left.\frac{\partial R}{\partial \alpha_j}\right|_{\alpha_j^{(0)}} \rho_{ij}\sigma_i\sigma_j
\end{aligned}
\tag{12.49}
$$

Thus, Equation (12.49) relates the uncertainties of the input parameters to the uncertainty of the response.

We usually denote the derivative

$$
\left.\frac{\partial R}{\partial \alpha_i}\right|_{\alpha_i^{(0)}} = S_i
$$

as a sensitivity element or as *sensitivity*. This equation relates the input parameter α_i to the response R. The higher the sensitivity value, the more important the input parameter in question. Equation (12.49) can also be written as:

$$
\operatorname{var} R(\boldsymbol{\alpha}) = \sum_{i=1}^{k}\sum_{j=1}^{k} S_i\, S_j\, \operatorname{cov}(\alpha_i\, \alpha_j)
\tag{12.50}
$$

One of the interesting cases is the uncorrelated one. In this case, the uncorrelated matrix elements $\rho_{ij} = 0$ for $i \neq j$. This uncorrelated case happens when the input parameters are independent of each other and their measurements are also independent.

Consider a case in which two independent input parameters α_i and α_j are both influenced by the same factors. For example, let both input parameters α_i and α_j be sensitive to temperature and pressure. If both input parameters are measured at the same time over a period of time, we expect the uncertainties σ_i of α_i and σ_j of α_j to be correlated. Usually, there will be no correlation if the measurement of each of these input parameters was performed independently. However, if part of the uncertainty in parameter α_i is due either to instrumentation or to measurement technique, and if the same instruments or techniques were also used to measure parameter α_j, we do expect that there will be a correlation between the uncertainties of these two parameters.

In order to determine if the uncertainties of the parameters are uncorrelated, we need to show that the correlation elements ρ_{ij} $(i \neq j)$ are zero or very small. Consider that two parameters α_i and α_j are each measured N times. Let $\alpha_{i,k}$ be the value of the kth measurement of the input parameter α_i and $\alpha_{j,k}$ be the kth measurement of the α_j input parameter. As a result, the correlation matrix elements are given by:

$$
\rho_{i,j} = \frac{\sum_{k=1}^{N} (\alpha_{i,k} - \alpha_i^{(0)})(\alpha_{j,k} - \alpha_j^{(0)})}{\left[\sum_{k=1}^{N} (\alpha_{i,k} - \alpha_i^{(0)})^2\ \sum_{k=1}^{N} (\alpha_{j,k} - \alpha_j^{(0)})^2\right]^{1/2}}
\tag{12.51}
$$

where $\alpha_i^{(0)}$ and $\alpha_j^{(0)}$ are calculated from Equation (12.44). For this uncorrelated case, Equation (12.50) will have the form:

$$\text{var } R(\alpha) = \sum_{i=1}^{k} S_i^2 \text{ var } (\alpha_i)$$

$$= \sum_{i=1}^{k} S_i^2 \, \sigma_i^2 \tag{12.52}$$

or

$$\sigma_R = \sqrt{\left(\sum_{i=1}^{k} S_i^2 \sigma_i^2 \right)} \tag{12.53}$$

Namely, the standard deviation of the response is equal to the square root of the sum of the variances of the input parameters weighted by the square value of the sensitivity.

For instance, consider a response R, which is:

$$R = \alpha_1 + \alpha_2 \tag{12.54}$$

Its input parameters are $\alpha_1 \pm \sigma_1$ and $\alpha_2 \pm \sigma_2$. Its sensitivities are $\partial R / \partial \alpha_1 = 1$ and $\partial R / \partial \alpha_2 = 1$. So:

$$\sigma_R = \sqrt{(\sigma_1^2 + \sigma_2^2)} \tag{12.55}$$

for

$$R = 5\alpha_1 + \alpha_2 \tag{12.56}$$

giving

$$\sigma_R = \sqrt{(25\sigma_1^2 + \sigma_2^2)} \tag{12.57}$$

There are cases in which there is an advantage in using relative relations. Let us divide both sides of Equation (12.52) by R^2, then multiply and divide the RHS by α. We get:

$$\frac{\sigma_R^2}{R^2} = \left(\frac{\sigma_R}{R} \right)^2 = \sum_{i=1}^{k} \left(\frac{\partial R / R}{\partial \alpha_i / \alpha_i} \right)_{\alpha_i^{(0)}}^2 \left(\frac{\sigma_i}{\alpha_i} \right)^2 \tag{12.58}$$

So the relative uncertainty of this response is σ_R / R, as determined by the relative uncertainties of the input parameters σ_i / α_i $(i = 1, \ldots, k)$ and by the relative sensitivity value $(\partial R / R)/(\partial \alpha_i / \alpha_i)$.

As an example, consider the following simple problem. The uncertainty in measuring the radius of a circle is 1%. What is the uncertainty of the circle area squared? The area of the circle is $R = \pi D^2 / 4$ and the input parameter of α is the diameter D. Thus, we have:

$$\frac{\partial R}{\partial \alpha} = \frac{\pi D}{2} \tag{12.59}$$

and

$$\frac{\partial R/R}{\partial \alpha/\alpha} = \frac{\pi D}{2} \frac{\alpha}{R} = \frac{\pi D}{2} \frac{D}{\pi D^2} \frac{4}{} = 2 \tag{12.60}$$

So

$$\frac{\sigma_R}{R} = \sqrt{(2^2 \times 0.01^2)} = 0.02 \tag{12.61}$$

Thus, the relative uncertainty of the area is 2%.

Another example is:

$$R = \alpha_1 \, \alpha_2 \tag{12.62}$$

$$\frac{\partial R/R}{\partial \alpha_1/\alpha_1} = 1 \qquad \frac{\partial R/R}{\partial \alpha_2/\alpha_2} = 1 \tag{12.63}$$

so

$$\frac{\sigma_R}{R} = \sqrt{\left[\left(\frac{\sigma_1}{\alpha_1} \right)^2 + \left(\frac{\sigma_2}{\alpha_2} \right)^2 \right]} \tag{12.64}$$

Or, we have

$$R = \frac{\alpha_1}{\alpha_2} \tag{12.65}$$

In this case

$$\frac{\partial R/R}{\partial \alpha_1/\alpha_1} = 1 \quad \text{and} \quad \frac{\partial R/R}{\partial \alpha_2/\alpha_2} = -1 \tag{12.66}$$

and

$$\left(\frac{\sigma_R}{R} \right)^2 = \sqrt{\left[\left(\frac{\sigma_1}{\alpha_1} \right)^2 + \left(\frac{\sigma_2}{\alpha_2} \right)^2 \right]} \tag{12.67}$$

The standard deviation of an average can also be calculated in this manner. Given a set of parameters X_i ($i = 1, \ldots, N$) and their uncertainties σ_i ($i = 1, \ldots, N$), the average value of X_i is:

$$\overline{X} = \sum_{i=1}^{N} \frac{X_i}{N} \tag{12.68}$$

and

$$S_i = \frac{\partial \overline{X}}{\partial X_i} = \frac{1}{N}$$

$$\sigma(\overline{X}) = \sqrt{\left(\sum_{i=1}^{N} S_i^2 \sigma_i^2 \right)} = \sqrt{\left(\sum_{i=1}^{N} \frac{\sigma_i^2}{N^2} \right)} \tag{12.69}$$

If all the uncertainties σ_i are the same and $\sigma_i = \sigma$, for Equation (12.69) we will have:

$$\sigma(\overline{X}) = \sqrt{\left(\sum_{i=1}^{N} \frac{\sigma_i^2}{N^2}\right)} = \sqrt{\left(\frac{N\sigma^2}{N^2}\right)} = \frac{\sigma}{\sqrt{N}} \tag{12.70}$$

Namely, the uncertainty of the average is equal to the uncertainty of the measured value X_i divided by the square root of the number of parameters. So we see that what we should expect is that the larger the number of parameters, the smaller the uncertainty of the average.

12.10.1   Determining the sensitivities

We have seen that sensitivities are, in matter of fact, the weights of the uncertainties of the input parameters, and, as a result, they play an important role in the uncertainty analysis of the response. For those cases in which the response is given as a function of the input parameters, a sensitivity is just the derivative of the response with respect to a given input parameter.

However, in many cases, the response is not a direct function of the input parameters. For instance, consider a case in which a function ϕ is calculated from:

$$\phi(\boldsymbol{\alpha}) = H\ \phi(\boldsymbol{\alpha}) + S(\boldsymbol{\alpha}) \tag{12.71}$$

where $\boldsymbol{\alpha}$ is the input parameter vector and H is some operator. The function $S(\boldsymbol{\alpha})$, if equal to zero, makes Equation (12.71) a homogeneous one. The response might be:

$$R(\boldsymbol{\alpha}) = B\ \phi(\boldsymbol{\alpha}) \tag{12.72}$$

where B is a function or an operator. In such cases, the sensitivity of the response, as given in Equation (12.72) with respect to a given input parameter, cannot be determined directly in an analytical way.

In most of the realistic cases, Equation (12.71) and (12.72) are solved numerically, which makes analytical derivatives impossible. In such cases, two methods are possible for deriving the sensitivities: one is the 'adjoint method' (some aspects of the 'adjoint method' will be described later and can also be found in Ronen (1990); the other is known as 'brute force'.

Given a response calculated for a nominal set of input parameters $R(\boldsymbol{\alpha}^{(0)})$ where $\boldsymbol{\alpha}^{(0)} = \alpha_1^{(0)}, \ldots, \alpha_k^{(0)}$, another value for that response is calculated when one of the input parameters $\alpha_i^{(0)}$ is changed to $\alpha_i^{(0)} + \delta\alpha_i$. The new input parameter vector is $\boldsymbol{\alpha} = \alpha_1^{(0)}, \ldots, \alpha_i^{(0)} + \delta\alpha_i, \ldots, \alpha_k^{(0)}$ and the new response is $R(\boldsymbol{\alpha})$. The sensitivity S_i is given by:

$$S_i = \left.\frac{\partial R}{\partial \alpha_i}\right|_{\alpha_i^{(0)}} \simeq \frac{R(\boldsymbol{\alpha}) - R(\boldsymbol{\alpha}^{(0)})}{\delta\alpha_i} \tag{12.73}$$

So in order to obtain the k values of all the sensitivities we need to calculate the response $(k + 1)$ times, once for the nominal values $\boldsymbol{\alpha}^{(0)}$ and k times using Equation (12.73).

If a better accuracy is needed for calculating the derivative we need to expand the response into a Taylor series. We have $R(\boldsymbol{\alpha}^{(0)})$, $R(\boldsymbol{\alpha}^+)$ and $R(\boldsymbol{\alpha}^-)$, where $\boldsymbol{\alpha}^{(0)}$ are the nominal values of the input parameters and $\boldsymbol{\alpha}^+ = (\alpha_1^{(0)}, \ldots, \alpha_i + \delta\alpha_i, \ldots, \alpha_k^{(0)})$ and $\boldsymbol{\alpha}^- = (\alpha_1^{(0)}, \ldots, \alpha_i - \delta\alpha_i, \ldots, \alpha_k^{(0)})$. Namely, in $\boldsymbol{\alpha}^+$ we change each time each one of the input parameters by $+\delta\alpha_i$ $(i = 1, \ldots, k)$, and in $\boldsymbol{\alpha}^-$ we change each time each one of the input parameters by $-\delta\alpha_i$ $(i = 1, \ldots, k)$. Using the Taylor expansion up to the second order we have:

$$R(\boldsymbol{\alpha}^+) = R(\boldsymbol{\alpha}^{(0)}) + \left.\frac{\partial R}{\partial \alpha_i}\right|_{\alpha_i^{(0)}} \delta\alpha_i + \left.\frac{\partial^2 R}{\partial \alpha_i^2}\right|_{\alpha_i^{(0)}} \delta\alpha_i^2 \tag{12.74}$$

and

$$R(\boldsymbol{\alpha}^-) = R(\boldsymbol{\alpha}^{(0)}) - \left.\frac{\partial R}{\partial \alpha_i}\right|_{\alpha_i^{(0)}} \delta\alpha_i + \left.\frac{\partial^2 R}{\partial \alpha_i^2}\right|_{\alpha_i^{(0)}} \delta\alpha_i^2 \tag{12.75}$$

Subtracting Equation (12.75) from Equation (12.74) we have:

$$R(\boldsymbol{\alpha}^+) - R(\boldsymbol{\alpha}^-) = 2\left.\frac{\partial R}{\partial \alpha_i}\right|_{\alpha_i^{(0)}} \delta\alpha_i \tag{12.76}$$

or

$$\left.\frac{\partial R}{\partial \alpha_i}\right|_{\alpha_i^{(0)}} = \frac{R(\boldsymbol{\alpha}^+) - R(\boldsymbol{\alpha}^-)}{2\delta\alpha_i}. \tag{12.77}$$

So the sensitivity given by Equation (12.77) is accurate up to the second order. However, in the case in which the sensitivities are represented by Equation (12.77), we need to calculate the response $(2k + 1)$ times, once for the nominal value, k times for $\boldsymbol{\alpha}^+$ and k times for $\boldsymbol{\alpha}^-$.

The 'brute force' method for calculating sensitivities, as presented by Equation (12.73) or (12.77), is very easy to implement. However, it requires calculating the response $(k + 1)$ or $(2k + 1)$ times. This might pose a problem, if the number of input parameters k is very big and the computation of the response requires a long computation time. In such a case, the 'adjoint method' is preferable.

With respect to the number of input parameters, it should be noted that, in any problem that we are solving, the number of important input parameters is not very big. Namely, the number of input parameters with large sensitivities is not very big, since otherwise the uncertainty of the response will be too large.

Consider an important input parameter for which

$$S_i = \frac{\partial R/R}{\partial \alpha_i/\alpha_i} \geq 1$$

and that we have N important input parameters. Also assume that all of them have the same relative uncertainty $\delta\alpha_i/\alpha_i = \sigma$ $(i = 1, \ldots, N)$. The contribution of the uncertainties of these input parameters to the uncertainty of the response, while assuming $S_i \geq 1$, will be:

$$\left(\frac{\sigma_R}{R}\right)^2 = \sum_{i=1}^{N} S_i^2 \sigma_i^2 \geq \sum_{i=1}^{N} \sigma^2 = N\sigma^2 \tag{12.78}$$

or

$$\frac{\sigma_R}{R} \geq \sqrt{(N)}\sigma \tag{12.79}$$

We can see from Equation (12.79) that if we have more than 100 important input parameters with $\frac{\partial R/R}{\partial \alpha_i/\alpha_i} \geq 1$, and if these parameters are known within about $\pm 5\%$, then the uncertainty in the response will be greater than 50%. Thus, large numbers of important input parameters, which are not known very accurately, make the calculation of the uncertainty of the response quite meaningless. Therefore, usually in meaningful problems, the number of important input parameters will not be large. Another advantage of the 'brute force' method is that, when calculating or solving the problem $(k + 1)$ or $(2k + 1)$ times, it enables us to calculate many responses related to $\phi(\alpha)$ of Equation (12.71) and, as a result, to calculate their uncertainties.

12.10.2   Vector notation

When considering the uncertainties of several responses, there is an advantage in using vector notation. In vector notation, the input parameters vector is α, given by Equation (12.28), and the vector of the sensitivities is given by:

$$\mathbf{S} = \begin{pmatrix} S_1 \\ \vdots \\ S_R \end{pmatrix} \tag{12.80}$$

and

$$\mathbf{S}^T = (S_1, \ldots, S_k) \tag{12.81}$$

The uncertainty matrix $\mathbf{C}$ is the covariance-variance matrix, given by its elements:

$$C_{ij} = \begin{matrix} \mathrm{cov}\ (\alpha_i \alpha_j) & i \neq j \\ \mathrm{var}\ (\alpha_i) & i = j \end{matrix} \tag{12.82}$$

The determinant of the uncertainty matrix is $|\mathbf{C}|$ and the PDF is given by:

$$f(\alpha) = (2\pi)^{-k/2} |\mathbf{C}|^{-1/2} \exp\left[-1/2(\alpha - \alpha_0)^T \mathbf{C}^{-1}(\alpha - \alpha_0)\right] \tag{12.83}$$

Using the vector notation, the variance of the response, given by Equation (12.50), can be presented as:

$$\mathrm{var}\ (R) = \mathbf{SCS}^T \tag{12.84}$$

also known as the 'sandwich rule'.

12.11   Non-linear aspects of uncertainty analysis

The basis for the sensitivity approach to uncertainty analysis is the expansion of the response into a Taylor series, due to the changes in the input parameters. In our analysis, only first order terms were considered, as expressed in Equation (12.33).

However, for non-linear problems, first order terms, which are a linearization of the response with respect to the input parameters, might not be accurate enough. In such a case, second order terms might be necessary. In a second order uncertainty analysis, we have the form (Ronen 1990):

$$\text{var } (R) = \sum_{i,j=1}^{k} S_i S_j \text{ cov } (\alpha_i \alpha_j) + \frac{1}{4} \sum_{i,j,n,m=1}^{k} Q_{ij} Q_{nm} \tag{12.85}$$

$$[\text{cov } (\alpha_i \alpha_n) \text{ cov } (\alpha_j \alpha_m) + \text{cov } (\alpha_i \alpha_m) \text{ cov } (\alpha_j \alpha_n)]$$

where Q_{ij} is the second order sensitivity:

$$Q_{ij} = \frac{\partial^2 R}{\partial \alpha_i \, \partial \alpha_j}\bigg|_{\alpha^{(0)}} \tag{12.86}$$

If the input parameters are uncorrelated, Equation (12.85) will be:

$$\text{var } (R) = \sum_{i=1}^{k} S_i^2 \text{ var } (\alpha_i) + \frac{1}{4} \sum_{i,j=1}^{k} Q_{ij} \text{ var } (\alpha_i) \text{ var}(\alpha_j) \tag{12.87}$$

12.12 Uncertainty analysis for several responses

In this analysis, only one response has been considered up to this point. However, the more practical situation is the one in which there are several responses. In the analysis presented in this section, we consider treating several responses simultaneously using vector and matrix notation. Our analysis is restricted to linear problems or linear approximations.

Consider the response vector which contains N responses:

$$\mathbf{R}^T = (R_1, \ldots, R_N) \tag{12.88}$$

The uncertainty matrix of these responses is $\mathbf{C}_R$ and is given by:

$$(\mathbf{C}_R)_{nm} = \langle \delta \mathbf{R}_n \, \delta \mathbf{R}_m \rangle \tag{12.89}$$

or in vector notation:

$$\mathbf{C}_R = \langle (\delta \mathbf{R} \, \delta \mathbf{R}^T) \rangle \tag{12.90}$$

The uncertainty matrix is a square of the order $k \times k$, where k is the number of input parameters.

The sensitivities of the responses to each of the input parameters are the relevant partial derivatives:

$$(\mathbf{S})_{ni} = \partial R_n / \partial \alpha_i \tag{12.91}$$

The matrix $\mathbf{S}$ is the sensitivity matrix and it is a rectangular matrix of order $N \times k$, where N is the number of responses. Now consider the first order approximation:

$$\mathbf{R}(\alpha_0 + \delta\alpha) = \mathbf{R}(\alpha_0) + \delta\mathbf{R} \cong \mathbf{R}(\alpha_0) + \mathbf{S} \, \delta\alpha \tag{12.92}$$

Within such a first order approximation or in cases in which the responses are linear functions of the input parameters, the required uncertainty matrix of these responses is:

$$
\begin{aligned}
\mathbf{C}_R &= \langle \delta \mathbf{R} \ \delta \mathbf{R}^T \rangle = \langle \mathbf{S} \ \delta \boldsymbol{\alpha} \ (\mathbf{S} \ \delta \boldsymbol{\alpha})^T \rangle \\
&= \langle \mathbf{S} \ \delta \boldsymbol{\alpha} \ \delta \boldsymbol{\alpha}^T \ \mathbf{S}^T \rangle = \mathbf{S} \langle \delta \boldsymbol{\alpha} \ \delta \boldsymbol{\alpha}^T \rangle \mathbf{S}^T \\
&= \mathbf{S} \mathbf{C} \mathbf{S}^T
\end{aligned}
\tag{12.93}
$$

Equation (12.93) follows the same format, the 'sandwich rule', as Equation (12.84).

12.13   Data adjustment

In this section, we present the formalism for data adjustment and for obtaining the uncertainty matrix of the adjusted parameters. The approach used here relies, to a large extent, on the approach suggested by Wagschal (1984) as appeared in the work by Ronen (1990).

12.13.1   *No correlations between input parameters and responses*

Consider a case in which we measured the responses. The measured response vector will be denoted as $\mathbf{R}_0$, where:

$$
\mathbf{R}_0 = (R_{01}, \ldots, R_{0n})^T
\tag{12.94}
$$

The uncertainty matrix of the measured response will be denoted by $\mathbf{C}_{R_0}$ and given by:

$$
\mathbf{C}_{R_0} = \langle \delta \mathbf{R}_0 \ \delta \mathbf{R}_0^T \rangle
\tag{12.95}
$$

The measured responses $\mathbf{R}_0$ deviate from the calculated responses $\mathbf{R}$ and the deviation is given by:

$$
\mathbf{d} = \mathbf{R} - \mathbf{R}_0
\tag{12.96}
$$

The uncertainty matrix associated with the deviations is:

$$
\mathbf{C}_d = \langle \delta(\mathbf{R} - \mathbf{R}_0) \ \delta(\mathbf{R} - \mathbf{R}_0)^T \rangle = \mathbf{C}_R + \mathbf{C}_{R_0}
\tag{12.97}
$$

For the cases where the uncertainties in the measured responses are smaller than the uncertainties in the calculated responses, we can use this fact to improve the uncertainties of the input parameters and their values. The corrected input parameters, which take into account the fact that the measured response uncertainties are smaller than the calculated ones, are known as the *adjusted* input parameters.

Let us denote the adjusted input parameter vector as $\boldsymbol{\alpha}'$. The responses calculated with the adjusted input parameters will be:

$$
\mathbf{R}' = \mathbf{R}(\alpha').
\tag{12.98}
$$

The new deviation $\mathbf{y}$ is given by:

$$
\mathbf{y} = \mathbf{R}' - \mathbf{R}_0
\tag{12.99}
$$

and it is smaller than **d**.

The change in the input parameters, i.e. the difference between the adjusted and unadjusted input parameters, is given by:

$$\mathbf{x} = \boldsymbol{\alpha}' - \boldsymbol{\alpha}_0 \tag{12.100}$$

If the adjusted input parameters $\boldsymbol{\alpha}'$ and the adjusted responses $\mathbf{R}'$ are the 'true' parameters and responses, the *a priori* PDF for the input parameters and responses is:

$$f \approx \exp[-\,Q(\mathbf{x}, \mathbf{y})] = \exp[-1/2(\boldsymbol{\alpha}_0 - \boldsymbol{\alpha}')^T \mathbf{C}^{-1}(\boldsymbol{\alpha}_0 - \boldsymbol{\alpha}') + (\mathbf{R}_0 - \mathbf{R}')^T \mathbf{C}_{R_0}^{-1}(\mathbf{R}_0 - \mathbf{R}')] \tag{12.101}$$

Thus, adjustment is obtained for the values **x** and **y**, when minimizing the following quadratic form:

$$Q(\mathbf{x}, \mathbf{y}) = \mathbf{x}^T \mathbf{C}^{-1} \mathbf{x} + \mathbf{y}^T \mathbf{C}_{R_0}^{-1} \mathbf{y} \tag{12.102}$$

However, the vector **y** is a function of the vector **x**. In our first order approximation:

$$\mathbf{R}' = \mathbf{R}(\alpha') = \mathbf{R}(\alpha_0) + \mathbf{Sx} \tag{12.103}$$

Subtracting $\mathbf{R}_0$ from both sides of Equation (12.103), we have:

$$\mathbf{y} = \mathbf{d} + \mathbf{Sx} \tag{12.104}$$

Thus, in the adjustment procedure, we require a minimum of Q(**x**, **y**) of Equation (12.102) subject to the constraint represented by Equation (12.104). The method of obtaining a conditional minimum is by using the Lagrange multipliers, in order to minimize P(**x**, **y**), which is defined as:

$$P(\mathbf{x}, \mathbf{y}) = Q(\mathbf{x}, \mathbf{y}) + 2\boldsymbol{\lambda}^T(\mathbf{Sx} - \mathbf{y}) \tag{12.105}$$

where $2\boldsymbol{\lambda}$ is the N-dimensional vector of Lagrange multipliers. The extremum of P(**x**, **y**) is obtained by requiring the derivatives of P(**x**, **y**), with respect to **x** and **y**, to be equal to zero, such that:

$$\frac{\partial P}{\partial \mathbf{x}} = 2(\mathbf{C}^{-1}\mathbf{x} + \mathbf{S}^T \boldsymbol{\lambda}) = 0 \tag{12.106}$$

and

$$\frac{\partial P}{\partial \mathbf{y}} = 2(\mathbf{C}_{R_0}^{-1}\mathbf{y} - \boldsymbol{\lambda}) = 0 \tag{12.107}$$

From Equations (12.106) and (12.107), we have:

$$\mathbf{x} = -\mathbf{CS}^T \boldsymbol{\lambda} \tag{12.108}$$

and

$$\mathbf{y} = \mathbf{C}_{R_0} \boldsymbol{\lambda} \tag{12.109}$$

The values of the Lagrange multipliers are obtained by using the condition of Equation (12.104). Thus:

$$\mathbf{y} - \mathbf{Sx} = (\mathbf{C}_{R_0} + \mathbf{SCS}^T)\boldsymbol{\lambda}$$
$$= \mathbf{C}_d\boldsymbol{\lambda} = \mathbf{d} \tag{12.110}$$

Therefore, we have:

$$\boldsymbol{\lambda} = \mathbf{C}_d^{-1}\mathbf{d} \tag{12.111}$$

$$\mathbf{C}_{\lambda} = \langle \delta\boldsymbol{\lambda}\ \delta\boldsymbol{\lambda}^T \rangle = \mathbf{C}_d^{-1}\langle \delta\mathbf{d}\ \delta\mathbf{d}^T \rangle \mathbf{C}_d^{-1} = \mathbf{C}_d^{-1}\mathbf{C}_d\mathbf{C}_d^{-1} = \mathbf{C}_d^{-1} \tag{12.112}$$

$$\mathbf{x} = -\mathbf{CS}^T\mathbf{C}_d^{-1}\mathbf{d} \tag{12.113}$$

and

$$\mathbf{y} = \mathbf{C}_{R_0}\mathbf{C}_d^{-1}\mathbf{d} \tag{12.114}$$

The adjusted input parameters have a smaller uncertainty than the original uncertainty of the input parameters.

The uncertainty matrix of the adjusted input parameters is:

$$\mathbf{C}' = \langle \delta\boldsymbol{\alpha}'\ \delta\boldsymbol{\alpha}'^T \rangle \tag{12.115}$$

Since $\boldsymbol{\alpha}' = \boldsymbol{\alpha}_0 + \mathbf{x}$, Equation (12.115) will be:

$$\mathbf{C}' = \langle \delta\boldsymbol{\alpha}\ \delta\boldsymbol{\alpha}^T \rangle + \langle \delta\mathbf{x}\ \delta\mathbf{x}^T \rangle + \langle \delta\boldsymbol{\alpha}\ \delta\mathbf{x}^T \rangle + \langle \delta\mathbf{x}\ \delta\boldsymbol{\alpha}^T \rangle \tag{12.116}$$

From Equation (12.113), we have:

$$\delta\mathbf{x} = -\mathbf{CS}^T\mathbf{C}_d^{-1}\delta\mathbf{d} = -\mathbf{CS}^T\mathbf{C}_d^{-1}\ \delta\mathbf{R} \tag{12.117}$$

Since $\delta\mathbf{R} = \mathbf{S}\ \delta\boldsymbol{\alpha}$, Equation (12.117) is either:

$$\delta\mathbf{x} = -\mathbf{CS}^T\mathbf{C}_d^{-1}\mathbf{S}\ \delta\boldsymbol{\alpha}$$

or

$$\delta\mathbf{x} = -\mathbf{CS}^T\delta\boldsymbol{\lambda} \tag{12.118}$$

Now, using Equation (12.112), we have:

$$\langle \delta\mathbf{x}\ \delta\mathbf{x}^T \rangle = \mathbf{CS}^T\langle \delta\boldsymbol{\lambda}\ \delta\boldsymbol{\lambda}^T \rangle \mathbf{SC}$$
$$= \mathbf{CS}^T\mathbf{C}_{\lambda}\mathbf{SC} = \mathbf{CS}^T\mathbf{C}_d^{-1}\mathbf{SC} \tag{12.119}$$

$$\langle \delta\mathbf{x}\ \delta\boldsymbol{\alpha}^T \rangle = -\mathbf{CS}^T\mathbf{C}_d^{-1}\mathbf{S}\langle \delta\boldsymbol{\alpha}\ \delta\boldsymbol{\alpha}^T \rangle$$
$$= -\mathbf{CS}^T\mathbf{C}_d^{-1}\mathbf{SC} \tag{12.120}$$

and

$$\langle \delta\boldsymbol{\alpha}\ \delta\mathbf{x}^T \rangle = -\langle \delta\boldsymbol{\alpha}\ \delta\boldsymbol{\alpha}^T \rangle \mathbf{S}^T\mathbf{C}_d^{-1}\mathbf{SC}$$
$$= -\mathbf{CS}^T\mathbf{C}_d^{-1}\mathbf{SC} \tag{12.121}$$

Thus, Equation (12.116) will be:

$$\mathbf{C}' = \mathbf{C} - \mathbf{C}\mathbf{S}^T\mathbf{C}_d^{-1}\mathbf{S}\mathbf{C} \tag{12.122}$$

In order to calculate the uncertainty matrix of the adjustment responses, i.e. the calculations of the responses with the adjustment input parameters, we get [using Equation (12.122)]:

$$\mathbf{C}_{R'} = \mathbf{S}\mathbf{C}'\mathbf{S}^T = \mathbf{S}\mathbf{C}\mathbf{S}^T - \mathbf{S}\mathbf{C}\mathbf{S}^T\mathbf{C}_d^{-1}\mathbf{S}\mathbf{C}\mathbf{S}^T \tag{12.123}$$

Using Equation (12.65), Equation (12.95) will be reduced to:

$$\mathbf{C}_{R'} = \mathbf{C}_R - \mathbf{C}_R\mathbf{C}_d^{-1}\mathbf{C}_R \tag{12.124}$$

The consistency of the data used in this adjustment is usually checked by the χ^2 test. The input data is considered to be consistent if the χ^2 per degree of the adjustment is near unity. We also expect the individual adjusted data to be consistent with the individual evaluated data, i.e. within or of the order of one standard deviation. A large value for the χ^2 per degree of freedom, or significant differences between the available and adjusted input parameters, would suggest some inconsistency in the input data or some problems with the applications of the adjusted procedure, i.e. the omission of important parameters. This may invalidate the expectation that the adjustment improves the calculation of responses not included in the adjustments. The sources of such inconsistencies should be identified, understood and corrected.

The adjusted input parameters, with their associated uncertainty, can be used to calculate other responses of the system, besides those considered in Equations (12.123) and (12.124). Furthermore, these adjusted input parameters can be used in different cases or different systems, in which all or part of the adjusted input parameters are used.

Consider that with such adjusted input parameters, we may calculate the uncertainties of a new response. The responses r'_i are elements of the vector $\mathbf{r}'$ with L components:

$$\mathbf{r}' = (r'_1, \ldots, r'_L)^T \tag{12.125}$$

and

$$r'_i = r_i(\boldsymbol{\alpha}') \tag{12.126}$$

The sensitivity matrix elements for these new responses are:

$$(\mathbf{S}_{\mathbf{r}'})_{ij} = \partial r'_i / \partial \alpha'_j \tag{12.127}$$

The sensitivity matrix $\mathbf{S}_{\mathbf{r}'}$ is rectangular and of order $L \times K$. The uncertainty matrix for our new responses is obtained using the 'sandwich rule' and Equations (12.122) and (12.127). Thus:

$$\mathbf{C}_{r'} = \mathbf{S}_{\mathbf{r}'}\mathbf{C}'\mathbf{S}_{\mathbf{r}'}^T = \mathbf{S}_{\mathbf{r}'}\mathbf{C}\mathbf{S}_{\mathbf{r}'}^T - \mathbf{S}_{\mathbf{r}'}\mathbf{C}\mathbf{S}_{\mathbf{r}'}^T\mathbf{C}_d^{-1}\mathbf{S}_{\mathbf{r}'}^T\mathbf{C}\mathbf{S}_{\mathbf{r}'}^T \tag{12.128}$$

where $\mathbf{C}_{r'}$ is the uncertainty matrix of the new responses, calculated with the adjusted input parameter.

12.13.2 *Correlations between input parameters and response measurements*

In the derivation so far, we have assumed no correlation between the input parameters and the response measurements. In other words, the experimental devices for the responses are different from those of the input parameters, which is usually the case. However, there might be a situation in which the same experimental devices are used for measuring both the input parameters and the responses. Other correlations might also be induced.

For such a case in which there is a correlation between the responses and the input parameter measurements, Equation (12.97) would no longer be valid. The uncertainty matrix $\mathbf{C}_d$ would have the form:

$$
\begin{aligned}
\mathbf{C}_d &= \langle \delta(\mathbf{R} - \mathbf{R}_0) \, \delta(\mathbf{R} - \mathbf{R}_0)^T \rangle \\
&= \langle \delta\mathbf{R} \, \delta\mathbf{R}^T \rangle + \langle \delta\mathbf{R}_0 \, \delta\mathbf{R}_0^T \rangle - \langle \delta\mathbf{R} \, \delta\mathbf{R}_0^T \rangle - \langle \delta\mathbf{R}_0 \, \delta\mathbf{R}^T \rangle
\end{aligned}
\tag{12.129}
$$

The last two terms on the LHS of Equation (12.129) are those in which the correlations between responses and input parameters are expressed. Since $\delta\mathbf{R} = \mathbf{S} \, \delta\boldsymbol{\alpha}$, Equation (12.129) will have the form:

$$
\mathbf{C}_d = \langle \delta\mathbf{R} \, \delta\mathbf{R}^T \rangle + \langle \delta\mathbf{R}_0 \, \delta\mathbf{R}_0^T \rangle - \mathbf{S}\langle \delta\boldsymbol{\alpha} \, \delta\mathbf{R}_0^T \rangle - \langle \delta\mathbf{R}_0 \, \delta\boldsymbol{\alpha}^T \rangle \mathbf{S}^T
\tag{12.130}
$$

The response-input parameter uncertainty matrix is:

$$
\mathbf{C}_{\alpha R} = \langle \delta\boldsymbol{\alpha} \, \delta\mathbf{R}_0^T \rangle
\tag{12.131}
$$

and its transpose:

$$
\mathbf{C}_{\alpha R}^T = \langle \delta\mathbf{R}_0 \, \delta\boldsymbol{\alpha}^T \rangle
\tag{12.132}
$$

When there is no correlation between input parameters and response measurements, Equations (12.131) and (12.132) are equal to zero and Equation (12.97) is valid. Thus, rewriting Equation (12.130) we have:

$$
\mathbf{C}_d = \mathbf{C}_R + \mathbf{C}_{R_0} - \mathbf{S}\mathbf{C}_{\alpha R} - \mathbf{C}_{\alpha R}^T \mathbf{S}^T
\tag{12.133}
$$

As a result of the correlations between input parameters and responses, the quadric form of Equations (12.101) and (12.102) is no longer valid. The new expression is:

$$
Q(\mathbf{x}, \mathbf{y}) = (\mathbf{x}^T, \mathbf{y}^T) \begin{pmatrix} \mathbf{C} & \mathbf{C}_{\alpha R}^{-1} \\ \mathbf{C}_{\alpha R} & \mathbf{C}_{R_0} \end{pmatrix} \begin{pmatrix} \mathbf{x} \\ \mathbf{y} \end{pmatrix}
\tag{12.134}
$$

Evaluating Equation (12.134), gives:

$$
Q(\mathbf{x}, \mathbf{y}) = \mathbf{x}^T \mathbf{T}\mathbf{x} - 2\mathbf{x}^T \mathbf{T}\mathbf{C}_{\alpha R}\mathbf{C}_{R_0}^{-1}\mathbf{y} + \mathbf{y}(\mathbf{C}_{R_0}^{-1} + \mathbf{C}_{R_0}^{-1}\mathbf{C}_{\alpha R}^T \mathbf{T}\mathbf{C}_{\alpha R}\mathbf{C}_{R_0}^{-1})\mathbf{y}
\tag{12.135}
$$

where

$$
\mathbf{T} = (\mathbf{C} - \mathbf{C}_{\alpha R}\mathbf{C}_{R_0}^{-1}\mathbf{C}_{\alpha R}^T)^{-1}
\tag{12.136}
$$

In order to obtain the vectors $\mathbf{x}$ and $\mathbf{y}$, we minimize the functional $P(\mathbf{x}, \mathbf{y})$ of Equation (12.105). However, the term $Q(\mathbf{x}, \mathbf{y})$ is the one given in Equation (12.135). Thus:

$$\frac{\partial P}{\partial \mathbf{x}} = 2(\mathbf{Tx} - \mathbf{TC}_{\alpha R}\mathbf{C}_{R_0}^{-1}\mathbf{y} + \mathbf{S}^T\boldsymbol{\lambda}) = 0 \tag{12.137}$$

and

$$\frac{\partial P}{\partial \mathbf{y}} = 2[-\mathbf{C}_{R_0}^{-1}\mathbf{C}_{\alpha R}^T\mathbf{Tx} + (\mathbf{C}_{R_0}^{-1} + \mathbf{C}_{R_0}^{-1}\mathbf{C}_{\alpha R}^T\mathbf{TC}_{\alpha R}\mathbf{C}_{R_0}^{-1})\mathbf{y} - \boldsymbol{\lambda}] = 0 \tag{12.138}$$

Multiplying Equation (12.138) by $\mathbf{C}_{R_0}$ gives:

$$-\mathbf{C}_{\alpha R}^T\mathbf{Tx} + \mathbf{y} + \mathbf{C}_{\alpha R}^T\mathbf{TC}_{\alpha R}\mathbf{C}_{R_0}^{-1}\mathbf{y} = \mathbf{C}_{R_0}\boldsymbol{\lambda} \tag{12.139}$$

From Equation (12.137), we have:

$$\mathbf{Tx} = \mathbf{TC}_{\alpha R}\mathbf{C}_{R_0}^{-1}\mathbf{y} - \mathbf{S}^T\boldsymbol{\lambda} \tag{12.140}$$

Substituting Equation (12.140) in Equation (12.139), we have:

$$-\mathbf{C}_{\alpha R}^T\mathbf{TC}_{\alpha R}\mathbf{C}_{R_0}^{-1}\mathbf{y} + \mathbf{C}_{\alpha R}^T\mathbf{S}^T\boldsymbol{\lambda} + \mathbf{y} + \mathbf{C}_{\alpha R}^T\mathbf{TC}_{\alpha R}\mathbf{C}_{R_0}^{-1}\mathbf{y} = \mathbf{C}_{R_0}\boldsymbol{\lambda} \tag{12.141}$$

or

$$\mathbf{y} = (\mathbf{C}_{R_0} - \mathbf{C}_{\alpha R}^T\mathbf{S}^T)\boldsymbol{\lambda} \tag{12.142}$$

Multiplying Equation (12.140) from the left by $\mathbf{T}^{-1}$, we have:

$$\mathbf{x} = \mathbf{C}_{\alpha R}\mathbf{C}_{R_0}^{-1}\mathbf{y} - \mathbf{T}^{-1}\mathbf{S}^T\boldsymbol{\lambda} \tag{12.143}$$

Substituting Equations (12.125) and (12.129) in Eq. (12.143), we get:

$$\mathbf{x} = \mathbf{C}_{\alpha R}\mathbf{C}_{R_0}^{-1}\mathbf{C}_{R_0}\boldsymbol{\lambda} - \mathbf{C}_{\alpha R}\mathbf{C}_{R_0}^{-1}\mathbf{C}_{\alpha R}^T\mathbf{S}^T\boldsymbol{\lambda} - \mathbf{CS}^T\boldsymbol{\lambda} + \mathbf{C}_{\alpha R}\mathbf{C}_{R_0}^{-1}\mathbf{C}_{\alpha R}^T\mathbf{S}^T\boldsymbol{\lambda} \tag{12.144}$$

Thus:

$$\mathbf{x} = (\mathbf{C}_{\alpha R} - \mathbf{CS}^T)\boldsymbol{\lambda} \tag{12.145}$$

The Lagrange multipliers, which were obtained by using Equation (12.111), are still valid. Thus, with the substitution of the Lagrange multipliers, Equation (12.111) into Equations (12.142) and (12.145), we get:

$$\mathbf{y} = (\mathbf{C}_{R_0} - \mathbf{C}_{\alpha R}^T\mathbf{S}^T)\mathbf{C}_d^{-1}\mathbf{d} \tag{12.146}$$

and

$$\mathbf{x} = (\mathbf{C}_{\alpha R} - \mathbf{CS}^T)\mathbf{C}_d^{-1}\mathbf{d} \tag{12.147}$$

The variation in $\mathbf{x}$ is:

$$\begin{aligned} \delta\mathbf{x} &= (\mathbf{C}_{\alpha R} - \mathbf{CS}^T)\mathbf{C}_d^{-1}\delta\mathbf{R} \\ &= (\mathbf{C}_{\alpha R}\mathbf{CS}^T)\mathbf{C}_d^{-1}\mathbf{S}\,\delta\boldsymbol{\alpha} \end{aligned} \tag{12.148}$$

Using Equation (12.116), we can calculate the uncertainty matrix of the adjusted input parameters. For:

$$\langle \delta \mathbf{x} \, \delta \mathbf{x}^T \rangle = (\mathbf{C}_{\alpha R} - \mathbf{C} \mathbf{S}^T)\langle \delta \boldsymbol{\lambda} \, \delta \boldsymbol{\lambda}^T \rangle(\mathbf{C}_{\alpha R}^T - \mathbf{S} \mathbf{C})$$
$$= (\mathbf{C}_{\alpha R} - \mathbf{C} \mathbf{S}^T)\mathbf{C}_d^{-1}\mathbf{C}_{\alpha R}^T - (\mathbf{C}_{\alpha R} - \mathbf{C} \mathbf{S}^T)\mathbf{C}_d^{-1}\mathbf{S} \mathbf{C} \tag{12.149}$$

$$\langle \delta \mathbf{x} \, \delta \boldsymbol{\alpha}^T \rangle = (\mathbf{C}_{\alpha R} - \mathbf{C} \mathbf{S}^T)\mathbf{C}_d^{-1}\mathbf{S}\langle \delta \boldsymbol{\alpha} \, \delta \boldsymbol{\alpha}^T \rangle$$
$$= (\mathbf{C}_{\alpha R} - \mathbf{C} \mathbf{S}^T)\mathbf{C}_d^{-1}\mathbf{S} \mathbf{C} \tag{12.150}$$

$$\langle \delta \boldsymbol{\alpha} \, \delta \mathbf{x}^T \rangle = \langle \delta \boldsymbol{\alpha} \, \delta \boldsymbol{\alpha}^T \rangle \mathbf{S}^T \mathbf{C}_d^{-1}(\mathbf{C}_{\alpha R}^T - \mathbf{S} \mathbf{C})$$
$$= \mathbf{C} \mathbf{S}^T \mathbf{C}_d^{-1}\mathbf{C}_{\alpha R}^T - \mathbf{C} \mathbf{S}^T \mathbf{C}_d^{-1}\mathbf{S} \mathbf{C} \tag{12.151}$$

Substituting Equations (12.149)–(12.151) in Equation (12.116), we obtain:

$$\mathbf{C}' = \mathbf{C} + \mathbf{C}_{\alpha R}\mathbf{C}_d^{-1}\mathbf{C}_{\alpha R}^T - \mathbf{C} \mathbf{S}^T \mathbf{C}_d^{-1}\mathbf{C}_{\alpha R}^T - (\mathbf{C}_{\alpha R} - \mathbf{C} \mathbf{S}^T)\mathbf{C}_d^{-1}\mathbf{S} \mathbf{C}$$
$$+ (\mathbf{C}_{\alpha R} - \mathbf{C} \mathbf{S}^T)\mathbf{C}_d^{-1}\mathbf{S} \mathbf{C} + \mathbf{C} \mathbf{S}^T \mathbf{C}_d^{-1}\mathbf{C}_{\alpha R}^T - \mathbf{C} \mathbf{S}^T \mathbf{C}_d^{-1}\mathbf{S} \mathbf{C} \tag{12.152}$$
$$= \mathbf{C} - \mathbf{C} \mathbf{S}^T \mathbf{C}_d^{-1}\mathbf{S} \mathbf{C} + \mathbf{C}_{\alpha R}\mathbf{C}_d^{-1}\mathbf{C}_{\alpha R}^T$$

The first two terms of the RHS of Equation (12.152) represent the uncertainty matrix of the adjusted input parameters, when there is no correlation between the measured responses and the measured input parameters, i.e. these two terms are those of Equation (12.122).

The uncertainty matrix obtained in Equation (12.152) can be used to calculate the uncertainty in responses which are functions of input parameters. Let us denote the matrix $\mathbf{S}_{R'}$ as the sensitivity matrix of the responses to be calculated. The uncertainty matrix of these responses $\mathbf{C}_{R'}$ will be given by using the 'sandwich rule' with the uncertainty matrix $\mathbf{C}'$ given in Equation (12.152). Thus:

$$\mathbf{C}_{R'} = \mathbf{S}_{R'}\mathbf{C}'\mathbf{S}_{R'}^T = \mathbf{S}_{R'}\mathbf{C} \mathbf{S}_{R'} - \mathbf{S}_{R'}\mathbf{C} \mathbf{S}^T \mathbf{C}_d^{-1}\mathbf{S} \mathbf{C} \mathbf{S}_{R'}^T + \mathbf{S}_{R'}\mathbf{C}_{\alpha R}\mathbf{C}_d^{-1}\mathbf{C}_{\alpha R}^T \mathbf{S}_{R'}^T \tag{12.153}$$

When calculating the original responses, we have:

$$\mathbf{S}_{R'} = \mathbf{S} \tag{12.154}$$

and Equation (12.153) will be:

$$\mathbf{C}_{R'} = \mathbf{C}_R = \mathbf{S} \mathbf{C} \mathbf{S}^T - \mathbf{S} \mathbf{C} \mathbf{S}^T \mathbf{C}_d^{-1}\mathbf{S} \mathbf{C} \mathbf{S}^T + \mathbf{S} \mathbf{C}_{\alpha R}\mathbf{C}_d^{-1}\mathbf{C}_{\alpha R}^T \mathbf{S}^T$$
$$= \mathbf{C}_R - \mathbf{C}_R\mathbf{C}_d^{-1}\mathbf{C}_R + \mathbf{S} \mathbf{C}_{\alpha R}\mathbf{C}_d^{-1}\mathbf{C}_{\alpha R}\mathbf{S}^T \tag{12.155}$$

12.14 References

Navon, Y. & Ronen, Y. 1988, Applications of generalized bias operators in reactor theory, *Nucl. Sci. Eng.*, vol. 100, p. 125.

Ronen, Y., Cacuci, D.G. & Wagschal, J. J. 1981, Determination and application of generalized bias operators using an inverse perturbation approach, *Nucl. Sci. Eng.*, vol. 77, p. 426.

Ronen, Y. & Dubi, A. 1983, A probabilistic rejection test for multivariable sensitivity analysis, *J. of Math. Phys.* vol. 24, p. 2800.

Ronen, Y. 1990, *Uncertainty Analysis*, CRC Press, Boca Raton, Florida, USA.

Wagschal, J.J. 1984, private communication.

13 Artificial neural networks – unlikely but effective tools in analytical chemistry

13.1 Introduction

Analytical chemistry techniques require high precision, traceability, and repeatability in order to vouch for the required accuracy in the results. The rise in the amounts of data available from modern multi-channel instrumentation led to the creation of the chemometrics discipline, combining complex mathematical manipulation and analysis of the chemical data to arrive at the desired results. However, the increase in the volume of the instrumentation data available from advanced analytical techniques has posed difficult processing and modeling issues.

Artificial neural network (ANN) is a branch of computational intelligence in which data is used to 'train' a network of simple computing elements called 'neurons' to recognize patterns, similar to the learning ability of our biological neural networks in our brains. During the early 1950s the ANN technique was considered as the 'new bright future' of artificial intelligence (AI), but the simple network architectures used at that time were proven to be deficient, and interest in this ANN waned until more efficient ANN architectures and training algorithms were developed. Since its resurgence in the mid 1980s, more than one hundred thousand papers on ANN have been published – in all fields of science, engineering, medical, industrial and commercial activities and in almost every conceivable area where computers are used to model and classify data.

In 1991, a review titled *Neural networks: A new method for solving chemical problems or just a passing phase?* appeared in the journal *Analytica Chimica Acta* (Zupan 1991). A decade passed, and today it is evident that it was not a passing phase – more than twenty thousand titles are found in databases such as *Web of Science* and *IEEE Explore*, containing for combinations of keywords 'artificial and neural and networks and chemistry'. However, ANN is considered one of the more heuristic and less rigorous artificial intelligence (AI) techniques, thus it is not surprising that ANN techniques are not mentioned in analytical chemistry textbooks nor in most chemometrics books.

There were several researchers that did try to apply ANN modeling in analytical chemistry, and reviews were published during the 1990s. Zupan published a book in 1993 describing ANN to chemists (Zupan 1993), and a tutorial on some ANN architectures in analytical chemistry in 1995 (Zupan 1995). The number of papers with connection to analytical chemistry grew steadily – from six papers in 1992, through 18 in 1993, 26 in 1994, to about 60 papers in 2002. In a chapter in a book published in 2000 reviewing applications of ANN in chemistry (Peterson 2000), out of almost 300 references, about 50 papers describe applications of ANN in analytical chemistry. Thus the question arises, who needs a review of ANN in analytical chemistry?

There are several answers to this question. The first is that in the four years since Peterson's review, many new applications of ANN modeling deal with subjects related to analytical chemistry, or of interest to analytical chemists, and about 300 new ones were published in the last three years or were not included in Peterson's reference list. The second answer appears in the introduction to the review book (Lipkowitz 2000) 'Neural networks are mathematical models for discerning complex relationships between input variables and a numerical outcome.... Neural networks calculations have the advantage of being very fast and inexpensive. Software for these calculations runs on inexpensive personal computers' – and analytical chemists like these attributes.

The third reason is more complex. As noted above, chemometricians do not consider ANN modeling to be valid enough, especially when the number of input variables is large – hundreds to thousands are usually found in analytical instrumentation spectra analysis. Because of the fear of over-fitting a model with a large number of parameters to a small number of calibration examples, they recommend the time-honored techniques of Principal Components Analysis (PCA), Partial Least Square (PLS) or other matrix manipulating techniques that drastically reduce the number of inputs to the model. Thus most analytical chemists regard ANN modeling as a hard-to-train black box, without the capability to explain its results, and thus lacking the robustness and validity demanded by the application. As I hope to show in this chapter, these attitudes are over-conservative. There are robust ways of training large-scale ANN models, and by analyzing them a small set of relevant inputs can be identified. As in the title of a recent article (Lawrence 1997), 'Neural networks are harder to over-train than you think'.

This review is written from a somewhat unusual perspective. I am neither an analytical chemist nor a computer science practitioner. I am a chemical engineer by training, and my expertise is in modeling and controlling of industrial processes. I do have, however, friends and colleagues that are analytical chemists. Since I started using in 1990 large-scale ANN for modeling and classification, I have helped analytical chemists in applying ANN modeling to solve seemingly difficult or intractable problems in their data. Following the published chemical literature and chemical conferences proceedings, I am struck by the apparent dichotomy – most ANN papers are published by people working in, or contracted by, industry, while academic research papers are mostly publishing non-ANN chemometric techniques and solutions. It may be that the non-rigorous, heuristic nature of the ANN modeling does not seem respectable, or safe, for MSc or PhD theses in chemistry. Thus this review will include some of my (subjective) views based on my experiences, with a non-proportional inclusion of many of my publications. I find that there are still many myths and misconceptions about ANN that discourage analytical chemists from considering the use of ANN in modeling and understanding the data they generate. In this chapter I will try my best to dispel these misconceptions.

The acronyms that are used in the chapter are given in Appendix D; the references in the text and the tables name only the first author and the publication year, and are described as if a single person did the work.

13.2   Overview and goals

The review starts with a brief description of the historical development of the ANN since the 1940s, with its amazing ups and downs. Then the fundamentals of the ANN ideas and structures will be presented, along with the accepted terms used in the ANN papers. The ANN techniques most suitable for chemical data processing will be discussed, with their advantages and shortcomings.

Examples of ANN applications to the more prevalent types of analytical chemistry instrumentation and techniques will be given. The important questions of robustness, explainability and new knowledge extraction will be discussed. Some hybrid modeling applications, combining ANN with Fuzzy Logic (FL) and Genetic Algorithms (GA) will be briefly described, as well as the newly popular Support Vector Machines (SVM) classification algorithm. Several active areas in chemistry, where ANN modeling is widely used, such as Quantitative Structure Activity Relationship (QSAR), mining of bio-medical data and drug discovery will not be covered in this review. 'Artificial nose' sensor arrays are now developed as analytical chemistry tools, and are included in this review. Although English-language abstracts of papers written in non-English-language journals are now available in the literature databases, most of them are not included in the reference list.

The goal of this review is to get more analytical chemists involved in ANN applications, by using the available ANN software, commercial or free. If adventurous, they may consider writing their own programs, more geared to their specific applications. The availability of high-powered PCs and user-friendly programing languages should make this task quite easy.

The review is not intended to include a detailed mathematical description of the ANN theory, as many books on this subject are available, Zupan (1993, 2002) has published books specifically intended for chemists; Bishop's book on pattern recognition by ANN has detailed mathematical discussion on this subject (Bishop 1995). Paterson's chapter (Paterson 2002) reviews the ANN architectures, training and results analyzing equations that were in chemistry applications up to the year 2000. There are quite good review articles in the journals (Savozyl 1997).

For up-to-date information of ANN books and software, and on any ANN subject, the reader is referred to Warren Sarle's excellent web site of Frequently Asked Questions on ANN, monthly updated (Sarle 2004). Additional help can be found in the active comp.ai.neural-nets Internet discussion group. Papers on ANN application in chemistry are regularly published in the main analytical chemistry journals, *Analytical Chemistry* and *Analytica Chimica Acta*, as well as the *Journal of Chemical Information and Computer Sciences*, *Chemometrics and Intelligent Laboratory Systems*, and *Journal of Chemometrics*.

13.3   A brief history of artificial neural networks

When computers were beginning to be used in the early 1950s, an attempt was made to mimic the operation of the neurons in our brains for learning from examples, as humans do. Hebb found that the synaptic connection between the neurons could be modified during learning. The resulting neural networks, containing these learned connections are the basis of our memory. Thus interacting 'neurons' were simulated in the computer as 'perceptrons', with connection weights that were modified during presentation of examples, until arriving at a steady state. When presented with new examples, the perceptrons recognized, or classified correctly, these examples.

However, it was proved mathematically that these perceptrons could only classify linearly separable examples, and could not correctly classify simple XOR examples. Like a pricked bubble, all research in the ANN modeling ceased, and other forms of AI were investigated. In particular, another way of mimicking the human brain was explored, namely Expert Systems (ES). These are a collection of known rules based on previous human expertise that are triggered by external inputs. Evaluating the logically connected rules, in the form of 'if...then...'statements, the system gives advice that, hopefully, a human expert would have reached given the same input data. ES reached their highest popularity in the 1980s and early 1990s. They were gradually abandoned as it became evident that precise rule elicitation from experts is not a simple task, the number of rules in a complex situation is very large, and their 'truth maintenance' in evolving situations is a heavy resource-consuming effort.

In 1974, in a Harvard University PhD thesis, a more complex architecture of neural networks was shown to be trainable by a 'back-propagation of errors' between the ANN model outputs and the known targets (Werbos 1974, 1994). However, few people read PhD theses and these findings were ignored for another decade. In the meantime, several developments were made in other ANN models, Hopfield net, and Kohonen self-organized maps (SOM). The first has its roots in physics, the second in the brain structure theory. Only in 1985 the error back-propagation ideas were independently arrived at by MIT researchers, and applied in multi-layer perceptrons (MLP) (Rumelhard 1986). Since then, the number of people researching the ANN architectures, training algorithms and applications, and their published results, has soared to the levels mentioned in the introduction.

In contrast with the Expert Systems, whose results can be explained by tracing back the evoked rules and conditions, the ANN was, and still is, considered a 'black-box' with no explanation facility. This, coupled with the heuristic way used in determining the ANN architectures, training algorithms and parameters, led to separation from the conventional mathematical and statistical community engaged in modeling data. In fact, similar and even identical concepts are named differently in the statistical and the ANN communities' papers (Stegmann 1999).

Although many ANN architectures and training algorithms were developed in the 1990s, only few have survived the collective experience in applying them to real-life situations. It was estimated that 90% of the ANN applications in chemistry are done by MLP. My review of the published papers in the analytical chemistry area indicated that two main types that are actively used now are the single hidden layer MLP and SOM, and only these two will be described in the next section.

In recent years there are some new developments in hybrid ANNs, in which FL, and increasingly GA, are combined with the ANN, either in the preprocessing of its inputs or in the training algorithms, to achieve better results. Recently, the SVM algorithm has begun to compete with the ANN in classification applications.

13.4   Multi-layer perceptrons ANN

The basic unit of our brain is the neuron cell. It receives 'inputs' from other neurons through its dendrites, up to ten thousand threads connected to its cell body. The inputs are various amounts of neuro-transmitter chemicals passing through the synaptic gaps. If the combined amount of the neuro-transmitter passes a certain threshold, the neuron cell becomes active, producing neuro-transmitter chemicals through its axon 'output' in proportion to its activation level. The axon, in turn, may be connected to up to ten thousand dendrites of other neurons.

The efficiency of the neuro-transmitter passage from an axon to the dendrites through the synaptic gap is increased incrementally if both neurons, the sending and the receiving, are active at the same time. Thus our memory is formed, made of networks of neurons connected by the learned stronger or weaker synaptic efficiencies. This is a simplistic description, as there are also inhibitory types of connections between neurons, but it is suitable for understanding the artificial neural networks.

Thus the basic unit in ANN is the 'neuron' that has an activation function that determines its output relationship to the inputs it gets. Many types of transfer functions have been proposed and tried, but the surviving, most used one is the logistic sigmoid function:

$$\text{output} = \frac{1}{1 + e^{-\Sigma(\text{weighted inputs})}}$$

or its close relative, the hyperbolic tangent

$$\text{output} = \frac{e^{+\Sigma(\text{weighted inputs})} - e^{-\Sigma(\text{weighted inputs})}}{e^{+\Sigma(\text{weighted inputs})} + e^{-\Sigma(\text{weighted inputs})}}$$

The sigmoid function value is gradually increasing from zero when the sum of inputs is a large negative number, through 0.5 when the input sum is 0, approaching asymptotically to 1.0 when the input sum is a large positive number (Fig. 13.1). The corresponding values for the hyperbolic tangent are $[-1.0, 0, 1.0]$.

Many ANN architectures, the way the neurons are connected together in the network, were suggested. Again, the one most used is the fully connected, feed-forward, single hidden neuron layer MLP. It is formed by an input layer of fan-out units, all of which are connected to each neuron unit in a 'hidden' layer. The outputs of each of those neurons are connected to the inputs of the 'output' layer neuron. The resulting values of these neurons transfer functions are the ANN outputs (Fig. 13.2).

The qualifiers in the name of the MLP indicate that all neurons are connected in the forward direction between each layer, with no layer bypassing, lateral or backward connections, with only one hidden layer. The number of input units and output neurons are fixed by the data structure presented to the ANN, the number

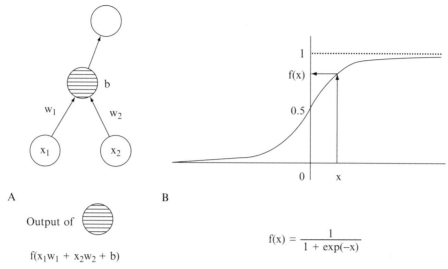

A   B

Output of ⊜

$f(x_1w_1 + x_2w_2 + b)$

$$f(x) = \frac{1}{1 + \exp(-x)}$$

Fig. 13.1 (a) A model of a neuron. (b) A sigmoid logistic function.

of hidden neurons is determined by the user, as the name 'hidden' derives from the fact that it is not connected to the outside world. Initially there was a feeling that multi-hidden layers may learn more complex models better, but it was mathematically proved that ANN with one single hidden layer is sufficient for approximating any complex function (Hornik, *et al.*, 1989). However, it was empirically found that there is an optimal number of hidden neurons for any data set. Too few do not learn well enough and too many learn too much, that is, they over-fit the model by learning the peculiarities of the training data and not generalizing for applicability to new data.

The most demanding task is the training process, in which the data set is presented repeatedly to the ANN, in one of two forms. In the 'supervised' mode, the data is in the form of input and targets. For example, the inputs may be an instrumental spectrum vector set, while the outputs may be the set of the compositions of a mixture from which the spectrum was measured. The task of the ANN is to learn the relationship between the spectra presented as inputs, and the composition of each mixture presented as its targets. The errors between the known training targets and the current outputs of the ANN are used to incrementally modify the connection weights so as to decrease the error when the training data will be presented again. When the error is decreased to a specified low limit, or the present number of data presentations, 'epochs', is reached, the generalization capacity of the trained ANN is measured by its success in correctly predicting the composition of a mixture that was not present in the training set, from its inputs (Fig. 13.3). Additional use of this ANN is to find out qualitative relationships between the inputs and outputs.

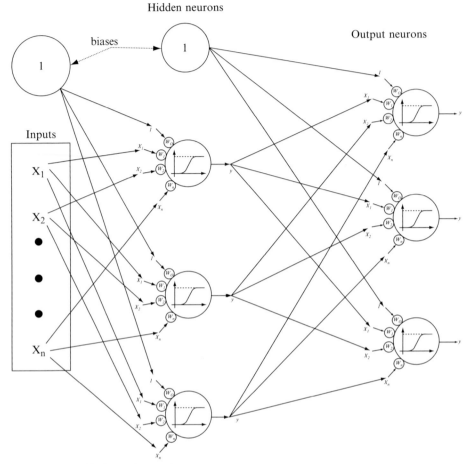

Fig. 13.2 The artifical neural network.

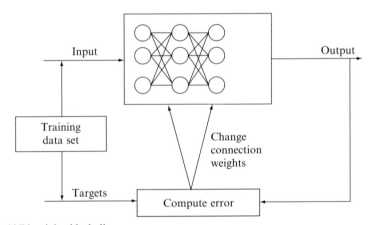

Fig. 13.3 ANN training block diagram.

The second type is the 'un-supervised' or auto-associative ANN, in which the same input vector is presented both as an input and as a target. The task of the AA-ANN is to learn the relationships within the spectra, and classify them into separate groups according to similar patterns in the data. Additional use of the AA-ANN is to check the validity of new data presented to the AA-ANN, by monitoring deviations between the actual inputs and the predicted inputs.

The basic error back-propagation ANN training algorithm works like this:

- the selected ANN is initialized by assigning small random values to the connection weights between the units of the ANN;
- an input vector is presented to the ANN, is multiplied by the input-hidden neurons connection weight matrix to form the inputs to the hidden neurons;
- each hidden neuron sums its inputs, and calculates its output using its activation function;
- the outputs of the hidden neurons are then multiplied by the hidden-output neurons connection weight matrix, to form the inputs to the output neurons;
- each output neuron sums its inputs and performs the activation function;
- the output neurons' values are then compared to the target (mixture concentration) values of the particular spectrum, and an error metric (usually the squared error) is calculated;
- this error metric is now used to modify the connection weights in a way that will reduce the error metric of the ANN by calculating the derivatives of each output neuron error as a function of each of its inputs, and adjusting the connection weight of each input in proportion to this error derivative;
- then a similar process is done in each hidden neuron, for adjusting the connection weight of each of its inputs – the formal equations are:

$$\text{error}_i = \text{output}_i - \text{target}_i$$

$$\text{delta}_j^{\text{out}} = (\text{target}_j - \text{output}_j){}^*\text{output}_j^*(1 - \text{output}_j)$$

This equation assumes that the activation function derivative can be calculated and in the case of the sigmoid function, the derivative is:

$$\text{der} = \text{function}^*(1 - \text{function})$$

$$\text{delta}_j^{\text{hid}} = (\Sigma\, \text{delta}_j^{\text{out}*}\, W_{kj}^{\text{out}}){}^*\, y_j^{\text{hid}\,*}(1 - y_j^{\text{hid}})$$

for all r number of examples,

$$\text{delta}\, W_{ij} = \eta^*\text{delta}_j^{\text{last}\,*}\text{output}_i^{\text{layer}-1} + \mu^*\text{delta}\, W_{ij}^{\text{layer}}(\text{previous})$$

where η and μ are the gain and momentum coefficients that are used to avoid undesired oscillations during the training process. Both proportional coefficients may be specified by the user, or can be modified during the training according to some error-dependent conditions. Trial and error, or previous experience, is used in assigning these coefficients, and it is claimed that their sum should be close to one.

This process is repeated for each training example, and when all the examples in the training set have been presented (called an 'epoch'), the same data set is presented again and again, until some stopping criteria is met – either the error metric is reduced below the desired limit, or the number of presentation epochs has exceeded the specified limit.

In practice there are several modifications. To avoid getting the same value if the input sum is zero, a 'bias' input with a constant value of one is added to the input layer and to the hidden layer, with adjustable connection weights similar to the other connection weights. To avoid undesired oscillations during the training process, a momentum part proportional to the most recent connection weight is added to the training equation.

Both proportional and momentum coefficients may be specified by the user, or can be modified during the training according to some error dependent conditions. Trial and error, or previous experience, is used in assigning these coefficients.

The training process is equivalent to optimizing a high-dimensional function with a potential complex error surface. Thus it is not surprising that standard function optimization methods such as Conjugate Gradient (CG), or Marquand–Leuvenberg (LM) are used to improve the training rate and efficiency. Still, local minima may be encountered, and either local minima escape algorithms can be employed, and if it fails to improve, a restart of the ANN training with different initial conditions, or a different ANN architecture, is tried.

Periodically during the training, the current ANN is usually presented with the testing examples, and the resulting predicting or classification error is evaluated. In some cases the training is stopped when the testing error begins to increase, as it may indicate 'over-training'. The generalization capacity is estimated by the ANN model error when presented with the validation examples, representing the case where an entirely new example set will be available.

13.5   The Kohonen self-organizing map ANN

A self-organizing map (SOM) is used for classification purposes, for the unsupervised situation when there is no a priori class assignment. An output 'map' is generated in a form of $n \times n$ grid points, where n is usually between 5 and 25. Each grid point is considered to be a neuron, with all the inputs in the example vectors connected to all neurons by connection weights. Again, the connection weights are initialized as small random numbers. When presented to the SOM, each neuron is activated by the sum of inputs times the connection weights. The neuron with the highest activation value is deemed the winner, and the connection weights leading to it are modified to increase its activation value, while the adjacent neurons are also modified in the same direction, but with smaller values. The process is repeated for many epochs, while the modification parameters and the range of adjacent neurons are gradually diminished. The expected result of the SOM training is clusters of winning neurons in different grid areas; the examples making these winning neurons are considered belonging to the same class.

The connection weight update equations are:

$$W_{ij}^{new} = W_{ij} + C^*(X_{ij} - W_{ij})$$

where W_{ij} is the connection weight from input j to output neuron i, X_{ij} is the input vector that caused neuron i to be the 'winning' neuron, that is the one with the highest value, C is the learning coefficient that depends on the identity of the neuron. If it is the winning one, C will be the largest allowed value. C will be decreased gradually in proportion to the distance from the winning neurons, even to having a negative value for the more distant neurons.

13.6   ANN modeling tasks

The ANN modeling tasks are as follows:

- Data collection, cleaning and preprocessing
- ANN model architecture selection
- ANN training and validation
- Model outputs post-processing and evaluation
- Knowledge extraction from the trained ANN.

13.6.1   *Data collection, cleaning and preprocessing*

Data collection, cleaning and preprocessing is often the most time-consuming phase of the ANN modeling. All inputs and targets have to be transferred directly from instrumentation or from other media, tagged and arranged in a matrix of vectors with the same lengths. Vectors with outliers and missing values have to be identified, and either corrected or discarded.

Correct preprocessing of the input and target values is essential for good ANN model training. The inputs have to be scaled into common range, so as to give equal importance to values in different ranges. For example, temperatures may be measured in a scale of hundreds, while concentrations may be measured in a fractional mole per liter. One of the common preprocessing methods is range scaling, in which the range between lowest and highest expected values is used as a scaling factor,

$$\text{range scaled input} = \frac{\text{input value} - \text{lowest input value}}{\text{highest input value} - \text{lowest input value}}$$

that results in 0–1 range of preprocessed inputs and outputs.

Another widely used method is auto-scaling of input values, in which each data column is zero-centered by subtracting the column mean, and unit normalized by dividing the result by the column standard deviation,

$$\text{auto-scaled input} = \frac{\text{input value} - \text{mean input column}}{\text{standard deviation of the input column}}$$

resulting in values usually in the range -3 to $+3$, i.e. three standard deviations from the mean. Sometimes the logarithm of the value replaces the actual value when their range spans several decades, and there are cases that require even more elaborate preprocessing methods (Mazzatorata, *et al.* 2002).

If the inputs of all examples are of the same scale, such as found in instrumentation spectra, a row-wise normalization can be done, in which all the input values (or their squared values) are summed and multiplied by a factor that normalizes the sum to one. Each factored value is the input to the ANN. This method is useful when classifying a 'shape' of spectra, even if the magnitude of the spectra may be different by an order of magnitude from each other because of different analyte concentrations.

The target values need a little more preprocessing than the inputs, as their value range has to conform to the ANN activation range, (0–1) for the sigmoid, (−1−+1) for the hyperbolic tangent. Thus the preprocessing procedure of the inputs has to be followed by these range adjustments. Experience has shown that slightly narrowing the target limits to the range of (0.1–0.9), or (−0.9−+0.9) helps to speed the training process. The reason is simple – recall that the connection weight adjustment depends on the derivative of the error. When the ANN outputs approach asymptotically to the target, the error derivative becomes small, and the connection weight adjustment process is slow. If the ANN output is allowed to overshoot the reduced target limit, the error derivative is changing sign and thus makes the training faster.

The above preprocessing techniques are valid for continuous values or simple binary ones. When ordinal, or multiple-class inputs or outputs are used, some coding should be employed, the simplest one is to assign a separate input or output to each class in a group. Thus an example with class 2 of five possible classes will have one as the second input of the group, the others are set to zero.

The preprocessing should also identify outliers. The outliers may compress other examples into a small range of values that will make it harder for the ANN to train. The identified outliers should be examined for data or instrumentation errors, and either corrected or discarded.

Another preprocessing task is the partitioning of the available examples into three subsets – one for the training, one for testing during the training for signs of overtraining, and one for validating the trained ANN model. This partitioning can be done randomly, by presetting the desired proportions of each subset, or by specifying an equal representation of each class or value range in each subset. When more robust validation is needed, or the number of examples is small, the leave-'n'-out method is used, where 'n' examples that are not used in the training are set aside for validation, and then repeating ANN model training with a different 'n' examples, until all examples are used for validation of an ANN model (Bauman, et al. 2002b).

13.6.2 ANN model architecture selection

This task is where the prior experience of the researcher is most needed. The most prevalent ANN architecture is the one-hidden-layer, fully connected feed-forward ANN. Still, the choice of the neuron activation function, the number of hidden neurons, the representation of the inputs and outputs and the training algorithm, are at the researcher's discretion.

The representation of the inputs, especially when the dimension of the input is large, such as is found in instrumentation spectra, is of critical importance. ANN with large numbers of inputs and outputs are regarded too complex, difficult to train, and liable to over-train if the ratio of training examples to the number of connection weights in the ANN model is less than 2. As shown later, I believe that these concerns are exaggerated.

The two most popular methods of input-dimension reduction are the PCA and the Fast Fourier Transform (FFT), although in recent years the wavelets transform is increasingly used. PCA transforms the data into a small number of composite, orthogonal vectors that describe best the variance in the data. FFT transforms time-dependent data into the frequency domain, giving the signal power and phase at a user selected number of frequencies. The wavelets transform tries to generate coefficients of a function with increasing frequencies whose sum will approximate the input values. The PCA vectors or the FFT and wavelets coefficients can be used as input to the ANN. The equations and subroutines for this preprocessing are readily available, and the dimension reduction effect is dramatic. One objection to the use of these transformations is that the relation between the original input identity and values to the ANN output is lost, and thus there may be less possibility to explain the ANN results, or 'new knowledge' extraction. Another objection is that future new data may contain important information in one of the inputs that is not represented in any composite vector.

The neuron activation function most used in the analytical chemistry application is the sigmoid, followed by the hyperbolic tangent. The number of hidden neurons is determined most of the time by trial-and-error, looking for the optimal number of hidden neurons in the testing error sense. In most cases this number is in the range of 5–15, depending on the selected training algorithm and its parameters.

13.6.3 *ANN training and validation*

As mentioned earlier, the classical error-back-propagation, while still used, can be made more efficient and fast by the conjugate-gradient or the ML algorithm. Other training algorithms are rarely used now.

For starting the training, random small connection weights are initially set. As the error surface may be quite complex, local minima may be encountered during the training, and several algorithms can be employed to get out of these local minima, searching for directions that provide possible escape to lower error regions (Fukuoka 1998). Alternatively, the ANN training may be restarted from a different set of random initial connection weights, hoping to avoid some local minima. Thus many results of ANN modeling are given as the average of several such ANN re-training, with the standard deviation of the ANN errors and the best ANN error values. There is an algorithm for calculating non-random initial connection weights, based on linear algebra processing of the training data that allows very efficient training of large-scale ANN modeling. Although published years ago (Guterman 1994), and proved successful in many large-scale modeling applications (Boger 1992, 1997a) including in analytical chemistry modeling (Boger 1994a, 1994b, 2003c), it remains largely ignored. The main equations of the derivation of this algorithm are given in Appendix A. This algorithm also provides a suggestion of how many hidden neurons should be tried in the ANN, although my personal recommendation is to start with five hidden neurons for training with real-world data, no matter how many inputs are there, and if not accurate enough, to try 4 or 6 hidden neurons ANN.

Once the ANN has been trained, there remains the task of validating the model and interpreting the model results. The validation is done by presenting the ANN with the validation set of examples, preprocessed by the same method that was used for the training and testing sets. Care should be taken when using any preprocessing technique, as the training input range may be smaller than some of the validation data ranges, and as it appears in the denominator, small values in it will result in a very large preprocessed input value that will saturate the trained hidden neurons. If the ANN results of this phase meet the required modeling error goals, the ANN is said to achieve generalization. If not, different ANN architectures, training algorithms or preprocessing techniques may be tried.

13.6.4   *Model outputs post-processing and evaluation*

The post-processing depends on the ANN goal. If used for prediction of some value, the outputs have to be transformed into their original units by reversing the preprocessing algorithm with the saved coefficients.

If classification is the ANN modeling goal, several possibilities exist. One may select the class according to the output neuron that is higher than a certain threshold limit, usually selected as 0.50. Alternatively, the neuron with the highest output value can be selected as indicating the class. As the values of the output neurons may be regarded as the posterior probabilities of the class density distribution, the second method may result in a rather reduced confidence in the classification results. In a two-class modeling, an estimate of the classification efficiency can be obtained by the oddly named Receiver Operator Characteristic (ROC) curve. The coordinates of this plot are the rate of the correct classification versus the rate of correct rejection as a function of the decision threshold of the output neuron, from 0 to 1. The higher the area under the curve, the better the classification. An alternative representation of the ROC curve is the selectivity vs. (1 − specificity), which is defined as:

selectivity = # of true positives/(# of true positives + # of false negatives)

the ability to classify correctly, i.e. the ratio of the found positive out of all positive examples

specificity = # of true positives/(# of true positives + # of false positives)

the ability to accept correctly, i.e. the ratio of the truly found positives to the found positives.

A more elaborate clustering mechanism is the examination of the outputs of the *hidden* neurons when all examples are presented to the trained ANN. Those examples that produce a similar pattern in the hidden neurons' outputs may be considered as belonging to the same class (Boger 1997b). If the ANN is of the auto-associative ANN type, this method can be used for unsupervised clustering (Boger 2001a, 2003b).

13.6.5 *Knowledge extraction from the trained ANN*

As mentioned before, one of the drawbacks of ANN modeling compared with chemometric techniques and ES is its apparent lack of explanation facility of its results. To overcome this drawback, many attempts are made to extract new knowledge, or at least to formulate some human-understandable rules. Sometime the ANN model architecture allows translating the trained connection weights into 'if-then' rules. This may happen if the inputs are binary yes/no, or one-out-of-M classes, and the hidden neuron outputs are close to one or zeros, as in many well-trained ANN (Guterman 1997). Recalling that the output of a neuron sigmoid activation function is greater than 0.5 when the input to the neuron is positive, one explores the conditions of the hidden neurons that make this sum positive or negative. For example, a set of connection weights to an output neuron, 0.94, 1.83, $-5.2 - 2.3$, for hidden neurons H1–H3 and the bias means that if the H3 output is 1 the output neuron value will be less than 0.5, no matter what are the outputs of H1 and H2. On the other hand, if H2 is zero, both H1 and H2 need to be 1 to have the sums of their connection weights overcome the negative bias connection value and make the output greater than 0.5. By checking similar conditions that make H1–H3 outputs ones or zeros as a function of the input state and their respective connection weights to each hidden neuron, a comprehensive set of if-then rules can be found to explain the ANN output state. Some improvement in the rule extraction process can be made by pruning the connection weights by adding a penalty term to the error used in adjusting the connection weights during the training. If this penalty term is proportional to the sum of absolute values of the connection weights, it leads to the selection of smaller connection weights, and those approaching zero can be safely deleted. The sparser ANN structure is easier to analyze, and has a better ratio of training examples to the number of connection weights.

A quasi-quantitative relationship between each input and each output may the Causal Indices (CI) (Baba 1990). The CI is calculated as the sum of the products of all 'pathways' between each input and output; these are easily calculated from the trained ANN connection weights, by the equation

$$CI = \Sigma \mathbf{W}_{kj} \times \mathbf{W}_{ji}$$

over all h hidden neurons. $\mathbf{W}_{kj}$ are the connection weights from hidden neuron j to output k, $\mathbf{W}_{ji}$ are the connection weights between input i to hidden neuron j.

The CI numerical results are relative; that is, they indicate the magnitude and sign of a change in an output when a particular input is changed. Their advantage is that they do not depend on a particular input vector, but on the connection weight set that represents all the training input vectors. This is also one of their limitations, as a local situation may be overlooked in the global representation. Although somewhat heuristic, this method is more reliable than the local sensitivity checks, as it is based on the whole ANN trained on all the available states. It has to be remembered, however, that the interactions between inputs are not readily visible.

Examining the CI for each output as a function of the inputs' number reveals the direction (positive or negative) and the relative magnitude of the relationship of the inputs on the particular output. Large positive CI (input, output) indicates an increase of the output value when the input value is increased, and vice versa. The CI method seems to give good relationships, but suffers from the fact that it has a global-single variable behavior, based on the ANN connection weights that reflect the overall behavior, disregarding some local or combined variables' peculiarities – see the section on integration with other AI techniques.

One of the more important forms of knowledge extraction is the identification of the more relevant inputs. When identified, they may be treated as a reduced input set for re-training another ANN model. In many cases this leads to a better model accuracy, and improves the ratio of training examples to the number of connection weights. Several methods can be employed for such identification, most of them based on sensitivity analysis that can be made either by calculating the complex output derivative relative to each input, juggling each input and observing the change in the output, or systematically removing each input for re-training and observing the change in the ANN modeling errors. The ANN pruning mentioned can sometimes eliminate all connection from an input.

More complex mathematical correlations are sometimes employed to identify redundant, co-linear inputs, or inputs with little information content, The use of GA to select subsets of inputs, and evolve the selection from the better trained ANN, may achieve the desired input reduction (Leardi 2002).

The method I usually employ is based on the calculation of the relative contribution of each input to the variance of the hidden neurons – once the whole data set is presented to the trained ANN. The rationale is that if an input does not contribute much to any hidden neuron activity, it is because either it is almost constant in the data, or that the training assigned to it small connection weights to each hidden neuron, thus making it almost constant, no matter how it varies in the data. In both cases it can be safely discarded from the re-training of the next reduced ANN. The equations leading to these results are given in Appendix B (Guterman 1997). For example, this algorithm was applied to the modeling of the cetane number of diesel oil (the equivalent of the octane number of gasoline) from IR spectra. It reduced the number of wavelengths from the original 600 to 12 in 7 ANN re-trainings, while improving the prediction accuracy of the cetane number (Boger 2003b, Appendix C).

13.7 Integration with other AI techniques

Several AI techniques that are in general use, including Fuzzy Logic and Genetic Algorithms, have been integrated with ANN modeling of analytical chemistry. Fuzzy Logic transforms any numerical input into a set of membership values in predefined classes. Thus, for example, a pH value of 5 would be transformed to 90% membership in the acid class, 10% membership in the neutral class, 0% in the basic class.

Sometimes this transformation is useful in the CI knowledge extraction. When modeling the solubility product of a salt in a complex mixture, the negative value obtained by CI analysis of the trained ANN indicated correctly that the salt solubility decreased with the increasing acidity. However, it was known that at high values of acidity the trend was reversed, possibly because of the formation of a salt–acid complex. To find this local feature, the mixture concentration values were fuzzified into membership values in the low, medium and high classes, and the ANN re-trained with those values as inputs. The CI analysis then correctly indicated a negative CI for the low and medium acidity, and a positive CI for the high acidity (Boger 2001b).

Genetic algorithms are optimization techniques mimicking the Darwinian 'survival of the fittest'. The parameters of a known function are placed side by side in a 'chromosome' vector. A large number of those vectors are generated with random values (within their allowed ranges) of the variables. Randomly selected pairs of 'chromosomes' are cleaved in a random position and the parts re-attached crosswise. The next generation of chromosomes is selected in some way proportional to the evaluated function value, keeping the number of chromosomes constant. To avoid being stuck in local minima, a randomly selected parameter in the chromosome is modified, with a low probability, a 'mutation'.

GA have been integrated with ANN modeling in two ways. One, more prevalent in analytical chemistry applications, is the selection of the inputs to a model, be it ANN or other chemometric model. The accuracy of the model is used as the evaluating function, and the GA usually converges to a good solution. The second way of using GA is for determining the architecture of the ANN, including the number of hidden neurons in the chromosome containing the input variable. The third way is during the ANN training, when the connection weights are generated by the GA. The input selection by GA gives modeling results that are generally equivalent to the input reduction algorithm results described in Appendix C. The selection of the ANN architecture by GA, while possible, was shown to be much more computationally demanding (Boger 2000).

The latest AI classification technique that is used is the Support Vector Machine (SVM). This technique tries to find a few examples that partition a two-class space with the maximum separation between the classes. If the space is non-linearly separable, a kernel function transforms the data into a higher-dimensional space that may be separable. If not possible, a small number of mis-classified examples is allowed, with penalty, in the search for the Support Vectors. The SVM has the advantage that the number of inputs does not increase the computational burden, and thus is suitable for the modeling of chemical instrumentation spectra. It has the disadvantage that the computational burden increases with the third power of the number of training examples.

13.8   Review of recent ANN applications in analytical chemistry

In this section papers of ANN applications in analytical chemistry are grouped by their instrumentation types. A 'general' list give some references to useful papers describing the ANN and other related techniques used, with no particular analytical chemistry area.

Several common topics can be found in the referenced papers. The first is the ANN comparison with the more conventional techniques such as PCA, PLS, SIMCA, etc. In all these papers the ANN modeling results are better, or at least equal to the other model's results. The second is that the ANN modeling is more prevalent in the analytical instrumentation areas that give more complicated, less sharply defined spectra or outputs, such as infra-red spectra or electrophoresis applications. Both topics are interrelated, as the more easily posed problems can be solved by the more accepted and accessible linear techniques such as PCA. Only when the problem is more complicated, ANN modeling is tried, usually the 'standard' feed-forward error back-propagation architecture.

An interesting, rather recent, application is the use of the ANN modeling to optimize the analytical technique itself, when the operating parameters are not easily selected for best results. This can be viewed as a kind of useful knowledge extraction, similar to the selection of the reduced number of inputs to achieve better spectral analysis.

It is expected that the ANN application in analytical chemistry will become more widely used, as the computing power available to the researcher and computer literacy increase. In addition, ANN techniques are also getting used more widely in other chemistry-related areas, not covered in this review, such as Quantitative Structure Activity Relationship (QSAR) research for drug discovery, large database mining, and microchip based systems.

13.8.1 *Tutorials, reviews and theory*

This section includes references that may help the reader to get more insights on ANN technology and related artificial intelligence algorithms such as Expert Systems (ES), Fuzzy Logic (FL), Genetic Algorithms (GA) and Support Vector Machines (SVM), as applied in analytical chemistry. It includes tutorials, reviews and books, and some journal articles that highlight specific subjects of interest. An Internet site address contains a compilation of all aspects of ANN, updated monthly (Sarle 2004). There is also an active Internet user group, comp.ai.neural-net, that may be used for getting answers on specific questions.

First author	Year	Type	Comments
Adler	1993		ES
Agatonovic-Kustrin	2000	tutorial	
Al-Khalifa	2003		SVM
Alsberg	2000		input importance
Baba	1990		causal index
Baumann	2002a		variable selection
Baumann	2002b		model validation
Blank	1993		data processing
Boger	1997a		input reduction
Boger	2000		GA
Boger	2003		auto-associative classification
	2000		
Bonastre	2001		distributed ES
Bos	1993c	review	data processing
Bourquin	1997	tutorial	
Bourquin	1998b		pitfalls
Byvatov	2003		SVM
Chen	2002a		prior knowledge, evolving algorithms
Chretien	2002		chemometrics
Danzer	1995		multi-dimensional calibration

Table (*Continued*)

First author	Year	Type	Comments
Daszykowski	2003	auto-associative tutorial	low dimensional visualization
Darks	1998	tutorial	
Defernez	1999		overfitting
Egan	2001		ANN optimization
Fraud	2002		methodology to explain ANN classification
Forsstrom	1995	tutorial	
Fukuoka	1998		avoid false local minima.
Guterman	1997		non-random initial connection weights
Harrington	2000		detect errors and non-linearities
Hervas	2001		pruning
Hopke	2003	review	chemometrics
Klawun	1995		strategies to improve the prediction capability
Lavigne	2001		selective receptors
Lavine	1999a	tutorial	GA
Leardi	2002		chemometrics: from classical to genetic algorithms
Lerner	1999		
Li	1999a	review	feature extraction
Liebman	1994		intelligent instruments
Lo	1995		multivariate calibration
Mazzatorata	2002		pre-processing
Milne	1997		mathematics in chemistry
Morris	1991		ES
Mukesh	1997		introduction to ANN in process chemistry
Olden	2002		variable contributions
Papadopoulos	2001		confidence estimation
Peris	1994		ES
	1997		
	1998		
Persaud	1998b	review	EN
Peterson	2000	review	ANN and their use in chemistry
Place	1994	review	
Place	1995	review	
Raj	1999	review	input reduction
Rius	1998	review	reliability
Sarle	2004	weekly updated web page	frequently asked questions ftp://ftp.sas.com/pub/neural/FAQ.html
Shao	2003		wavelets
Smits	1993		drift correction
Shukla	1999		global training algorithm
Spining	1994		'black box'
Stefan	1999		sensor arrays
Sumpter	1994	review	spectroscopy
Svozil	1997		introduction to ANN
Tafeit	1999	review	medical laboratory
Thodberg	1996	review	Bayesian ANN
Tominaga	1999		comparison of classification techniques
Tetko	2002		introduction to associative NN
Tolle	2002		pharmacokinetic data sets

(*Continues*)

Table (*Continued*)

First author	Year	Type	Comments
Zhang	2002a	review	overlapping peaks
Zupan	1993	book	neural networks for chemists – an introduction
Zupan	1999	book	neural networks in chemistry and drug design

13.8.2 *Gas, liquid and ion chromatography*

Two main subjects of interest are evident in the papers dealing with chromatography aspects of analytical chemistry – predicting the retention indices (RI) of compounds, and using ANN modeling for optimizing the operating parameters of the columns to get better separation and identification. In some cases, the ANN model is used for experimental design.

First author	Year	Technology	Application area	Comments
Agatonovic-Kustrin	1998a	HPLC	amiloride, methychlothiazide	optimization
Agatonovic-Kustrin	1998b	HPLC		optimization
Agatonovic-Kustrin	1999b	HPLC	diuretics	
Agatonovic-Kustrin	2000a	HPLC	pharmaceutical	tutorial
Aires-de-Sousa	2002a	GC	chirality	RI
Angerosa	1999	GC	olive oil	
Araujo	2002	GC-MS	catalytic activity	RI
Booth	1997	HPLC	aromatic acids, amides	
Branca	2003	GC-MS	perfumes	
Bruchmann	1993	GC	terpenes	
Bruzzoniti	1998	RPLC, electrophoresis	carboxylic acids	RI
Bucinski	2002	HPLC	flavonoids	
Bylund	1997	HPLC	pharmaceutical R,S-oxybutynin chloride	optimization
Bylund	2003	LC-MS	pharmaceutical lactate dehydrogenase	
Chen	2000a	chromatogram	noise filtration	Elman recurrent ANN
Chen	2001a	GC/MS	wine aroma	
Cordella	2003	Anion-ex.chrom.	sugars in honey	
de Lucas	2001	GC, HPLC	diesel particle emission	
Debeljak	2001	chromatography		SOM
Diaz	2001	HPLC	nitrophenol pesticides	experimental design
Dieterle	2003b	HPLC	urinary nucleosides	LVQ
Divjak	1998	ion-chromatography	transition metals	experimental design
Doble	2003	GC-MS	gasoline grade	
Fatemi	2002	GC		RI, Kovats
Francelin	1993	GC	vegetable oils	
Frenich	2001	HPLC	pesticides water soil	
Galvan	1996	GC	kinetic data	
Garcia-Parilla	1997	HPLC	wine vinegar	
Garcia-Valls	2001	HPLC	rare earths	

Table (*Continued*)

First author	Year	Technology	Application area	Comments
Gennaro	2003	LC		optimization experimental design
Gorodkin	2001	HPLC	metabolites	pruning
Guillaume	2000	HPLC	p-hydroxybenzoic esters	optimization
Guo	2000	GC	dioxane	
Guo	2000	GC	alcohols	RI
Guo	2001	GC	alcohols	RI
Hadjiiski	1999	GC	air pollutants	
Havel	1999b	ion chromatography		RI
Havlis	2001	HPLC	DNA	optimization
Heberger	2002	GC		RI input selection
Hilder	2000	anion IX	inorganic anions	
Horimoto	1997	GC	milk, bacteria	RI
Hu	1996	GC		
Jakus	1993	GC		RI
Jakus	1993	GC		RI
Jalali-Heravi	1998	FID		
Jalali-Heravi	2000a	GC-MS	MS simulation	
Jalali-Heravi	2000b	FID		
Jalali-Heravi	2000c	GC	different adsorbents	RI
Jalali-Heravi	2001b	GC, RI	noncyclic and monocyclic terpenes	
Jalali-Heravi	2002c	GC		Kovats RI
Kaliszan	2003	HPLC		optimization
Kurbatova	1999	GC	alkyladamantanes	
Kushalappa	2002	GC	potato diseases	
Lee	1998	GC	fatty acids	
Lleti	2003 2000	GC-MS	proteinaceuos binders	SOM
Loukas	2001	LC	RI	RBF
Madden	2001	anion-chromatography		RI
Magelssen	1997	GC	araclors	
Marengo	2002	GC-MS	wines	
Marengo	1998	HPLC	pesticides	optimization
Metting	1996	HPLC		optimization
Micic	1998	GC-MS	organics in geological	
Moberg	2000	LC-MS	estriol/estrogens, buprofen/urine morphine/codeine	
Mulholland	1995	IC		detector selection
Mulholland	1996	IC		ES
Ni	1997	HPLC	polymerization	
Onjia	2002	HPLC	phenols	
Pappa-Louisi	2000	HPLC	adenosine	optimization
Patnaik	1996	chromatography	bio-process	optimization, ES

(*Continues*)

Table (*Continued*)

First author	Year	Technology	Application area	Comments
Petritis	2002	reversed-phase LC	peptides	elution times
Petritis	2003	LC-MS	peptide, bacteria MS/MS	
Pompe	1997	GC	organics	RI
Qi	2000a	GC	naphta	RI wavelets
Qi	2000b	GC LC	amines	
Questier	2002	GC	Maillard reaction products	SOM
Rowe	1994	GC	peak classification	
Ruan	2002	IR, GC/MS	deoxynivalenol	
Sackett	1996	GC	VOC in breath	
Shan	2002	HPLC		optimization
Shen	2003	Py-GC-MS		tablet production
Shih	2000	PZ detectors for HPLC, GC	various	
Siouffi	2002	GC, LC		
Song	1998	GC	fatty acids	fuzzy ARTMAP
Song	2002	IX chromatograph	protein	RI SVM
Srecnik	2002a	ion chromatograph	anion	RI
Srecnik	2002b	ion chromatograph	anion	RI
Steinhart	2000	GC	odor	optimization
Sutter	1997	GC		RI
Suzuki	2001	chromatograph	arylalkylcarbinols	
Tandler	1995	GC/MS	alkene	
Taylor	2002	GC/MS	metabolomics	ES
Tham	2002	RP-HPLC	amino acids, RI	GA
Tong	1999	MS, GC		
Urruty	2002	GC/MS	strawberry aroma	
Wan	2000	GC/MS	aromatic carbamates	SOM
Wang	2003	chromatography		process
Welsh	1996	HPLC	drug fingerprinting	
Xiang	2002	HPLC	N-Benzylideneanillines	RBF
Xie	1995	HPLC		
Yan	1998	GC		RI CC
Yan	2000a	GC	alkylbenzenes	
Yan	2001	GC	alkylbenzenes	RI
Yang	2002a	LC GC/MS	cetane number diesel	
Yang	2002b	HPLC	nucleosides markers	RI
Yin	2001c	GC	alkanes	wavelets
Zhang	1997a	GC	wine	
Zhang	1999	GC	alkyl-benzenes	RI
Zhang	2001	GC	naphta	RI
Zhao	1999	HPLC	bile acid	

13.8.3 *Sensor arrays*

Chemical analysis by sensor arrays is gaining popularity as part of the move from specific reagents or specific peaks to pattern recognition by non-specific sensors. The generic name Electronic Nose (EN) is used for such arrays, most of them based on the electrical conductivity change of metal oxide semiconductors related to the oxidation-reduction reactions with the gas mixtures.

First author	Year	Technology	Application area	Comments
Aleïxandre	2002	MOS	CO, NO$_2$, organics	PNN
Alexander	1998	tin-oxide	beer	
Al-Khalifa	2003	thermal microsensor		SVM
Barko	1997	piezo-electro	VOC	
Barshick	1998		fuel odor	
Beling	1998	sensor	odor	SOM, ART
Blonder	1997	QCM	antibodies	
Boilt	2002	tin-oxide	fruit	GA
Boger	1997a	tin-oxide	organic gas mixture	
Boger	2002a	tin-oxide	organic gas mixture	input reduction
Boger	2002b	tin-oxide	organic gas mixture	input reduction
Boger	2002c	tin-oxide	organic gas mixture	input reduction
Boger	2003	tin-oxide	organic gas mixture	input reduction
Boger	2003	tin-oxide, IR	chemical warfare agents	
Boger	2003	tin-oxide, IR	fuel, organic gas mixture	
Boger	2003	tin-oxide, IR	CWA	
Bolarin	2002	tin-oxide, IR	essential oils	ensemble
Branca	2003		perfumes	GC-MS
Brezmes	1997	tin-oxide	aroma	
Bull	1993	sensor array	fuel cell gas	
Carrapiso	2001	EN	pig fatty acids	
Cerrato-Oliveros	2002	tin-oxide	olive oil	
Chang	1991	piezoelectric crystals	odor	
Chong	2000	tin-oxide	hydrocarbons in air	
Choi	2002	tin-oxide	organic gas mixture	RBF
Di Natale	1995	QMB	organic gas mixture	
Di Natale	1996	glass sensors	heavy metal ions	
Di Natale	1997	EN		SOM
Di Natale	1999a	QMB metalloporphyrin	organic gas mixture	
Di Natale	1999b	EN	blood in urine	
Di Natale	2000	EN, tongue	wine	
Dickert	1999	QMB	xylene isomers	
Dieterle	2002	polymer sensor	R22, R134a refrigerants	
Distante	2000	tin-oxide	organic gases	SOM
Distante	2002	tin-oxide	organic gases, humidity	
Dutta	2003a	EN	tea quality	
Dutta	2003b	EN	tea quality	
Eklov	1999a	MOS and catalyst	H2, ethylene	

(*Continues*)

Table (*Continued*)

First author	Year	Technology	Application area	Comments
Eklov	1999b	MOS and catalyst	H2, ethylene, NH3	
Eklov	1999c	gas sensors		variable selection
Fu	2002	EN	breath diagnosis	
Gallardo	2003	electronic tongue	K+, NH4+ in water	
Galdikas	2000	tin-oxide	meat freshness	
Garcia-Gonzalez	2003	MOS	olive oil	
Gardner	1992	MOS	wine, beers	
Gardner	1999	MOS	breath analysis	
Gardner	2000a	MOS	health diagnosis	
Gardner	2000b	EN	bacteria potable water	
Getino	1998	MOS	soil VOC	
Gibson	1997	EN	bacteria	
Gonzalez-Martin	2001	MOS	vegetable oil	
Grove	2001	SAW	solvent vapors	
Guernion	2001	EN	bacteria in urine	
Hanaki	1996	quartz resonator	aroma	
Henkel	2002	QMB	CO2 and H2O in gas mixtures	
Hobson	2003	EN	temperature correction	
Hoffheins	1992	simplified chemical sensor arrays		
Hong	1996	semiconductor	aroma	
Hsueh	1999	sensor array	redox active substances	
Hudon	2000	EN	odor intensity	
Joo	2001	tin-oxide		RBF
Jarvis	2003	tin-oxide	gases	
Kato	2000	tin-oxide	gases	FFT
Keller	1994 1999	tin-oxide	chemical vapors	
Kim	2001	MOS	VOC	
Kunt	1997	micro-hot-plate	ethanol methanol	
Lau	2000	SAW	organics	SOM
Lee	2000	tin-oxide	VOC	
Lee	2001	tin-oxide	VOC	
Lee	2003a	micro-hot-plate	combustible gases	
Legin	1998	ET	Fe , U	
Li	1999b	MOS	hydrocarbons	counter-propagation
Liden	2000	EN	metabolites	
Llobet	1997	MOS	ethanol, toluene and	
Lu	2000	EN	oxylene alcohol and petroleum vapors	
Mandenius	2000	EN	metabolites	review
Martin	2001	tin-oxide	NO2, CO	
McAlernon	1999	QMB	organic gases	
McCarrick	1996	MOS	jet fuel	
Mohamed	2002 2003	EN	breath diagnosis	

Table (*Continued*)

First author	Year	Technology	Application area	Comments
Moon	2002	tin-oxide	H2, CO	
Nakamoto	2001	sensor array	odor	
Nakamoto	2002	sensor array	odor	
Nanto	1996a	QMB	alcohols	
Nanto	1996b	sol-gel QMB	alcohols	
Nazarov	2003	laser spectroscopy	Cs in water	
Niebling	1995	EN	organic gas mixture	
Oliveros	2002 1997 1999	MOS	olive oil	
Paulsson	2000	MOS	alcohol in breath ethyl	
Penza	2002	SAW	acetate, ethanol, acetone	
Penza	2003	SAW	methanol/2-propanol	
Persaud	1998a	sensor array	odor	Sammon
Ping	1997	EN	odor	
Pinto	2001	EN		review
Prado	1999a	MOS	VOC	feature extraction
Prado	1999b	MOS	cheese classification	
Prado	2000	MOS	VOC	reliability
Qin	2001	sensor array	environmental monitoring gas mixtures	
Qin	2001	tin-oxide		
Reibel	2000	SAW	gas mixtures	
Rodriguez-Mendez	2000	Langmuir-Blodgett films	VOC	
Roppel	2001	sensor array	fault checking	
Rudnitskaya	2001	PVC and solid-state array	inorganic pollutants	
Ryman-Tubb	1992	sensor array	wine	
Sawant	2002	tin-oxide	odors	
Sberveglieri	1998	MOS	milk	
Schnurer	1999	EN	spoilage volatiles	
Shaffer	1999	SAW	chemical warfare agents	PNN
Shin	2000	MOS	bacteria in water	
Shu	1994	sensor array	taste	SOM
Sinesio	2000	EN	tomato quality	
Stetter	1992	sensor array	grain quality	
Taurino	2003	EN	VOC	
Vazquez	2003	EN	anchovies	
Vlasov	1998	EN	non-selective	
Wang	2002a	QMB	benzene, toluene	
Wen	2002	sensor array	Freon	
Wilson	2000	tin-oxide	odors	
Yan	2000b	sensor array	ethanol gasoline	
Yang	2000	SAW	organic gases, beer	
Yea	1997	tin oxide	inflammable gas	FL
Yea	1999	sensor array	inflammable gases	FL
Zhou	2000a	EN	environmental	review

13.8.4   *Electrophoresis*

Although similar to chromatography, electrophoresis is a good example for the utility of ANN models, as the optimal operating parameters of the electrophoresis methods are essential for getting good results. Most of the papers on this subject are by Havel and his co-workers.

First author	Year	Technology	Application area	Comments
Bocaz-Beneventi	2002a	electrophoresis		experimental design
Bocaz-Beneventi	2002b	electrophoresis	serum potassium	
Cancalon	1999	electrophoresis	citrus juice	
Casamento	2003	electrohoresis	explosives	
Ceppi	1999	electrophoresis	humic substances	
Dobiasova	2002	electrophoresis	sorbic acid, wine	
Dohnal	1998	electrophoresis	naloxon, naltrexon, nalorfin	
	1999			
Dohnal	2003	electrophoresis	amino acids enantiomers	experimental design
Dohnal	2002	electrophoresis	optical isomers	experimental design
Farkova	1999	electrophoresis microemulsion electrokinetic		experimental design
Fatemi	2003	chromatography	benzene derivatives migration indices	
Gajdosova	2003	electrophoresis micellar liquid	humic acids	
Garcia-Alvarez-Coque	1997	chromatography		experimental design
Garcia-Ruiz	2003	electrophoresis	phenylpyrazoles	optimization
Grus	1997	electrophoresis	proteins	
Grus	2001	electrophoresis	proteins	
Havel	1998	electrophoresis		optimization
Havel	1999a	micellar chromatography		optimization
Herber	2001	2-D electrophoresis 2-D polyacrylamide gel	tear proteins of people with diabetics	
Jacobsen	2001	electrophoresis	gliadin extracted from wheat	
Jalah–Heravi	2002	electrophoresis	alkyl- and alkenylpyridines	
Jalali–Heravi	2001a	electrophoresis	sulfonamides	
Jensen	1997	iso electrophoresis	potatoes	
Jimenez	1997a	micellar LC		
Jimenez	1997b	micellar LC	dihydropyridines	

Table (*Continued*)

First author	Year	Technology	Application area	Comments
Kalny	2001	electrophoresis	Pd	
Klempfl	2002	electrokinetic chromatography		optimization
Latorre	2001	electrophoresis	amino acid derivatives	
Li	2000	electrophoresis		optimization
Li	2002	electrophoresis micellar electrokinetic	medicines	
Liu	2002	capillary chromatography	plant hormones	LM
Liu	2002	electrophoresis	Chinese medication	
Liu	2002	micellar electrokinetic	plant hormones	optimization
Malovana	2001	electrophoresis	esmolol	
Malovana	2002	electrophoresis	oximes, CW agents	
Marti	1999	electrophoresis	metal cyanide	optimization
Muzikar	2002	electrophoresis	metal crown-ethers	
Muzikar	2003	electrophoresis	sulphate	optimization
Pacheco	2001	electrophoresis	U + humic acids	
Pacheco	2002	electrophoresis	humic acids	
Pacheco	2003	electrophoresis	humic acids + xenobiotics	
Pazourek	2000	electrophoresis	polyphenols in wines	
Pokorna	1999	electrophoresis	galanthamine, anti-Alzheimer	
Pokorna	2000	electrophoresis	humic acids	
Polaskova	2002	electrophoresis	pharmaceutical	
Regeniter	2003	electrohoresis	immunofixation	
Reichova	2002	electrophoresis	adamantane anti-viral	
Ruanet	2001	electrophoresis	gliadin in wheat	
Sanchez	2002	electrophoresis, MALDI	Rh(III)	
Sanders	2003	electrophoresis-microchip	DNA	optimization
Schrim	2001	electrophoresis, UV	pentosan polysulphate	QA
Siouffi	2002	GC, LC electrophoresis	meat proteins	
Skarpeid	1998	isoelectric electrophoresis	thyreostatic drugs	
Vargas	1998a	electrophoresis UV	beta-cyclodextrin	
Vargas	1998b	electrophoresis UV	anti-Parkinson drugs	
Vashi	2001	electrophoresis	chlorophenols	RBF
Wang	2003	electrophoresis	arylpropionic acids	optimization
Wolbach	2001	electrophoresis	proteins	
Woodward	2001	electrophoresis flourophore electrokinetic	metabolic substrates	
Zakaria	2003	chromatography	aromatic bases	optimization

13.8.5 *Electrical sensors*

This section deals with signals generated by electrical properties sensors, such as pH, polarographs, conducting polymers, etc. One important class, conductivity change of semi-conducting materials, is described separately in the sensor array section. As in other sections, the ANN modeling is sometimes used for optimizing the operating parameters or for experiment design.

First author	Year	Technology	Application area	Comments
Bachmann	2000	amperometric biological sensors	carbamate, organo-phospherous	
Basheer	2000	pH	bacteria	tutorial
Bastarache	1997	pH	water	
Bos	1990b	ion-selective	Ca, K,	
Bessant	1999	dual pulse voltammetry	ethanol, fructose and glucose	
Bessant	2000	dual pulse voltammetry	ethanol, fructose and glucose	
Bessant	2003	dual pulse voltammetry	ethanol, fructose and glucose	
Brodnjak–Voncina	1999	potentiometric titration	wines	
Brodnjak–Voncina	2002	potentiometric titration	river water	
Byun	2002	conducting polymer	organics	SVD
Cabanillas	2000	polarograph	atrazine-simazine, terbutryn-prometryn	
Chan	1997a	voltametric	Cu, Pb, Zn, Cd	
Chen	1999	organic thin films	NO2, NO	
Chen	2000b	organic thin films	NO2, NO	
Cladera	1997	voltammetric		
Cordella	2003	pulsed amperometeric	sugars in honey	anion-exchange chromatography
Cukrowska	2000	pulse polarograph	hydrogen evolution	experimental design
Cukrowska	2001	pulse polarograph	adenine and cytosine	experimental design
Cukrowski	2000	polarograph	metal-ligand	experimental design
Cukrowski	2001	polarograph	metal-ligand	experimental design
Davey	1998	conductance	biomass	
De Carvalho	2000	voltammetry	catechol and hydroquinone	
De Souza	1999a	polypyrrole array	odor	
De Viterbo	2001	pH	metallic complex	
Ehrman	1996	pH	water chemistry	
Garcia-Villar	2001	ion-selective array	lysine	
Emilio	2001	pH	photocatalytic reaction of EDTA with TiO2	
Geeraerd	1998	PH	bacteria, atrazine-simazine and terbutryn-prometryn	

Table (*Continued*)

First author	Year	Technology	Application area	Comments
Guiberteau	2001	polarograph	captan-captafol and folpet	
Guwy	1997	pH	anaerobic digester	
Hajmeer	1997	pH	bacteria growth	
Hajmeer	1998	pH	bacteria growth	
Hitzmann	1994	pH-FIA	penicillin	
Kuznetsov	1999	pH	optical membrane	
Lastres	1997	stripping voltammetry	Cu, Zn.	
Lee	2003b	conducting polymers	VOC	
Lubal	2000a	potentiometric	U-humic acids	
Lubal	2000b	potentiometric	W-tartarate	
MacDonald	2002	electrochemical impedance	corrosion	
Mongelli	1997	pH	blood	ES
Moya-Hernandez	2002	pH		
Palacios-Santander	2003	electro-chemical signals	Tl+, Pb2+	FT
Pishvaie	2000	pH		
Pradova	2002	electro-chemical sensors		review
Pavlou	2000	conductive polymers	bacteria Helicobacter pylori	GA
Racca	2001	pH	paleolimnology	
Raj	1999	electrochemistry		review, input reduction
Richards	2002a	voltammetric	ethanol, fructose and glucose,	GA
Richards	2002b	electroanalysis	data analysis	
Richards	2003	voltammetric	ethanol, fructose and glucose,	
Saini	2000	voltammetry	organics	patent
Sarabia	2002	polarograph	Ti/Pb	accuracy, review
Schepers	2000	pH	bacterial growth	
Seker	2003	potential sweep	D-glucose oxidation	
Shamsipur	2002	pH	anthanilic, nicotinic, picolinic, sulfanilic acids	
Taib	1996	pH	optical sensor	
Talaie	1996a	conductive polymer	Ca, Na, Li ions	
Talaie	1996b	conductive polymer	formate	
Talaie	1996c	conductive polymer	pH	
Talaie	1999	cyclic voltammetry, resistometry, gramometry	polypyrrole electrode.	

(*Continues*)

Table (*Continued*)

First author	Year	Technology	Application area	Comments
Talaie	2001	conductive polymer	pH	
Van Can	1998	pH	penicillin	
Van der Linden	1989	ion selective	Ca, K	
Vlasov	1997	glass electrodes	heavy metals	
Winquist	1998	tongue, voltammetry	milk	
Woodward	1996	dielectric spectroscopy	metabolites	optimization
Xing	1997	Berberine selective electrode		
Yang	2001	RF capacitance	oil in water	
Zampronio	2002	potentiometric flow injection	orange juice	
Zuppa	2002	conducting polymer	acetonitrile, methanol, propanol, acetone, butanol, water	SOM

13.8.6 *Spectroscopy*

One of the largest areas of ANN modeling applications, especially in spectra without sharply defined specific peaks, as in infrared (IR). A major aim is to reduce the input vector length in order to improve the ratio of examples to the number of connection weights in the ANN, to avoid over-fitting.

First author	Year	Technology	Application area	Comments
Absalan	2001	spectrophotometry	Se in the presence of Te	
Agatonovic-Kustrin	1999a	IR	ranitidine-HCl	input reduction
Agatonovic-Kustrin	2000c	IR, XRD	terbutaline sulphate	
Agatonovic-Kustrin	2000b	XRD	ranitidine-HCl	
Agatonovic-Kustrin	2001a	diffuse reflectance FTIR	entiometric ibuprofen	
Agatonovic-Kustrin	2000	IXRD	terbutaline sulphate	
Agatonovic-Kustrin	2001	input reduction		
Alam	1994	NIR	resin	
Al-Amma	2000	ICP-AES	ICP-MS	input reduction
Allanic	1992	fluorescence		
Alsberg	1998	FTIR		input importance
Alsberg	1998	FTIR	bacteria	
Alsberg	1998	FTIR	bacteria	input importance
Anderson	1999	ICP	food (potatoes)	classification
Andreu	2002	fluorescence	bilirubin	
Andrews	1994	fluorescence	fuel	
Arakawa	2003a	IR microcalorimetry	bacteria (anthrax)	
Arakawa	2003b	IR microcalorimetry	bacteria	
Babri	2000	NIR	hydroxyl	
Balamurugan	2003	UV	drugs	

Table (*Continued*)

First author	Year	Technology	Application area	Comments
Becker	2002	FIA	ethanol, lactate	optimization
Bell	1992	IMS		
Bell	1993	IMS		
Bell	1999	IMS		
Bell	2000	IMS	high temperature	
Benjathapanun	1997	UV-Vis	wastewater contaminants	
Benoudjita	2004	spectrophotometer		variable selection
Berry	1997	optical	heavy metals	
Bertran	1999	reflectance NIR	water in ampicillin trihydrate	
Bertran	2000	NIR	vegetable oil	
Bessant	2001	synchronous scan fluorescence	insulation oils	
Bhandare	1993	mid-IR	blood glucose	
Bhandare	1994a	mid-IR	blood glucose	
Bhandare	1994b	FTIR	blood glucose	
Blanco	1995	optical	Cu, Fe, Co	
Blanco	1996	optical	benzylamine-butylamine	
Blanco	1997	optical	tanning bath colors	
Blanco	1998	IR	pharmaceutical industry	
Blanco	1999a	enzymatic spectroscopy	ethanol methanol	
Blanco	1999b	reflectance IR	particle size effects	
Blanco	2000	NIR	acrylic fibers	
Blanco	2001a	NIR	bitumens	
Blanco	2001b	NIR	geographical origin of petroleum crudes	
Blanco	2001c	circular dichoism UV-Vis	entiomers	
Blanco	2002	NIR	finishing oil	
Boger	1993	IR	organic	
Boger	1994a	IMS	platinum group, organic	
Boger	1994b	IMS	bromine, HF	
Boger	1994c	IMS	organic	
Boger	1996	UV	distillation	
Boger	2003	IR	fuel, organic	input reduction
Boke	2004	FTIR	calcium minerals	
Borggaard	1992			minimal ANN
Bos	1991a	XRF		GA
Bos	1993	XRF	Fe-Cr-Ni	
Bratz	1994	UV-Vis		input reduction
Briandet	1996	NIR	instant coffee	
Bro	1996	UV-Vis	food enzyme	
Brown	1998	NIR	classification	
Bryjak	2000	optical	glucoamylase	
Burden	1997	electronic absorption	PAH	
Cabrol-Bass	1995	IR, MS		
Camara	2003	spectrophotometric	dextropropoxyphene, dipyrone	
Capitan-Vallvey	2000	fluorescence	flufenamic, mefenamic meclofenamic acids	input selection by SOM

(Continues)

Table (*Continued*)

First author	Year	Technology	Application area	Comments
Carrieri	1995	IR	chemical warfare agents	
Carlin	1994	IR	water in fish products	
Catasus	1995	ICP-AES	Fe, Zn in Mg matrix	
Centner	2000	NIR	multivariate calibration	
Chan	1997b	optical	Na, K	
Charef	2000	UV-Vis, pH	wastewater COD	
Chen	1995	NIR	wheat classification	
Chen	1996	Vis-NIR	poultry	
Chen	2001b	NIR	drug content	
Chen	2002b	FIA	Ir(III)	optimization
Chen	2003	IMS		wavelets
Citrovic	1997	spectroscopy		
Cleva	1997	IR	structure determination	
Cullen	2000	kinetic-spectrography	Ga Ni	
Dacosta	2001	NIR	triglycerides in plasma	
Da Silva	1998	UV	chlorophenols	rule base
Dane	1999	XRF	layered materials	
Dane	2001	guided microwave spectroscopy	moisture in tobacco	
Daniel	1997a	Raman	explosives	
Daniel	1997b	Raman	explosives	
de Cerqueira	2001	IR	N in wheat, BRIX sugar cane	
Debska	1999	IR	spectra interpretation	
Debska	2000a	IR		
Debska	2000b	IR	molecular sub-structure	pruning
De Souza	1997	Mossbauer		
De Souza	1999a	Mossbauer		GA, FL and ANN
De Souza	1999b	Mossbauer		review
De Souza	2001a	Mossbauer, XRF		
De Souza	2001b	Mossbauer, XRD, thermogravimetry, SEM	particulates	
De Souza	2001c	Mossbauer	iron in particulates	
De Souza	2002a	Mossbauer		review
De Souza	2002b	Mossbauer, XRF	elemental analysis of particulates	
Despagne	1998a	NIR	calibration	
Despagne	1998b	NIR	calibration	
Despagne	2000	NIR	water content	
Dickert	1994	optical	cholestric liquid crystals	
Dieterle	2001	reflective interference	ethanol butanol	
Dieterle	2003a	surface plasmon resonance	refrigerants	
Dieterle	2003c	surface plasmon resonance	organic gases	GA, input selection
Ding	1999	NIR		
Dolenko	2000	Raman	water media	
Dolmatova	1997	IR	paper coating	
Dolmatova	1998	IR	modified starches	

Table (*Continued*)

First author	Year	Technology	Application area	Comments
Dolmatova	1999	mid-IR	starches from corn, rice and potatoes	
Duponchel	1999a	IR	instrument calibration	
Duponchel	1999b	IR	instrument calibration	
Duponchel	2000	IR	instrument calibration	
Edelman	2002	FTIR	optical tongue, tannin	
Egan	1995	Raman	environmental contaminants	
Egan	2001	Raman		ANN optimization, GA
Ehrentreich	2001	IR	phenolic groups	selected wavelengths
Ehrentreich	1999	IR	naphthalene compounds	
Eiceman	2001a	IMS	chemical class	
Eiceman	2001b	IMS		
Estienne	2001	Raman		
Facchin	1999	EDXRF	Pb, S	
Ferreira	2001	biosensor FIA	alcohol fermentation	
Ferreira	2003	biosensor FIA	lactose	
Ferrer	1999	fluorescence	PAH water	CC SOM
Fidencio	2002	diffuse reflectance NIR	organic matter in soil	
Frake	1998	NIR	lactose particle	SOM
Gang	2000	UV-Vis	Pd(II)	
Garcia-Baez	1999	luminescence	binary and ternary PCB mixtures	
Gasteiger	1993	IR		
Gasteiger	1993	spectrography		
Gemperline	1997	NIR	hydroxyl in cellulose esters	
Gerdova	2002	Raman		
Goicoechea	2001	spectrophotometric	phenylephrine, diphenhydramine, naphazoline methylparaben	
Goicoechea	2002a	spectrophotometric	chlorpheniramine, naphazoline, dexamethasone	
Goicoechea	2002b	spectrophotometric		
Goodacre	1998b	FTIR, Raman	bacteria	
Goodacre	2003a	FTIR		
Goodacre	2003b	Raman	honey	
Gordon	1998	FTIR-PAS	fungi	
Gordon	2000	spectrophotometry	turbidity	
Hammer	2000	FTIR	trichloroethylene	
Harrington	1994b	IMS	volatile organics	CC
Havel	2003	spectroscopy	U-complex	
Havel	2002a	UV-Vis	U-oxalate	
He	2001	UV	oil in water	
He	2003a	optical	metal ions	
He	2003b	UV	oil in water	
Herrdor	2001	ICP-AES	tea classification	

(*Continues*)

Table (*Continued*)

First author	Year	Technology	Application area	Comments
Hernandez-Caraballo	2003	FAAS	Zn, Cu, Fe in wine	
Hervas	1998	chemoluminescence	trimeprazine, methotrimeprazine	GA
Hervas	2000	spectrography	L-Glycine	optimization
Hitzmann	1997	FIA	glucose, urea	
Huang	2003	short-wave NIR	salt and humidity in salmon	
Irudayaraj	2002	FTIR photo-acoustic	bacteria on food	
Irudayaraj	2003	FTIR	sugar in honey	
Izquierdo	1998	spectroscopy	Nb, Ta	
Jacobsen	1993	IR	carrageenans	local minima
Jasper	1994	NIR	fiber identification	
Jimenez-Prieto	1999	colorimetric	amino acids	
Johnson	1997	optical	VOC	
Karpas	1993	IMS		
Keller	1994	optical	dyes in liquids	
Kim	2002	vis-NIR	fruit	
King	2003	optical fiber, FT	ethanol in water	
Klawun	1994	IR		
Klawun	1995	multispectral		
Klawun	1996a	IR		
Klawun	1996b	IR, MS		
Koehler	2000	FTIR	acetone, SF6	
Kolehmainen	2003	IMS	fermentation rates	SOM
Kompany-Zareh	2002a	colorimetry	Hg	
Kompany-Zareh	2002b	colorimetry	Fe	
Kuzmanovski	2001	IR	urinary calculi	
Larsen	1992	plasma x-ray		
Lewis	1994	FT Raman, NIR	wood classification	
Li	1994	transition emission	complex fluorides	ANN parameters
Li	1999c	FTIR	toxic organics in air	
Li	2000a	fluorescence	dyes	
Lichan	1994	FT Raman	olive oil	
Lin	1994	colorimetry		
Long	1990	UV-Vis	pharma ingredients	
Lopez-Cueto	2000	spectrophotometry	non-linear	
Lopez-Molinero	2003	ICP-AES		SOM
Lozano	1995b	UV-Vis	Ca, Mg in water	SOM
Lubal	2001a	spectrography	Pd-chloroacetate	
Lubal	2001b	spectrography	Cu complexes	SOM
Luo	1996	XRF	geology	
Luo	1997	XRF	Cu alloys	
Luo	1998	XRF	Cu alloys	
Luo	2000a	XRF	matrix effect	
Luo	2000b	XRF	Cr,Fe,Ni	
Luo	2000c	IR	protein	
Lyons	2001	time domain reflectometry	contaminants in water	
Lyons	2002	time domain reflectometry	contaminants in water	

Table (*Continued*)

First author	Year	Technology	Application area	Comments
Magallanes	1998	EDX	steels	optimization
Magellanes	2001	hydride-ICP-AE	Ge	
Majcen	1995	oxide	total color difference	
Maquelin	2000	Raman	microbiological	
Maquelin	2002a	Raman	microbiological	
Maquelin	2002b	Raman, IR	microbiological	
Maquelin	2003	Raman, IR	microbiological	
Marigheto	1998	Raman, mid-IR	edible oils	
Marjoniemi	1994	optical	chromium tanning	review
Marshall	1998	AES		
Mayfield	1999	IR	organophosphorus	
McAvoy	1989	fluorescence	tryptophan and tyrosine	
McAvoy	1992	fluorescence		
Mello	1999	NIR reflectance	N in wheat	
Meyer	1992	IR		
Meyer	1993	IR		reduced inputs
Mittermayr	1994	UV		
Morgan	1994	plasma spectroscopy		
Muller	1997	FIA	enzymes	
Nascimento	1997	laser diffraction	particle size	
Nagata	2001	XRF		
Nazarov	1999	IMS		
Ni	1999	spectrophotometry	anti-oxidants	
Ni	2000	spectrophotometry	acetaminophen and phenobarbital	
Ni	2002	spectrography	metal ions electroplating	review
Olmos	1994	gamma spectroscopy		
Otto	1992	XRF, ICP-AES	fuzzy and NN	ES
Padin	2001	atomic spectroscopy	potatoes	
Penchev	2001	IR		GA
Penchev	1999	IR	variable selection	
Pereira	2001	spectrography	Co	
Perez-Trujillo	2003	ICP-MS	wine	
Plumb	2002	opacity	drug coating	
Polster	1995	FIA	penicillin	
Poppi	1998	NIR	polymers	pruning
Powell	1994	FTIR	sulphate in oligosaccharides	process
Raimundo	2003	optode	Cu, Zn, Hg in water	
Rantanen	2001	IR	moisture	
Rao	2000		octane number	review
Ren	1998	NIR	metronidazole	
Roggo	2003	NIR	sugar beet	McNamara's test
Roughooputh	1999	Raman	cocaine, heroin and explosives	
Ruan	2002	IR, GC/MS	deoxynivalenol	SOM
Ruckebusch	1999	mid-IR	bovine hemoglobin	
Ruckebusch	2001a	mid-IR	bovine hemoglobin	
Ruckebusch	2001b	mid-IR	bovine hemoglobin	SOM
Ruisanchez	1996	X-ray	mineralogical	

(*Continues*)

Table (*Continued*)

First author	Year	Technology	Application area	Comments
Ruisanchez	1997	UV-Vis	Ca, Mg in water	SOM
Rustoiu-Csvdari	2002	optical	Pb(II) in the presence of Cu(II) in river water	optimization
Safavi	2001a	spectrophotometry	Se and Te	
Safavi	2001b	spectrophotometry	Fe and V	
Safavi	2003	optode	pH	
Salles	1994	Mossbauer	sulphates, sulfides, sulfites, silicates	
Salles	1995	Mossbauer	crystal structure	
Saldhana	1999	UV-Vis		
Sante	1996	optical	chicken	
Saurina	1999	spectrography	aniline	
Saurina	1997	spectrography	amino acids	
Saurina	1998	spectrography	amino acids	
Schulze	1994	Raman, UV resonance	neurotransmitter	
Schulze	1995	Raman, UV resonance	neurotransmitter	
Selzer	2000	IR	organics	
Schrim	2001	electrophoresis, UV	pentosan polysulphate	QA
Simhi	1996	FTIR	blood	
Shaw	1999	Raman	biotransformation	
Sigman	1994	UV, IR	ionization potentials	
Simpson	1993	laser desorption – IMS	polymeric material	
Suah	2003a	optode	pH	optimization
Suah	2003b	optode	pH	optimization
Sun	2000a	FIA	Pd	optimization
Sun	2000b	spectrophotometry FIA	Pd	optimization
Sun	2000c	FIA	U in ore	
Sun	2003	spectrophotometry FIA	methylene blue	
Tanabe	1992	IR		
Tanabe	2001	IR	functional groups	
Tchistiakov	2000	IR	mixtures	input reduction
Tenhagen	1998	ICP-AES, IMS,	hashish classification	SOM
Thaler	1993	Raman	graphite transition	
Thodberg	1996	IR	meat	Bayesian
Thompson	2003	FTIR photo-acoustic	bacteria	
Tokutaka	2002	XPS	periodic table	
Udelhoven	2000a	reflectance spectroscopy	sediments	
Udelhoven	2000b	FTIR	bacteria	RBF
Udelhoven	2003	spectroscopy data		software
Visser	1994	IR	recognition	
Volmer	2001	FTIR	urinary calculi	
Walczak	1994	X-ray		input selection, ART
Wang	1998	IR	lub oil	
Wang	1999	optical	wheat	
Wang	2000	micro-PIXE	aerosols	
Wang	2001	microwave FIA	Ru	optimization
Wang	2002b	NIR	soybean	
Weigel	1992	IR	aromatic substitution	
Wessel	1994	IMS		
Wessel	1996	IMS		

Table (*Continued*)

First author	Year	Technology	Application area	Comments
Wienke	1994	UV-Vis, IR	phenantroline, optical glass	ART
Winson	1997	FTIR	metabolites	
Wiskur	2003	colorimetry	tartarate, maleate	
Wolf	2001	two-dimensional fluorimetry	biological pollutants	
Workman	1996	spectroscopy		review
Wu	1996	NIR	drugs	SOM
Wythoff	1990	IR		input reduction
Xiang	2002	NIR	rhubarbs	Hopfield
Yang	1999a	FTIR	alcohol mixtures	SOM
Yang	1999b	FTIR	alcohol mixtures	input reduction
Yang	1999c	Raman	wood classification	
Yang	2003a	FT Raman	bacteria on food	
Yang	2003b	FTIR	methyl chlorides mixture	
Yin	2001a	UV	amino acids	input selection
Yin	2001b	UV	Vitamin-B	wavelet ANN
Zeng	2001	microwave FIA	Ru	
Zhang	1994	optical	fermentation	
Zhang	1997b	fluorescence	dyes	hidden node pruning
Zhang	1998	fluorescence		experiment design
Zhang	2001	ICP-AES	serum	SOM
Zhang	2002b	IR	gas mixture	GA
Zhang	2002c	ICP-AES		review
Zheng	1996	IMS	volatile organics	CC
Zhou	2000b	FIA	Al in steel	optimization
Zupan	1995	optical		

13.8.7 *Mass spectroscopy*

In mass spectra analysis there is no uncertainty about peak location, but the applications of ANN modeling are usually in pattern recognition of 'soft' targets, such as food and drug products and bacteria. The latter were extensively researched by Goodacre and Goodfellow and their co-workers, using Curie-point pyrolysis to generate the inputs to the mass spectrometer (PyMS). In recent years this method was replaced by matrix assisted laser desorption ionization (MLADI).

First author	Year	Technology	Application area	Comments
Alberti	2003	laser ablation MS	tellurium, sulfur	wavelets
Alsberg	1997	PyMS	milk, olive oil	wavelets
Alsberg	1997	PyMS	milk, olive oil	fuzzy, wavelets
Alsberg	1997	PyMS	milk, olive oil	wavelets, clustering
Anklam	1997	PyMS	cocoa butters	
Belic	1997	MS		tutorial
Bloch	1999	MS MALDI	wheat	
Bloch	2001	MS MALDI	barley rye	
Chroma-Keull	2000	MALDI	ostazine black V-B dye	

(Continues)

Table (*Continued*)

First author	Year	Technology	Application area	Comments
Chun	1993a	PyMS	bacteria	
Chun	1993b	PyMS	bacteria	
Chun	1997	PyMS	bacteria	
Darsey	1999	MS	isomeric-isobalic classification	
Dumas	2002	MAB-TOFMS	anabolic steroids in cattle urine	optimization
Eghbaldar	1997 1998	MS	structural feature recognition	
Freeman	1994	PyMS	bacteria	
Freeman	1995	PyMS	bacteria	
Geng	2000	Py chromatogram	herbal medicine	
Ghassempour	2002	PyMS	penicillin hydrolysis	
Goodacre	1993a	PyMS	bacteria	
Goodacre	1993b	PyMS	bacteria	
Goodacre	1993c	PyMS	olive oil	
Goodacre	1994a	PyMS	bacteria	
Goodacre	1994b	PyMS	recombinant protein expression, bacteria	
Goodacre	1994c	PyMS	metabolite fermentation	
Goodacre	1994d	PyMS	protein mixtures	
Goodacre	1995	PyMS	bacteria fermentation	
Goodacre	1996a	PyMS	oral bacteria	
Goodacre	1996b	PyMS	drift correction	
Goodacre	1996c	PyMS	seed identification	
Goodacre	1996d	PyMS	bacteria	
Goodacre	1996e	PyMS	animal cell lines	
Goodacre	1996f	PyMS		tutorial
Goodacre	1997a	PyMS	orange juice	
Goodacre	1997b	PyMS	milk	
Goodacre	1997c	PyMS	instrument calibration	
Goodacre	1997d	PyMS	bacteria	
Goodacre	1998a	PyMS	bacteria	
Goodacre	1998b	PyMS, FTIR, Raman	bacteria	
Goodacre	1999	PyMS,	beverages	GA
Goodacre	2000	PyMS, FTIR	biomarker	
Goodacre	2002	ESI-MS	oils	
Goodacre	2003a	ESI-MS	oils	
Goodfellow	1997	PyMS	microbiology	
Harrington	1994a	MS		
Harrington	1998	MS	chlorinated bi-phenols	
Harrington	2002	MS	bacteria	
Harrington	1993	PyMS		
Jacobsen	2001	MLADI-TOF-MS	gliadin	
Jalali-Heravi	2000a	MS	noncyclic alkanes and alkenes	
Kang	1998a	PyMS	bacteria	
Kang	1998b	PyMS	clavulanic acid production	

Table (*Continued*)

First author	Year	Technology	Application area	Comments
Kenyon	1997	PyMS	bacteria	
Lavine	1999b	PyMS	skin fibroblasts	GA
Lohninger	1991	MS		
Lohninger	1992	MS		
Marsalek	2002	MALDI-TOFMS	amino acids	
McGovern	1999	PyMS	protein expression	
McGovern	2002	PyMS, FTIR	gibberellic acid production	
Montanarella	1995	PyMS, IR, NMR	wines	
Murphy	2003	Laser MS	particle analysis	
Nilsson	1996	PyMS	pencillium species	
Nord	1998	SIMS		variable contribution
Peres	2002	PyMS	cheese	
Phares	2003	particle MS		ART
Salter	1997	PyMS	olive oil	
Sanchez	2002	electrophoresis, MALDI	Rh(III)	
Sanni	2002	ToF SIMS	protein films	
Sebastian	2002	PyMS	feed of lambs	
Sedo	2003	MALDI –TOF	peptides	
Shen	2003	Py-GC-MS	tablet production	
Sisson	1995	PyMS	bacteria	
Song	1999	ATOFMS	aerosol	
Song	2001	MS	single particle gasoline diesel	
Sorensen	2002	MS	wheat	
Sorensen	2002	MALDI-TOF-MS	gliadin	
Tas	1994	MS		pattern recognition review
Taylor	1998	PyMS	bacteria	GA
Timmins	1997	PyMS	bacteria	
Timmins	1998a	PyMS, FTIR	bacteria	
Timmins	1998b	PyMS, FTIR	yeast	
Tong	1999	MS, GC		
Wan	1999	MS	olive oil	RBF
Werther	1994	MS	classification	

13.8.8   *Nuclear magnetic resonance*

Applications of ANN in the NMR, or its more 'politically correct' Magnetic Resonance Imaging (MRI) name, are a branch of image processing. This is more actively researched in the medical imaging area, where the main issue is the feature extraction algorithm. However, some analytical chemistry applications are reported.

First author	Year	Application area	comments
Aires-de-Souza	2002b		
Alkorpela	1997	pig liver metabolism	
Amendolia	1998	alditols	
Anker	1992		
Ball	1993		
Basu	2003	cetane number	
Binev	2004a	organics	auto-associative ANN
Binev	2004b	organics	
Forshed	2002	4-aminophenol paracetamol	Bayesian regularized
Freeman	2003	finding ideas for new experiments	
Kalelkar	2002	combinatorical chemistry	SOM
Kapur	2004	base oils	
Kvasnicka	1992	monosubstituted benzenes	
Montanarella	1995	wines	
Sharma	1994	polypeptides and proteins	
Svozil	1995	alkanes	
Vaananen	2002	hydrocarbons	

13.8.9 *Miscellaneous*

This section includes various applications with technologies other than those described in the previous sections.

First author	Year	Technology	Application area	Comments
Aires-de-Sousa	1996	amino acids	wine	
Angulo	1998	kinetics	vanadium	
Baba	1990		water	causal index
Balestrieri	2001	chemical analysis	cheese classification	
Baxter	1999		water quality	
Baxter	2001		water quality	
Berger	1996		process control sulfur	
Boger	1992	nitrate	wastewater	input reduction
Bos	1992		food (cheese)	
Bourquin	1998a		pharmaceutical, dosage	advantages
Brodnjak-Voncina	2002		river water	SOM
Burtis	1996		clinical	
Carson	1995	bacteria	image classification	
Charef	2000	pH, UV-Vis	wastewater COD	
Chretien	2000		QSAR	data mining
Damiani	2000	stereospecific analysis	triacylglycerols	
Da Silva	2002a		gas in transformer oil	
Da Silva	2002b		gas in transformer oil	
Deshpande	1997		on-line property prediction	
Dobrowski	2003	carbonation	sugar processing	
Du	1994	pH	metal leaching sludge	
Fernandez-Caceres	2001	metal analysis	tea origin	
Garcia	1999		DO in water	
Gerbi	1998		vinegars origin	

Table (*Continued*)

First author	Year	Technology	Application area	Comments
Gonzalez	2002		soft drinks	
Hajek	1995	gas analyzers	coal gasifiers	
Havel	2002b		equilibrium	experimental design
Jeyamkondan	2001		bacterial growth	
Karul	1999		lake water	
Kateman	1993a		analytical chemistry	black-box
Kateman	1993b		analytical chemistry	
Kocjancic	1997	fatty acids	olive oils	SOM
Kuzmanovski	2003		whewellite, weddellite, carbonate apatite	
Li	1993			SOM
Lopes	1999	bacterial metabolites	cheese classification	
Lozano	1995a		metabolic energy	
Ma	2000	volumetric	ceramics classification	
Magallanes	2003		SO2, environmental	
Marini	2003	chemical analysis	rice bran oil	
Matias	2001	Rutherford backscattering	thin films	
Meisel	1997		butter	
Mo	1996		Pb	
Morales	2001	acetification	wines, vinegars	
Noble	1997	rRNA	bacteria Chesapeake Bay	
Noble	2000	lipid profiles	bacteria sea	
Nott	1996	on-line	bakers' yeast production	
Parilla	1999	chemical analysis	wine phenolic ageing	
Patnaik	1999		bio-process	
Perez-Margarine	2004	chemical analysis	wine classification	
Pletnev	2002	complexing	metal ions	SOM
Prieto	1993		microbiology	
Remola	1996		archeological	
Riesenberg	1999		fermentation	process
Schleiter	1999		water	
Schlimme	1997		butter	
Slot	1997	Maillard		optimization
Sreekrishnan	1995		wastewater	bacteria
Steyer	1997		anaerobic	
Sun	1997	trace metal	wines	
Syu	1998a		wastewater	
Syu	1998b		SO2	
Teissier	1996	CO2	champagne fermentation	process
Thompson	2001	polarized light	mineral identification	
Tusar	1992			
Valcarcel	2001		selectivity	
Vieira	2000	Rutherford backscattering	Ge in Si	
Vieria	2001a	Rutherford backscattering	NiTaC	
Vieria	2001b	Rutherford backscattering	Ge in Si	

(*Continues*)

Table (*Continued*)

First author	Year	Technology	Application area	Comments
Viera	2002	Rutherford backscattering	sapphire	
Veltri	2001	image	biomarkers	
Wang	1996		overfitting	
Wailzer	20001	aroma	pyrazines	
Weinstein	1994		drug discovery	
Wesolowski	2001	chemical properties	rapeseed and soybean oils	reduced inputs
Wesolowski	2002	thermo-gravimetry	rapeseed and soybean oils	
Wesolowski	2002	thermo-gravimetry	fish-oils	
Wilcox	1995	bicarbonate	wastewater, anaerobic digestion	
Wunderlich	1992	thermal analysis		
Yan	2001		spearmint essences	SOM
Yu	1998	ORP	SBR wastewater	
Zhou	1995		bioprocess	
Zupan	1991			
Zupan	1999			
Zupan	1997			SOM
Zupan	1994	fatty acids	olive oils	
Zupan	2002	fatty acids	olive oils	SOM

13.9   References

Absalan, G., Safavi, A., & Maesum, S. 2001, Application of artificial neural networks as a technique for interference removal: kinetic-spectrophotometric determination of trace amounts of Se(IV) in the presence of Te(IV), *Talanta*, vol. 55, no. 6, pp. 1227–1233.

Adler, B., Schutze, P., & Will, J. 1993, Expert system for interpretation of x-ray diffraction, *Analytica Chimica Acta*, vol. 271, pp. 287–291.

Adler, B., Bruckner, G., & Winterstein, M. 1994, A multivariate sensor system for controlling of wastewater streams, *Chemische Technik*, vol. 46, no. 2, pp. 77–86.

Agatonovic-Kustrin, S., Zecevic, M., Zivanovic, L., & Tucker, I. G. 1998a, Application of artificial neural networks in HPLC method development, *Journal of Pharmaceutical and Biomedical Analysis*, vol. 17, no. 1, pp. 69–76.

Agatonovic-Kustrin, S., Zecevic, M., Zivanovic, L., & Tucker, I. G. 1998b, Application of neural networks for response surface modeling in HPLC optimization, *Analytica Chimica Acta*, vol. 364, no. 1–3, pp. 265–273.

Agatonovic-Kustrin, S., Tucker, I. G., & Schmierer, D. 1999a, Solid state assay of ranitidine HCl as a bulk drug and as active ingredient in tablets using DRIFT spectroscopy with artificial neural networks, *Pharmaceutical Research*, vol. 16, no. 9, pp. 1477–1482.

Agatonovic-Kustrin, S., Zecevic, M., & Zivanovic, L. 1999b, Use of ANN modelling in structure-retention relationships of diuretics in RP-HPLC, *Journal of Pharmaceutical and Biomedical Analysis*, vol. 21, no. 1, pp. 95–103.

Agatonovic-Kustrin, S. & Beresford, R. 2000a, Basic concepts of artificial neural network (ANN) modeling and its application in pharmaceutical research, *Journal of Pharmaceutical and Biomedical Analysis*, vol. 22, no. 5, pp. 717–727.

Agatonovic-Kustrin, S., Wu, V., Rades, T., Saville, D., & Tucker, I. G. 2000b, Ranitidine hydrochloride x-ray assay using a neural network, *Journal of Pharmaceutical and Biomedical Analysis*, vol. 22, no. 6, pp. 985–992.

Agatonovic-Kustrin, S., Beresford, R., Razzak, M. 2000c, Determination of enantiomeric composition of ibuprofen in solid state mixtures of the two by DRIFT spectroscopy, *Analytica Chimica Acta*, vol. 417, no. 1, pp. 31–39.

Agatonovic-Kustrin, S. & Alany, R. 2001a, Application of diffuse reflectance infrared Fourier transform spectroscopy combined with artificial neural networks in analysing enantiomeric purity of terbutaline sulphate bulk drug, *Analytica Chimica Acta*, vol. 449, no. 1–2, pp. 157–165.

Agatonovic-Kustrin, S., Rades, T., Wu, V., Saville, D., & Tucker, I. G. 2001b, Determination of polymorphic forms of ranitidine-HCl by DRIFTS and XRPD, *Journal of Pharmaceutical and Biomedical Analysis*, vol. 25, no. 5–6, pp. 741–750.

Aires-de-Sousa, J. 1996, Verifying wine origin – a neural-network approach, *Amer. J. Enol. Viticult.*, vol. 47, no. 4, pp. 410–414.

Aires-de-Sousa, J. & Gasteiger, J. 2002a, Prediction of enantiomeric selectivity in chromatography. Application of conformation-dependent and conformation-independent descriptors of molecular chirality, *Journal of Molecular Graphics and Modeling*, vol. 20, no. 5, pp. 373–388.

Aires-de-Sousa, J., Hemmer, M. C., & Gasteiger, J. 2002b, Prediction of H-1-NMR chemical-shifts using neural networks, *Analytical Chemistry*, vol. 74, no. 1, pp. 80–90.

Alam, M. K., Stanton, S. L., & Hebner, G. A. 1994, Near-infrared spectroscopy and neural networks for resin identification, *Spectroscopy*, vol. 9, no. 2, pp. 31–40.

Al-Ammar, A. S., & Barnes, R. M. 2000 Supervised cluster classification using the original n-dimensional space without transformation into lower dimension, *Journal of Chemometrics*, vol. 15 no. 1, pp. 49–67.

Al-Ammar, A. S. & Barnes, R. M. 2001, Supervised cluster classification using the original n-dimensional space without transformation into lower dimension, *Journal of Chemometrics*, vol. 15, no. 1, pp. 49–67.

Al-Khalifa, S., Maldonado-Bascon, S., & Gardner, J. W. 2003, Identification of CO and NO_2 using a thermally resistive microsensor and support vector machine, *IEE Proceedings: Science, Measurements and Technologies*, vol. 150, no. 1, pp. 11–14.

Alberti, M., Sedo, O., & Havel, J. 2003, Laser ablation synthesis and TOF mass spectrometric identification of tellurium, sulfur and mixed tellurium-sulfur clusters, *Polyhedron*, vol. 22, no. 18, pp. 2601–2605.

Alakorpela, M., Changani, K. K., Hiltunen, Y., Bell, J. D., Fuller, B. J., Bryant. D. J., Taylor-Robinson, S. D., & Davidson, B. R. 1997, Assessment of quantitative artificial neural-network analysis in a metabolically dynamic ex-vivo P-31 NMR pig-liver study, *Magnetic Resonance in Medicine*, vol. 38, no. 5, pp. 848.

Aleixandre, M., Sayago, I., Horrillo, M. C., Fernandez, M. J., Ares, L., Garcia, M., Santos, J. P., & Gutierrez, J. 2002, Sensor array for the monitoring of contaminant gases, *Proceedings of the International Workshop on New Developments on Sensors for Environmental Control, Lecce, Italy*, pp. 244–248.

Alexander, P. W., Di Benedetto, L. T., & Hibbert, D. B. 1998, A field-portable gas analyzer with an array of six semiconductor sensors. Part 2: identification of beer samples using artificial neural networks, *Field Analytical Chemistry and Technology*, vol. 2 no. 3, pp. 145–153.

Allanic, A. L., Jezequel, J. Y., & Andre, J. C. 1992, Application of neural networks theory to identify two-dimensional fluorescence spectra, *Analytical Chemistry*, vol. 64, pp. 2616–2622.

Alsberg, B. K., Goodacre, R., Rowland, J. J., & Kell, D. B. 1997, Classification of pyrolysis mass spectra by fuzzy multivariate rule induction-comparison with regression, K-nearest neighbor, neural and decision-tree methods, *Analytica Chimica Acta*, vol. 348, no. 1-3, pp. 389–407.

Alsberg, B. K., Woodward, A. M., Winson, M. K., Rowland, J. J., & Kell, D. B. 1998a, Variable selection in wavelet regression models, *Analytica Chimica Acta*, vol. 368, no. 1–2, pp. 29–44.

Alsberg, B. K., Wade, W. G., & Goodacre, R. 1998b, Chemometric analysis of diffuse reflectance-absorbance Fourier transform infrared spectra using rule induction methods: Application to the classification of Eubacterium species, *Applied Spectroscopy*, vol. 52, no. 6, pp. 823–832.

Alsberg, B. K. 2000, Parsimonious multiscale classification models, *Journal of Chemometrics*, vol. 14, no. 5–6, pp. 529–539.

Amendolia, S. R., Doppiu, A., Ganadu, M. L., & Lubinu, G. 1998, Classification and quantitation of H-1 NMR spectra of alditols binary mixtures using artificial neural networks, *Analytical Chemistry*, vol. 70, no. 7, pp. 1249–1254.

Anderson, K. A., Magnuson, B. A., Tschirgi, M. L., & Smith, B. 1999, Determining the geographic origin of potatoes with trace metal analysis using statistical and neural network classifiers, *Journal of Agricultural and Food Chemistry*, vol. 47, no. 4, pp. 1568–1575.

Andreu, Y., Ostra, M., Ubide, C., Galban, J., de Marcos, S., & Castillo, J. R. 2002, Study of a fluorometric-enzymatic method for bilirubin based on chemically modified bilirubin-oxidase and multivariate calibration, *Talanta*, vol. 57, no. 2, pp. 343–353.

Andrews, J. M., & Lieberman, S. H. 1994, Neural network approach to qualitative identification of fuels and oils from laser induced fluorescence spectra, *Analytica Chimica Acta*, vol. 285, pp. 237–246.

Angerosa, F., DiGiacinto, L., Vito, R., & Cumitini, S. 1996, Sensory evaluation of virgin olive oils by artificial neural network processing of dynamic head-space gas chromatographic data, *Journal of the Science of Food and Agriculture*, vol. 72, no. 3, pp. 323–328.

Angulo, R., Lopez-Cueto, G., & Ubide, C. 1998, The hexacyanomanganate(IV) hydrogen peroxide reaction. Kinetic determination of vanadium, *Talanta*, vol. 46, no. 1, pp. 63–74.

Anker, L. S., & Jurs, P. C. 1992, Prediction of Carbon-13 nuclear magnetic resonance chemical shifts by artificial neural networks, *Analytical Chemistry*, vol. 64, pp. 1157–1164.

Anklam, E., Bassani, M. R., Eiberger, T., Kriebel, S., Lipp, M., & Matissek, R. 1997, Characterization of cocoa butters and other vegetable fats by pyrolysis mass spectrometry, *Fresenius Journal of Analytical Chemistry*, vol. 357, no. 7, pp. 981–984.

Anstice, P. J. C. & Alder, J. F. 1999, Towards the prediction of the remaining lifetime of a carbon filter bed on hydrogen cyanide challenge using in-bed sensors and a neural network: preliminary findings, *Adsorption Science & Technology*, vol. 17, no. 9, pp. 771–781.

Arakawa, E. T., Lavrik, N. V., & Datskos, P. G. 2003a, Detection of anthrax simulants with microcalorimetric spectroscopy: Bacillus subtilis and Bacillus cereus spores, *Applied Optics*, vol. 42, no. 10, pp. 1757–1762.

Arakawa, E. T., Lavrik, N. V., Rajic, S., & Datskos, P. G. 2003b, Detection and differentiation of biological species using microcalorimetric spectroscopy, *Ultramicroscopy*, vol. 97, no. 1–4, pp. 459–465.

Araujo, A. S., Souza, C. D. R., Souza, M. J. B., Fernandes, V. J., & Pontes, L. A. M. 2002, Acid properties of ammonium exchanged AlMCM-41 with different Si/Al ratio, *Nanoporous Materials Iii*, vol. 141, pp. 467–472.

Baba, K., Enbut, I. & Yoda, M. 1990, Explicit representation of knowledge acquired from plant historical data using neural network, *Proceedings of the International Joint Conference on Neural Networks,* San Diego, vol. 1, pp. 155–160.

Babri, M., Rouhani, S., & Massoumi, A. 2000, Near infrared spectroscopy and artificial neural networks, *American Laboratory*, vol. 32, no. 7, pp. 28, 30, 32.

Bachmann, T. T., Leca, B., Vilatte, F., Marty, J.-L., Fournier, D., & Schmid, R. D. 2000, Improved multianalyte detection of organophosphates and carbamates with disposable multielectrode biosensors using recombinant mutants of Drosophila acetylcholinesterase and artificial neural networks *Biosensors & Bioelectronics*, vol. 15, no. 3–4, pp. 193–201.

Balamurugan, C., Aravindan, C., Kannan, K., Sathyanarayana, D., Valliappan, K., & Manavalan, R. 2003, Artificial neural network for the simultaneous estimation of multicomponent sample by UV spectrophotometry, *Indian Journal of Pharmaceutical Sciences*, vol. 65, no. 3, pp. 274–278.

Balestrieri, F., Damiani, F., & Marini, Domenico, 2001, Artificial neural networks to classify some dairy products, *Journal of Commodity Science*, vol. 40, no. 1, pp. 17–31.

Ball, J. W., & Jurs, P. C. 1993, Automated selection of regression models using neural networks for C NMR spectral predictions, *Analytical Chemistry*, vol. 65, pp. 505–512.

Baratti, R., Vacca, G., & Servida, A. 1995, Control of distillation columns via artificial neural networks, in Bulsari, A. B., & Kallio, S., eds. *Engineering Applications of Artificial Neural Networks*, Finnish Artificial Intelligence Society, pp. 13–16.

Barko, G., & Hlavay, J. 1997, Application of an artificial neural-network (ANN) and piezoelectric chemical sensor array for identification of volatile organic-compounds, *Talanta*, vol. 44, no. 12, pp. 2237–2245.

Barshick, S. A. 1998, Analysis of accelerants and fire debris using aroma detection technology, *Journal of Forensic Sciences*, vol. 43, no. 2, pp. 284–293.

Basheer, I. A. & Hajmeer, M. 2000, Artificial neural networks: fundamentals, computing, design, and application, *Journal of Microbiological Methods*, vol. 43, no. 1, pp. 3–31.

Bastarache, D., El Jabi, N., Turkkan, N., & Clair, T. A. 1997, Predicting conductivity and acidity for small streams using neural networks, *Canadian Journal of Civil Engineering*, vol. 24, no. 6, pp. 1030–1039.

Basu, B., Kapur, G. S., Sarpal, A. S., & Meusinger, R. 2003, A neural network approach to the prediction of cetane number of diesel fuels using nuclear magnetic resonance (NMR) spectroscopy, *Energy & Fuels*, vol. 17, no. 6, pp. 1570–1575.

Baumann, K., Albert, H., & von Korff, M. 2002a, A systematic evaluation of the benefits and hazards of variable selection in latent variable regression. Part I: Search algorithm, theory and simulations, *Journal of Chemometrics*, vol. 16, pp. 339–350.

Baumann, K., Albert, H., & von Korff, M. 2002b, A systematic evaluation of the benefits and hazards of variable selection in latent variable regression. Part II: practical applications, *Journal of Chemometrics*, vol. 16, pp. 351–360.

Becker, T. M. & Schmidt, H. L. 2000, Data model for the elimination of matrix effects in enzyme-based flow-injection systems, *Biotechnol.Bioeng.*, vol. 69, no. 4, pp. 377–384.

Belic, I. & Gyergyek, L. 1997, Neural network methodologies for mass spectra recognition, *Vacuum*, vol. 48, no. 7–9, pp. 633–637.

Beling, S., Blaser, G., Bock, J., Heinert, L., Traxler, M., & Kohl, D. 1998, Signal conditioning for semiconductor gas sensors being used as detectors in gas-chromatographs and similar applications, *Sensors and Actuators B-Chemical*, vol. 52, no. 1–2, pp. 15–22.

Bell, S. E., & Mead, W. C., 1992 Artificial intelligence and neural networks applied to ion mobility spectrometry, in *Computer Enhanced Analytical Spectroscopy*, Chapter 12.

Bell, S. E., Mead, W. C., Jones, R. D., Eiceman, G. A., & Ewing, R. G. 1993, Connectionist hyperprism neural-network for the analysis of ion mobility spectra – an empirical evaluation, *Journal of Chemical Information and Computer Sciences*, vol. 33, no. 4, pp. 609–615.

Bell, S., Nazarov, E., Wang, Y. F., & Eiceman, G. A. 1999, Classification of ion mobility spectra by functional groups using neural networks, *Analytica Chimica Acta*, vol. 394, no. 2–3, pp. 121–133.

Bell, S., Nazarov, E., Wang, Y. F., Rodriguez, J. E., & Eiceman, G. A. 2000, Neural network recognition of chemical class information in mobility spectra obtained at high temperatures, *Analytical Chemistry*, vol. 72, no. 6, pp. 1192–1198.

Benjathapanun, N., Boyle, W. J. O., & Grattan, K. T. V. 1997, Binary encoded 2nd-differential spectrometry using UV-VIS spectral data and neural networks in the estimation of species type and concentration, *IEE Proceedings: Science, Measurements and Technologies*, vol. 144, no. 2, pp. 73–80.

Benoudjita, N., Cools E., Meurens, M., & Verleysena, M. 2004, Chemometric calibration of infrared spectrometers: selection and validation of variables by non-linear models, *Chemometrics and Intelligent Laboratory Systems*, vol. 70, pp. 47– 53.

Berger, D., Landau, M. V., Herskowitz, M., & Boger, Z. 1996, Deep hydrodesulfurization of atmospheric gas oil – effects of operating conditions and modelling by artificial neural network techniques, *Fuel*, vol. 75, no. 7, pp. 907–911.

Berry, R. J., Harris, J. E., & Williams, R. R. 1997, Light-emitting diodes as sensors for colorimetric analyses, *Applied Spectroscopy*, vol. 51, no. 10, pp. 1521–1524.

Bertran, E., Blanco, M., Maspoch, S., Ortiz, M. C., Sanchez, M. S., & Sarabia, L. A. 1999, Handling intrinsic non-linearity in near-infrared reflectance spectroscopy, *Chemometrics and Intelligent Laboratory Systems*, vol. 49, no. 2, pp. 215–224.

Bertran, E., Blanco, M., Coello, J., Iturriaga, H., Maspoch, S., & Montoliu, I. 2000, Near infrared spectrometry and pattern recognition as screening methods for the authentication of virgin olive oils of very close geographical origins, *Journal of Near Infrared Spectroscopy*, vol. 8, no. 1, pp. 45–52.

Bessant, C. & Saini, S. 1999, Simultaneous determination of ethanol, fructose, and glucose at an unmodified platinum electrode using artificial neural networks, *Analytical Chemistry*, vol. 71, no. 14, pp. 2806–2813.

Bessant, C. & Saini, S. 2000, A chemometric analysis of dual pulse staircase voltammograms obtained in mixtures of ethanol, fructose and glucose, *Journal of Electroanalytical Chemistry*, vol. 489, no. 1–2, pp. 76–83.

Bessant, C., Ritchie, L., Saini, S., Pahlavanpour, B., & Wilson, G. 2001, Chemometric evaluation of synchronous scan fluorescence spectroscopy for the determination of regulatory conformance and usage history of insulation oils, *Applied Spectroscopy,* vol. 55, no. 7, pp. 840–846.

Bessant, C., & Richards, E. 2003, Neural networks for the calibration of voltammetric data, *Data Handling in Science and Technology*, vol. 23 (Nature-Inspired Methods in Chemometrics: Genetic Algorithms and Artificial Neural Networks), pp. 257–280.

Bhandare, P., Mendelson, Y., Peura, R. A., Janatsch, G., Krusejarres, J. D., Marbach, R., & Heise, H. M. 1993, Multivariate determination of glucose in whole-blood using partial least-squares and artificial neural networks based on midinfrared spectroscopy, *Applied Spectroscopy,* vol. 47, no. 8, pp. 1214–1221.

Bhandare, P., Mendelson, Y., Stohr, E., & Peura, R. A. 1994a, Comparison of multivariate calibration techniques for mid-IR absorption spectrometric determination of blood-serum constituents, *Applied Spectroscopy*, vol. 48, no. 2, pp. 271–273.

Bhandare, P., Mendelson, Y., Stohr, E., & Peura, R. A. 1994b, Glucose determination in simulated blood-serum solutions by Fourier-transform infrared-spectroscopy – investigation of spectral interferences, *Vibrational Spectroscopy*, vol. 6, no. 3, pp. 363–378.

Binev, Y., Corvo, M., & Aires-de-Sousa, J. O. 2004a [in press], The impact of available experimental data on the prediction of 1H NMR chemical shifts by neural networks, *Journal of Chemical Information and Computer Sciences*.

Binev, Y., & Aires-de-Sousa, J. O. 2004b [in press], Structure-based predictions of 1H NMR chemical shifts using feed-forward neural networks, *Journal of Chemical Information and Computer Sciences*.

Bishop, C. M. 1995, *Neural Networks for Pattern Recognition*, Clarendon Press.

Blanco, M., Coello, J., Iturriaga, H., Maspoch, S., & Redon, M. 1995, Artificial neural networks for multicomponent kinetic determinations, *Analytical Chemistry*, vol. 67, no. 24, pp. 4477–4483.

Blanco, M., Coello, J., Iturriaga, H., Maspoch, S., Redon, M., & Villegas, N. 1996, Artificial neural networks and partial least-squares regression for pseudo-first-order with respect to the reagent multicomponent kinetic-spectrophotometric determinations, *Analyst*, vol. 121, no. 4, pp. 395–400.

Blanco, M., Coello, J., Gene, J., Iturriaga, H., Maspoch, S., Canals, T., & Jumilla, J. 1997, Use of indirect multiple linear-regression for multicomponent dye analysis in a leather tanning bath, *J. Soc. Dyers Colour*, vol. 113, no. 11, pp. 311–316.

Blanco, M., Coello, J., Iturriaga, H., Maspoch, S., & Delapezuela, C. 1998, Near-infrared spectroscopy in the pharmaceutical industry, *Analyst*, vol. 123, no. 8, p. R135–R150.

Blanco, M., Coello, J., Iturriaga, H., Maspoch, S., & Porcel, M. 1999a, Simultaneous enzymatic spectrophotometric determination of ethanol and methanol by use of artificial neural networks for calibration, *Analytica Chimica Acta*, vol. 398, no. 1, pp. 83–92.

Blanco, M., Coello, J., Iturriaga, H., Maspoch, S., & Pages, J. 1999b, Calibration in nonlinear near-infrared reflectance spectroscopy – a comparison of several methods, *Analytica Chimica Acta*, vol. 384, no. 2, pp. 207–214.

Blanco, M., Coello, J., Iturriaga, H., Maspoch, S., & Pages, J. 2000, NIR calibration in non-linear systems: different PLS approaches and artificial neural networks, *Chemometrics and Intelligent Laboratory Systems*, vol. 50, no. 1, pp. 75–82.

Blanco, M., Maspoch, S., Villarroya, I., Peralta, X., Gonzalez, J. M., & Torres, J. 2001a, Determination of physical-properties of bitumens by use of near-infrared spectroscopy with neural networks – joint modeling of linear and nonlinear parameters, *Analyst*, vol. 126, no. 3, pp. 378–382.

Blanco, M., Maspoch, S., Villarroya, I., Peralta, X., Gonzalez, J. M., & Torres, J. 2001b, Geographical origin classification of petroleum crudes from near-infrared spectra of bitumens, *Applied Spectroscopy*, vol. 55, no. 7, pp. 834–839.

Blanco, M., Coello, J., Iturriaga, H., Maspoch, S., & Porcel, M. 2001c, Use of circular dichroism and artificial neural networks for the kinetic-spectrophotometric resolution of enantiomers, *Analytica Chimica Acta*, vol. 431, no. 1, pp. 115–123.

Blanco, M. & Pages, J. 2002, Classification and quantitation of finishing oils by near-infrared spectroscopy, *Analytica Chimica Acta*, vol. 463, no. 2, pp. 295–303.

Blank, T. B., & Brown, S. D. 1993, Data processing using neural networks, *Analytica Chimica Acta*, vol. 277, pp. 273–287.

Bloch, H. A., Kesmir, C., Petersen, M., Jacobsen, S., & Sondergaard, I. 1999, Identification of wheat-varieties using matrix-assisted laser desorption/ionisation time-of-flight mass-spectrometry and an artificial neural network, *Rapid Communications in Mass Spectrometry*, vol. 13, no. 14, pp. 1535–1539.

Bloch, H. A., Petersen, M., Sperotto, M. M., Kesmir, C., Radzikowski, L., Jacobsen, S., & Sondergaard, I. 2001, Identification of barley and rye varieties using matrix-assisted laser desorption/ionisation time-of-flight mass spectrometry with neural networks, *Rapid Communications in Mass Spectrometry*, vol. 15, no. 6, pp. 440–445.

Blonder, R., Bendov, I., Dagan, A., Willner, I., & Zisman, E. 1997, Photo chemically-activated electrodes – application in design of reversible immunosensors and antibody patterned interfaces, *Biosens.-Bioelectron.*, vol. 12, no. 7, pp. 627–644.

Bocaz-Beneventi, G., Latorre, R., Farkova, M., & Havel, J. 2002a, Artificial neural networks for quantification in unresolved capillary electrophoresis peaks, *Analytica Chimica Acta*, vol. 452, no. 1, pp. 47–63.

Bocaz-Beneventi, G., Tagliaro, F., Bortolotti, F., Manetto, G., & Havel, J. 2002b, Capillary zone electrophoresis and artificial neural networks for estimation of the post-mortem interval (PMI) using electrolytes measurements in human vitreous humour, *Int.J.Legal Med.*, vol. 116, no. 1, pp. 5–11.

Bocaz, G., Revilla, A. L., Krejci, J., & Havel, J. 1998, Characterization of a sapphire-epoxy coating for capillary electrophoresis, *Journal of Capillary Electrophoresis*, vol. 5, no. 5–6, pp. 165–170.

Boger, Z. 1992, Application of neural networks to water and waste-water treatment-plant operation, *ISA Transactions*, vol. 31, no. 1, pp. 25–33.

Boger, Z. 1993, Artificial neural networks for quantitative stationary spectroscopic measurements, *Proceedings of the 10th Israeli Conference on Artificial Intelligence, Computer Vision and Neural Networks, AICVNN-93*, Ramat-Gan, Israel, pp. 185–194.

Boger, Z. & Karpas, Z. 1994a, Application of neural networks for interpretation of ion mobility and x-ray-fluorescence spectra, *Analytica Chimica Acta*, vol. 292, no. 3, pp. 243–251.

Boger, Z. & Karpas, Z. 1994b, Use of neural networks for quantitative measurements in Ion Mobility Spectrometry (IMS), *Journal of Chemical Information and Computer Sciences*, vol. 34, no. 3, pp. 576–580.

Boger, Z., Guterman, H., & Segal, T. 1996, Application of large-scale artificial neural networks for modeling the response of a naphtha stabilizer distillation train, *Proceedings of the AIChE annual meeting*, Chicago. Available on line at http://www.che.wisc.edu/aiche/1996/10b-abstracts/24.html.

Boger, Z. 1997a, Experience in industrial plant model development using large-scale artificial neural networks, *Information Sciences – Applications*, vol. 101, no. 3–4, pp. 203–212.

Boger, Z., Ratton, L., Kunt, T. A., Mc Avoy, T. J., Cavicchi, R. E. & Semancik, S. 1997b, Robust classification of "Artificial Nose" sensor data by artificial neural networks, *Proceedings of the IFAC ADCHEM '97 Conference*, Banff, Canada, pp. 334–338.

Boger, Z. & Guterman, H. 1997c, Knowledge extraction from artificial neural networks models, *Proceedings of the IEEE International Conference on Systems Man and Cybernetics, SMC '97*, Orlando, Florida, pp. 3030–3035.

Boger, Z., & Weber, R. 2000, Finding an optimal artificial neural network topology in real-life modeling – two approaches *Proceedings of the Neural Computation 2000 Conference*, Berlin, Germany.

Boger, Z., 2001a, Hidden neurons as classifiers in artificial neural networks models, *Proceedings of the 5th Conference on Artificial Neural Networks and Expert Systems*, Dunedin, New-Zealand, pp. 4–7.

Boger, Z., 2001b, Connectionist methods and systems for knowledge discovery in real world applications, *Proceedings of the 5th Conference on Artificial Neural Networks and Expert Systems*, Dunedin, New Zealand, pp. 52–55.

Boger, Z., Semancik, S., & Cavicchi, R. E., 2002a, Artificial neural networks methods for identification of the most relevant inputs from conductometric microhotplate sensors, *Proceedings of the 9th International Meeting on Chemical Sensors*, Boston, pp. 190–191.

Boger, Z., Cavicchi, R. E. & Semancik, S. 2002b, Analysis of conductometric micro-sensor responses in a 36-sensor array by artificial neural networks modeling, *Proceedings of the 9th International Symposium on Olfaction and Electronic Nose*, Rome, 2002, Aracne Editrice S.r.l., Rome 2003, pp. 135–140. Also available at http://preprint.chemweb.com/analchem/0211001.

Boger, Z., Meier, D. C., Cavicchi, R. E., & Semancik, S. 2003a, Identification of the type and the relative concentration of chemical warfare agents using NIST conductometric microhotplate sensors, *Proceedings of the International Joint Conference on Neural Networks, IJCNN '03*, Portland, OR, pp. 1065–1070.

Boger, Z. 2003b, Finding patient cluster attributes using auto-associative ANN modeling, *Proceedings of the International Joint Conference on Neural Networks, IJCNN '03*, Portland, OR, pp. 2643–2648.

Boger, Z. 2003c, Selection of quasi-optimal inputs in chemometrics modeling by artificial neural network analysis, *Analytica Chimica Acta*, vol. 490, no. 1–2, pp. 31–40.

Boger, Z., Cavicchi, R. E., Meier, D.C., & Semancik, S. 2003d, Rapid identification of CW agents by artificial neural networks pruning of temperature programmed microsensor databases, *Sensor Letters*, vol. 1, pp. 86–92.

Boilot, P., Hines, E. L., Gongora, M. A., & Folland, R. S. 2003, Electronic noses inter-comparison, data fusion and sensor selection in discrimination of standard fruit solutions, *Sensors and Actuators B-Chemical*, vol. 88, no. 1, pp. 80–88.

Boke, H., Akkurt, S., Ozdemir, S., Gokturk, E. H., & Caner Saltik, E. N. 2004, Quantification of CaCO3-CaSO4·0.5H2O-CaSO4·2H2O mixtures by FTIR analysis and its ANN model, *Materials Letters*, vol. 58, no. 5, pp. 723–726.

Bonastre, A., Ors, R., & Peris, M. 2000, Monitoring of a wort fermentation process by means of a distributed expert system, *Chemometrics and Intelligent Laboratory Systems*, vol. 50, pp. 235.

Bonastre, A., Ors, R., Peris, M. 2001, Distributed expert systems as a new tool in analytical chemistry, *TRAC, Trends in Analytical Chemistry*, vol. 20, no. 5, pp. 263–271.

Booth, T. D., Azzaoui, K., & Wainer, I. W. 1997, Prediction of chiral chromatographic separations using combined multivariate regression and neural networks, *Analytical Chemistry*, vol. 69, no. 19, pp. 3879–3883.

Borggaard, C., & Thodberg, H. H., 1992, Optimal minimal neural interpretation of spectra, *Analytical Chemistry*, vol. 64, pp. 545–551.

Bos, M., Bos, A., & Van der Linden, W. E. 1990, Processing of signals from an ion-selective electrode array by a neural network, *Analytica Chimica Acta*, vol. 233 no. 1, pp. 31–39.

Bos, M., & Weber, H. T. 1991, Comparison of the training of neural networks for quantitative x-ray fluorescence spectrometry by a genetic algorithm and backward error propagation, *Analytica Chimica Acta*, vol. 247 no. 1, pp. 97–105.

Bos, A., Bos, M., & Van der Linden, W. E. 1992, Artificial neural networks as a tool for soft-modeling in quantitative analytical-chemistry – the prediction of the water-content of cheese, *Analytica Chimica Acta*, vol. 256, no. 1, pp. 133–144.

Bos, A., Bos, M., & van der Linden, W. E. 1993a, Artificial neural networks as multivariate calibration tool: modeling the Fe-Cr-Ni system in x-ray fluorescence spectroscopy, *Analytica Chimica Acta*, vol. 277, pp. 289–295.

Bos, M., Bos, A., & Vanderlinden, W. E. 1993b, Data-processing by neural networks in quantitative chemical-analysis, *Analyst*, vol. 118, no. 4, pp. 323–328.

Bos, M., Bos, A., & van der Linden, W. E. 1993c, Tutorial review: data processing by neural networks in quantitative chemical analysis, *Analyst*, vol. 118, pp. 323–328.

Bourquin, J., Schmidli, H., van Hoogevest, P., & Leuenberger, H. 1997, Basic concepts of artificial neural networks (ANN) modeling in the application to pharmaceutical development, *Pharm.Dev. Technol.*, vol. 2, no. 2, pp. 95–109.

Bourquin, J., Schmidli, H., van Hoogevest, P., & Leuenberger, H. 1998a, Advantages of Artificial Neural Networks (ANNs) as alternative modelling technique for data sets showing non-linear relationships using data from a galenical study on a solid dosage form, *Eur. J. Pharm. Sci.*, vol. 7, no. 1, pp. 5–16.

Bourquin, J., Schmidli, H., van Hoogevest, P., & Leuenberger, H. 1998b, Pitfalls of artificial neural networks (ANN) modelling technique for data sets containing outlier measurements using a study on mixture properties of a direct compressed dosage form, *Eur.J.Pharm.Sci.*, vol. 7, no. 1, pp. 17–28.

Branca, A., Simonian, P., Ferrante, M., Novas, E., & Negri, R. M. 2003, Electronic nose based discrimination of a perfumery compound in a fragrance, *Sensors and Actuators B-Chemical*, vol. 92, no. 1–2, pp. 222–227.

Bratz, E., Yacoubgeorge, E., Hecht, F. X., & Tiltscher, H. 1994, Spectrophotometric sensor devices working in a raster of a few wavelengths, *Fresenius Journal of Analytical Chemistry*, vol. 348, no. 8–9, pp. 542–545.

Brezmes, J., Ferreras, B., Llobet. E., Vilanova, X., & Correig, X. 1997, Neural-network-based electronic nose for the classification of aromatic species *Analytica Chimica Acta*, vol. 348, no. 1–3, pp 503–509.

Briandet, R., Kemsley, E. K., & Wilson, R. H. 1996, Approaches to adulteration detection in instant coffees using infrared-spectroscopy and chemometrics, *J.Sci.Food Agr.*, vol. 71, no. 3, pp. 359–366.

Bro, R. & Heimdal, H. 1996, Enzymatic browning of vegetables. Calibration and analysis of variance by multiway methods, *Chemometrics and Intelligent Laboratory Systems*, vol. 34, no. 1, pp. 85–102.

Brodnjak-Voncina, D., Dobcnik, D., Novic, M., & Zupan, J. 1999, Determination of concentrations at hydrolytic potentiometric titrations with models made by artificial neural networks, *Chemometrics and Intelligent Laboratory Systems*, vol. 47, no. 1, pp. 79–88.

Brodnjak-Voncina, D., Dobcnik, D., Novic, M., & Zupan, J. 2002, Chemometrics characterization of the quality of river water, *Analytica Chimica Acta*, vol. 462, no. 1, pp. 87–100.

Brown, C. W., & Lo, S-C. 1998, Chemical information based on neural network processing of Near-IR spectra, *Analytical Chemistry*, vol. 70, pp. 2983–2990.

Bruchmann, A., Zinn, P., & Haffer, C. M. 1993, Prediction of gas-chromatographic retention index data by neural networks, *Analytica Chimica Acta*, vol. 283, no. 2, pp. 869–880.

Bruzzoniti, M. C., Mentasti, E., & Sarzanini, C. 1998, Carboxylic acids: prediction of retention data from chromatographic and electrophoretic behaviours, *J.Chromatogr.B Biomed. Sci. Appl.*, vol. 717, no. 1–2, pp. 3–25.

Bryjak, J., Murlikiewicz, K., Zbicinski, I., & Stawczyk, J. 2000, Application of artificial neural networks to modeling of starch hydrolysis by glucoamylase, *Bioprocess.Eng*, vol. 23, no. 4, pp. 351–357.

Bucinski, A. &, Baczek, T. 2002, Optimization of HPLC separations of flavonoids with the use of artificial neural networks, *Polish Journal of Food and Nutrition Sciences*, vol. 11, no. 4, pp. 47–51.

Bull, D. R., Harris, G. J., & Ben Rashed, A. B. 1993, A connectionist approach to fuel cell sensor array processing for gas discrimination, *Sensors and Actuators, B-Chemical*, vol. 15, no. 1–3, pp, 151–61.

Burtis, C. A. 1996, Converging technologies and their impact on the clinical laboratory, *Clinical Chemistry*, vol. 42, no. 11, pp. 1735–1749.

Burden, F. R., Brereton, R. G., & Walsh, P. T. 1997, Cross-validatory selection of test and validation sets in multivariate calibration and neural networks as applied to spectroscopy, *Analyst*, vol. 122, no. 10, pp 1015–1022.

Bylund, D., Bergens, A., & Jacobsson, S. P. 1997, Optimisation of chromatographic separations by use of a chromatographic response function, empirical modelling and multivariate analysis, *Chromatographia*, vol. 44, no. 1–2, pp. 74–80.

Bylund, D., Samskog, J., Markides, K. E., & Jacobsson, S. P. 2003, Classification of lactate dehydrogenase of different origin by liquid chromatography-mass spectrometry and multivariate analysis, *Journal of the American Society for Mass Spectrometry*, vol. 14, no. 3, pp. 236–240.

Byun, H.-G., Kim, N.-Y., Persaud, K. C., Huh, J.-S., & Lee, D.-D. 2000, Application of adaptive RBF networks to odour classification using conducting polymer sensor array, *Proceedings of the 7th International Symposium on Olfaction and Electronic Noses*, pp. 121–126.

Byvatov, E., Fechner, U., Sadovski, J. & Schneider, G. 2003, Comparison of support vector machines and artificial neural network systems for drug/nondrug classification, *Journal of Chemical Information and Computer Sciences*, vol. 43, pp. 1882–1889.

Cabanillas, A. G., Diaz, T. G., Diez, N. M. M., Salinas, F., Burguillos, J. M. O., & Vire, J. C. 2000, Resolution by polarographic techniques of atrazine-simazine and terbutryn-prometryn binary mixtures by using PLS calibration and artificial neural networks, *Analyst*, vol. 125, no. 5, pp. 909–914.

Cabrol-Bass, D., Cachet, C., Cleva, C., Eghbaldar, A., & Forrest, T. P. 1995, Practical application of neuromimetic networks to (infrared and mass) spectroscopic data for structural elucidation, *Canadian Journal of Chemistry – Revue Canadienne de Chimie*, vol. 73, no. 9, pp. 1412–1426.

Camara, M. S., Ferroni, F. M., De Zan, M., & Goicoechea, H. C. 2003, Sustained modeling ability of artificial neural networks in the analysis of two pharmaceuticals (dextropropoxyphene and dipyrone) present in unequal concentrations, *Analytical and Bioanalytical Chemistry*, vol. 376, no. 6, pp. 838–843.

Cancalon, P. F. 1999, Analytical monitoring of citrus juices by using capillary electrophoresis, *Journal of AOAC International*, vol. 82, no. 1, pp. 95–106.

Capitan-Vallvey, L. F., Navas, N., del Olmo, M., Consonni, V., & Todeschini, R. 2000, Resolution of mixtures of three nonsteroidal anti-inflammatory drugs by fluorescence using partial least squares multivariate calibration with previous wavelength selection by Kohonen artificial neural networks, *Talanta*, vol. 52, no. 6, pp. 1069–1079.

Carlin, M., Kavli, T., & Lillekjendlie, B. 1994, A comparison of 4 methods for nonlinear data modeling, *Chemometrics and Intelligent Laboratory Systems*, vol. 23, no. 1, pp. 163–177.

Carrapiso, A. I., Ventanas, J., Jurado, A., & Garcia, C. 2001, An electronic nose to classify Iberian pig fats with different fatty acid composition, *Journal of the American Oil Chemists Society*, vol. 78, no. 4, pp. 415–418.

Carrieri, A. H., & Lim, P. I. 1995, Neural network pattern recognition of thermal-signature spectra for chemical defense, *Applied Optics*, vol. 34, no. 15, pp. 2623–2635.

Carson, C. A., Keller, J. M., Mcadoo, K. K., Wang, D. Y., Higgins, B., Bailey, C. W., Thorne, J. G., Payne, B. J., Skala, M., & Hahn, A. W. 1995, Escherichia-Coli O157-H7 restriction pattern-recognition by artificial neural network, *Journal of Clinical Microbiology*, vol. 33, no. 11, pp. 2894–2898.

Casamento, S., Kwok, B., Roux, C., Dawson, M., & Doble, P. 2003, Optimization of the separation of organic explosives by capillary electrophoresis with artificial neural networks, *Journal of Forensic Sciences*, vol. 48, no. 5, pp. 1075–1083.

Catasus, M., Branagh, W., & Salin, E. D. 1995, Improved calibration for inductively coupled plasma-atomic emission spectroscopy using generalized regression neural networks, *Applied Spectroscopy*, vol. 49, no. 6, pp. 798–807.

Centner, V., Verdu-Andres, J., Walczak, B., Jouan-Rimbaud, D., Despagne, F., Pasti, L., Poppi, R., Massart, D. L., & Denoord, O. E. 2000, Comparison of multivariate calibration techniques applied to experimental NIR data sets, *Applied Spectroscopy*, vol. 54, no. 4, pp. 608–623.

Ceppi, S. B., Velasco, M. I., Fetsch, D., & Havel, J. 1999, Fingerprints of humic acids by capillary zone electrophoresis, *Boletin de la Sociedad Chilena de Quimica*, vol. 44, no. 4, pp. 401–411.

Cerrato-Oliveros, M. C., Perez-Pavon, J. L., Garcia-Pinto, C., Fernandez-Laespada, M. E., Moreno-Cordero, B., & Forina, M. 2002, Electronic nose based on metal oxide semiconductor sensors as a fast alternative for the detection of adulteration of virgin olive oils, *Analytica Chimica Acta*, vol. 459, no. 2, pp. 219–228.

Chan, H., Butler, A., Falck, D. M., & Freund, M. S. 1997a, Artificial neural-network processing of stripping analysis responses for identifying and quantifying heavy-metals in the presence of intermetallic compound formation, *Analytical Chemistry*, vol. 69, no. 13, pp. 2373–2378.

Chan, W. H., Lee, A. W. M., Kwong, W. J., Liang, Y.-Z., & Wang, K.-M. 1997b, Simultaneous determination of potassium and sodium by optrode spectra and an artificial network algorithm, *Analyst*, vol. 122, pp. 657–661.

Chang, S.-M., Iwasaki, Y., Suzuki, M., Karube, I., & Muramatsu, H. 1991, Detection of odorants using an array of piezoelectric crystals and neural network pattern recognition', *Analytica Chimica Acta*, vol. 249, pp. 323–329.

Charef, A.; Ghauch, A., Martin-Bouyer, M. 2000, An adaptive and predictive control strategy for an activated sludge process, *Bioprocess Engineering*, vol. 23, no. 5, pp. 529–534.

Chen, C. W., Chen, D. Z., & Cao, G. Z. 2002a, An improved differential evolution algorithm in-training and encoding prior knowledge into feedforward networks with application in chemistry, *Chemometrics and Intelligent Laboratory Systems*, vol. 64, no. 1, pp. 27–43.

Chen, F., Hu, S. X., & Yu, M. H. 2000a, Using Elman recurrent neural-network trained in sections to filter noise of chromatographic spectrogram, *Chem.J.Chinese.Univ Chinese.*, vol. 21, no. 2, pp. 193–197.

Chen, G. X. & Harrington, P. D. 2003, Simplisma applied to 2-dimensional wavelet compressed ion mobility spectrometry data, *Analytica Chimica Acta*, vol. 484, no. 1, pp. 75–91.

Chen, J. C., Ju, Y. H., & Liu, C. J. 1999, Concentration determination of gas by organic thin-film sensor and back-propagation network, *Sensors and Actuators B-Chemical*, vol. 60, no. 2–3, pp. 168–173.

Chen, J. C., Liu, C. J., & Ju, Y. H. 2000b, Determination of the composition of NO2 and NO mixture by thin-film sensor and backpropagation network, *Sensors and Actuators B-Chemical*, vol. 62, no. 2, pp. 143–147.

Chen, H., Yu, Z. Y., & Zhu, G. B. 2001a, Recognition of the bouquet of Chinese spirits by artificial neural network analysis, *Journal of AOAC International*, vol. 84, no. 5, pp. 1579–1585.

Chen, X. G., Xu, H. P., Dong, L. J., Liu, H. T., Zeng, Y. B., & Hu, Z. D. 2002b, Application of artificial neural networks in multivariable optimization of an online microwave FIA system for catalytic kinetic determination of Iridium(III), *Analytical & Bioanalytical Chemistry*, vol. 373, no. 8, pp. 883–888.

Chen, Y. R., Delwiche, S. R., & Hruschka, W. R. 1995, Classification of hard red wheat by feedforward backpropagation neural networks, *Cereal Chemistry*, vol. 72, no. 3, pp. 317–319.

Chen, Y. R., Huffman, R. W., Park, B., & Nguyen, M. 1996, Transportable spectrophotometer system for online classification of poultry carcasses, *Applied Spectroscopy*, vol. 50, no. 7, pp. 910–916.

Chen, Y. X., Thosar, S. S., Forbess, R. A., Kemper, M. S., Rubinovitz, R. L., & Shukla, A. J. 2001b, Prediction of drug content and hardness of intact tablets using artificial neural-network and near-infrared spectroscopy, *Drug Development & Industrial Pharmacology*, vol. 27, no. 7, pp. 623–631.

Chong, K. H., & Wilson, D. M. 2000, Multilevel pattern recognition architectures for localization of mixed chemical/auditory stimuli, *Sensors and Actuators B-Chemical*, vol. 68, no. 1, pp. 58–68.

Chretien, J. R., Pintore, M., & Ros, F. 2000, Virtual high throughput screening (v-HTS), *Actualite Chimique*, no. 9, pp. 60–63.

Chretien, J. R. 2002, The state of the art for chemometrics in analytical chemistry, *Analytical and Bioanalytical Chemistry*, vol. 372, no. 4, pp. 511–512.

Chroma-Keull, H., Havlis, J., & Havel, J. 2000, Reactive dye Ostazine Black V-B: determination in the dye-bath and hydrolysis monitoring by matrix-assisted laser desorption/ionisation time of-flight mass spectrometry, *Rapid Communications in Mass Spectrometry*, vol. 14, no. 1, pp. 40–43.

Chun, J., Atalan, E., Ward, A. C., & Goodfellow, M. 1993a, Artificial neural network analysis of pyrolysis mass-spectrometric data in the identification of streptomyces strains, *FEMS Microbiology Letters*, vol. 107, no. 2–3, pp. 321–325.

Chun, J., Atalan, E., Kim, S. B., Kim, H. J., Hamid, M. E., Trujillo, M. E., Magee, J. G., Manfio, G. P., Ward, A. C., & Goodfellow, M. 1993b, Rapid identification of streptomycetes by artificial neural-network analysis of pyrolysis mass-spectra, *FEMS Microbiology Letters*, vol. 114, no. 1, pp. 115–119.

Chun, J. S., Ward, A. C., Kang, S. O., Hah, Y. C., & Goodfellow, M. 1997, Long-term identification of streptomycetes using pyrolysis mass spectrometry and artificial neural networks, *Zentralblatt für Bakteriologie – International Journal of Medical Microbiology Virology Parasitology and Infectious Diseases*, vol. 285, no. 2, pp. 258–266.

Cirovic, D. A. 1997, Feedforward artificial neural networks – applications to spectroscopy, *TRAC-Trends in Analytical Chemistry*, vol. 16, no. 3, pp 148–155.

Cladera, A., Alpizar, J., Estela, J. M., Cerda, V., Catasus, M., Lastres, E., & Garcia, L. 1997, Resolution of highly overlapping differential-pulse anodic-stripping voltammetric signals using multicomponent analysis and neural networks, *Analytica Chimica Acta*, vol. 350, no. 1–2, pp. 163–169.

Cleva, C., Cachet, C., Cabrol-bass, D., & Forrest, T. P. 1997, Advantages of a hierarchical system of neural-networks for the interpretation of infrared-spectra in structure determination, *Analytica Chimica Acta*, vol. 348, no. 1–3, pp. 255–265.

Cordella, C. B., Militao, J. S., Clement, M. C., & Cabrol-Bass, D. 2003, Honey characterization and adulteration detection by pattern recognition applied on HPAEC-PAD profiles. 1. Honey floral species characterization, *Journal of Agricultural and Food Chemistry*, vol. 51, no. 11, pp. 3234–3242.

Cukrowska, E., Cukrowski, I., & Havel, J. 2000, Application of artificial neural networks for analysis of highly overlapped and disturbed differential pulse polarographic peaks in the region of hydrogen evolution, *South African Journal of Chemistry – Suid-Afrikaanse Tydskrif Vir Chemie*, vol. 53, no. 3.

Cukrowska, E., Trnkova, L., Kizek, R., & Havel, J. 2001, Use of artificial neural networks for the evaluation of electrochemical signals of adenine and cytosine in mixtures interfered with hydrogen evolution, *Journal of Electroanalytical Chemistry*, vol. 503, no. 1–2, pp. 117–124.

Cukrowski, I. & Havel, J. 2000, Evaluation of equilibria with a use of artificial neural networks (ANN). I. Artificial neural networks and experimental design as a tool in electrochemical data evaluation for fully inert metal complexes, *Electroanalysis*, vol. 12, no. 18, pp. 1481–1492.

Cukrowski, I., Farkova, M., & Havel, J. 2001, Evaluation of equilibria with use of artificial neural networks (ANN). II. ANN and experimental design as a tool in electrochemical data evaluation for fully dynamic (labile) metal complexes, *Electroanalysis*, vol. 13, no. 4, pp. 295–308.

Cullen, T. F. & Crouch, S. R. 2000, Factors affecting the accuracy of multicomponent kinetic-spectrophotometric determinations based on multivariate calibration methods, *Analytica Chimica Acta*, vol. 407, no. 1–2, pp. 135–146.

Dacosta, P. A. & Poppi, R. J. 2001, Determination of triglycerides in human plasma using near-infrared spectroscopy and multivariate calibration methods, *Analytica Chimica Acta*, vol. 446, no. 1–2, pp. 39–47.

Da Silva, J. C. G. E. & Laquipai, M. C. P. O. 1998, Method for rapid screening of chlorophenols using a reduced calibration set of UV spectra and multivariate calibration techniques, *Analytical Letters*, vol. 31, no. 14, pp. 2549–2563.

Da Silva, I. N., De Souza, A. N., Hossri, R. M. C., & Hossri, J. H. C. 2000a, Intelligent system applied in diagnosis of transformer oil, *Eighth International Conference On Dielectric Materials, Measurements and Applications*, pp. 330–334.

Da Silva, I. N., de Souza, A. N., Hossri, J. H. C., & Zago, M. G. 2000b, Identification of the level of contamination and degradation of oil by artificial neural networks, *Proceedings of the IEEE International Symposium on Electrical Insulation*, pp. 275–279.

Damiani, P., Cossignani, L., Simonetti, M. S., Mattoscio, A., Neri, A., Balestrieri, F., Damiani, F., & Marini, D. 2000, Classification of virgin olive oils by artificial neural networks and statistical multivariate methods applied to the stereospecific analysis data of triacylglycerols, *Journal of Commodity Science*, vol. 39, no. 4, pp. 179–196.

Dane, A. D., van Sprang, H. A., & Buydens, L. M. C. 1999, A two-step approach toward model-free x-ray fluorescence analysis of layered materials, *Analytical Chemistry*, vol. 71, no. 20, pp. 4580–4586.

Dane, A. D., Rea, G. J., Walmsley, A. D., & Haswell, S. J. 2001, The determination of moisture in tobacco by guided microwave spectroscopy and multivariate calibration, *Analytica Chimica Acta*, vol. 429, no. 2, pp. 185–194.

Daniel, N. W., Lewis, I. R., & Griffiths, P. R. 1997a, Supervised and unsupervised methods of classification of Raman-spectra of explosives and non-explosives, *Mikrochimica Acta*, no. S14, pp. 281–282.

Daniel, N. W., Lewis, I. R., & Griffiths, P. R. 1997b, Interpretation of Raman-spectra of nitro-containing explosive materials. Part II: The implementation of neural, fuzzy, and statistical-models for unsupervised pattern-recognition, *Applied Spectroscopy*, vol. 51, no. 12, pp. 1868–1879.

Danzer, K. 1995, Calibration – A multidimensional approach, *Fresenius Journal of Analytical Chemistry*, vol. 351, no. 1, pp. 3–10.

Darsey, J. A., Lay, J. O., & Holland, R. D. 1999, Deconvolution of composite mass spectra by artificial neural networks, *Chimica Oggi – Chemistry Today*, vol. 17, no. 9, pp. 41–44.

Daszykowski, M., Walczak, B., & Massart, D. L. 2003, A journey into low-dimensional spaces with autoassociative neural networks, *Talanta*, vol. 59, no. 6, pp. 1095–1105.

Davey, C. L. & Kell, D. B. 1998, The influence of electrode polarisation on dielectric spectra, with special reference to capacitive biomass measurements. I. Quantifying the effects on electrode polarisation of factors likely to occur during fermentations, *Bioelectrochemistry and Bioenergetics*, vol. 46, no. 1, pp. 91–103.

Debska, B. & Guzowskaswider, B. 1999, Scankee – computer-system for interpretation of infrared-spectra, *Journal of Molecular Structure.*, vol. 512, no. Nov, pp. 167–171.

Debska, B. & Guzowskaswider, B. 2000, Fuzzy definition of molecular fragments in chemical structures, *Journal of Chemical Information and Computer Sciences*, vol. 40, no. 2, pp. 325–329.

Debska, B., Guzowskaswider, B., & Cabrol-bass, D. 2000, Automatic-generation of knowledge-base from infrared spectral database for substructure recognition, *Journal of Chemical Information and Computer Sciences*, vol. 40, no. 2, pp. 330–338.

De Carvalho, R. M., Mello, C., & Kubota, L. T. 2000, Simultaneous determination of phenol isomers in binary mixtures by differential pulse voltammetry using carbon fiber electrode and neural network with pruning as a multivariable calibration tool, *Analytica Chimica Acta*, vol. 420, no. 1, pp. 109–121.

De Cerqueira, E. O., de Andrade, J. C., Poppi, R. J., & Mello, C. 2001, Neural networks and its applications in multivariate calibration, *Quimica Nova*, vol. 24, no. 6, pp. 864–873.

De Lucas, A., Duran, A., Carmona, M., & Lapuerta, M. 2001, Modeling diesel particulate emissions with neural networks, *Fuel*, vol. 80, no. 4, pp. 539–548.

De Viterbo, V. D. & Belchior, J. C. 2001, Artificial neural networks applied for studying metallic complexes, *Journal of Computational Chemistry*, vol. 22, no. 14, pp. 1691–1701.

Debeljak, Z., Strapac, M., & Medic-Saric, M. 2001, Application of self-organizing maps for the classification of chromatographic systems and prediction of values of chromatographic quantities, *Journal of Chromatography A*, vol. 925, no. 1–2, pp. 31–40.

Defernez, M., & Kemsley, E. K. 1999, Avoiding overfitting in the analysis of high-dimensional data with artificial neural networks (ANNs), *Analyst*, vol. 124, no. 11, pp. 1675–1681.

Derks, E. P. P. A. & Buydens, L. M. C. 1998, Aspects of network training and validation on noisy data – Part 1. Training aspects, *Chemometrics and Intelligent Laboratory Systems*, vol. 41, no. 2, pp. 171–184.

Deshpande, P. B. 1997, Predict difficult-to-measure properties with neural analyzer, *Control Engineering*, July, pp. 55–56.

Despagne, F. & Massart, D. L. 1998a, Neural networks in multivariate calibration, *Chemometrics and Intelligent Laboratory Systems*, vol. 40, pp. 145–163.

Despagne, F., Walczak, B., & Massart, D. L. 1998b, Transfer of calibrations of near-infrared spectra using neural networks, *Applied Spectroscopy*, vol. 52, no. 5, pp. 732–745.

Despagne, F., Massart, D. L., & Chabot, P. 2000, Development of a robust-calibration model for non-linear in-line process data, *Analytical Chemistry*, vol. 72, no. 7, pp. 1657–1665.

De Souza, P. A., Jr., Garg, R., & Garg, V. K. 1997, Automation of the analysis of Mossbauer spectra, *Hyperfine Interactions*, vol. 112, no. 1–4, pp. 275–278.

De Souza, J. E. G., Neto, B. B., Dossantos, F. L., Demelo, C. P., Santos, M. S., & Ludermir, T. B. 1999a, Polypyrrole-based aroma sensor, *Synthetic Metals*, vol. 102, no. 1–3, pp. 1296–1299.

De Souza, P. A., Jr. 1999b, Automation in Mossbauer spectroscopy data analysis, *Laboratory Robotics and Automation*, vol. 11, no. 1, pp. 3–23.

De Souza, P. A., Jr., & Garg, V. K. 1999c, The challenge of an automatic Mossbauer analysis, *NATO Science Series, 3: High Technology*, vol. 66 *(Mossbauer Spectroscopy in Materials Science)*, pp. 359–372.

De Souza, P. A., Jr., Klingelhoefer, G., & Morimoto, T. 2001a, On-line and in situ identification of air pollution, *Proceedings of the International Conference on the Applications of the Mossbauer Effect*, pp. 487–490.

De Souza, Paulo A. Jr., Klingelhoefer, G., Guimaraes, A. F., Morimoto, T., de Queiroz, R. S.A. 2001b, Intelligent receptor modelling: a receptor modelling based on adaptive techniques, *Proceedings of the 94th Air & Waste Management Association's Annual Conference*, pp. 179–195.

De Souza, P. A., Jr., Klingelhoefer, G., Bernhardt, B., Schroeder, C., Guetlich, P., & Morimoto, T. 2001c, On-line and in situ characterization of iron phases in particulate matter, *Proceedings of the 94th Air & Waste Management Association's Annual Conference*, pp. 707–717.

De Souza, P. A., Garg, V. K., Klingelhoefer, G., Gellert, R., & Guetlich, P. 2002a, Portable and automatic Mossbauer analysis, *Hyperfine Interactions*, vol. 139/140, no. 1–4/1–4, pp. 705–714.

De Souza, P. A., Jr., Klingelhoefer, G., & Morimoto, T. 2002b, On-line and in situ identification of air pollution, *Proceedings of the International Conference on the Applications of the Moessbauer Effect*, pp. 487–490.

Diaz, T. G., Guiberteau, A., Ortiz, J. M., Lopez, M. D., & Salinas, F. 2001, Use of neural networks and diode-array detection to develop an isocratic HPLC method for the analysis of nitrophenol pesticides and related compounds, *Chromatographia*, vol. 53, no. 1–2, pp. 40–46.

Dickert, F. L., Haunschild, A., & Hofmann, P. 1994, Cholesteric liquid-crystals for solvent vapor detection – elimination of cross-sensitivity by band shape-analysis and pattern-recognition, *Fresenius Journal of Analytical Chemistry*, vol. 350, no. 10–11, pp. 577–581.

Dickert, F. L., Hayden, O., & Zenkel, M. E. 1999, Detection of volatile compounds with mass sensitive sensor arrays in the presence of variable ambient humidity, *Analytical Chemistry*, vol. 71, no. 7, pp. 1338–1341.

Dieterle, F., Nopper, D., & Gauglitz, G. 2001, Quantification of butanol and ethanol in aqueous phases by Reflectometric interference spectroscopy – different approaches to multivariate calibration, *Fresenius Journal of Analytical Chemistry*, vol. 370, no. 6, pp. 723–730.

Dieterle, F., Belge, G., Betsch, C., & Gauglitz, G. 2002, Quantification of the refrigerants R22 and R134a in mixtures by means of different polymers and reflectometric interference spectroscopy, *Analytical & Bioanalytical Chemistry*, vol. 374, no. 5, pp. 858–867.

Dieterle, F., Busche, S., & Gauglitz, G. 2003a, Growing neural-networks for a multivariate calibration and variable selection of time-resolved measurements, *Analytica Chimica Acta*, vol. 490, no. 1–2, pp. 71–83.

Dieterle, F., Mullerhagedorn, S., Liebich, H. M., & Gauglitz, G. 2003b, Urinary nucleosides as potential tumor-markers evaluated by learning vector quantization, *Artificial Intelligence in Medicine*, vol. 28, no. 3, pp. 265–279.

Dieterle, F., Kieser, B., & Gauglitz, G. 2003c, Genetic algorithms and neural networks for the quantitative analysis of ternary mixtures using surface-plasmon resonance, *Chemometrics and Intelligent Laboratory Systems*, vol. 65, no. 1, pp. 67–81.

Di Natale, C., Davide, F. A. M., Damico, A., Hierlemann, A., Mitrovics, J., Schweizer, M., Weimar, U., & Gopel, W. 1995, A composed neural-network for the recognition of gas-mixtures, *Sensors and Actuators B-Chemical*, vol. 25, no. 1–3, pp. 808–812.

Di Natale, C., Davide, F., Brunink, J. A. J., Damico, A., Vlasov, Y. G., Legin, A. V., & Rudnitskaya, A. M. 1996, Multicomponent analysis of heavy-metal cations and inorganic anions in liquids by a nonselective chalcogenide glass sensor array, *Sensors and Actuators B-Chemical*, vol. 34, no. 1–3, pp. 539–542.

Di Natale, C., Macagnano, A., Damico, A., & Davide, F. 1997, Electronic-nose modeling and data-analysis using a self-organizing map, *Measurement Science Technology*, vol. 8, no. 11, pp. 1236–1243.

Di Natale, C., Paolesse, R., Macagnano, A., Troitsky, V. I., Berzina, T. S., & Damico, A. 1999a, Pattern-recognition approach to the study of the interactions between metalloporphyrin Langmuir-Blodgett-films and volatile organic-compounds, *Analytica Chimica Acta*, vol. 384, no. 3, pp. 249–259.

Di Natale, C., Mantini, A., Macagnano, A., Antuzzi, D., Paolesse, R., & Damico, A. 1999b, Electronic nose analysis of urine samples containing blood, *Physiological Measurement*, vol. 20, no. 4, pp. 377–384.

Di Natale, C., Paolesse, R., Macagnano, A., Mantini, A., D'Amico, A., Ubigli, M., Legin, A., Lvova, L., Rudnitskaya, A., & Vlasov, Y. 2000, Application of a combined artificial olfaction and taste system to the quantification of relevant compounds in red wine, *Sensors and Actuators B-Chemical*, vol. 69, no. 3, pp. 342–347.

Ding, Q., Small, G. W., & Arnold, M. A. 1999, Evaluation of nonlinear model-building strategies for the determination of glucose in biological matrices by near-infrared spectroscopy, *Analytica Chimica Acta*, vol. 384, no. 3, pp. 333–343.

Distante, C., Artursson, T., Siciliano, P., Holmberg, M., & Lundstrom, I. 2000, Odour identification under drift effect, *Proceedings of the 7th International Symposium on Olfaction and Electronic Noses*, pp. 89–95.

Distante, C., Leo, M., Siciliano, P., & Persaud, K. C. 2002 On the study of feature extraction methods for an electronic nose, *Sensors and Actuators B-Chemical*, vol. 87 no. 2, pp. 274–288.

Divjak, B., Moder, M., & Zupan, J. 1998, Chemometrics approach to the optimization of ion chromatographic analysis of transition metal cations for routine work, *Analytica Chimica Acta*, vol. 358, no. 3, pp. 305–315.

Dobiasova, Z., Pazourek, J., & Havel, J. 2002, Simultaneous determination of trans-resveratrol and sorbic acid in wine by capillary zone electrophoresis, *Electrophoresis*, vol. 23, no. 2, pp. 263–267.

Doble, P., Sandercock, M., Du Pasquier, E., Petocz, P., Roux, C., & Dawson, M. 2003, Classification. of premium and regular gasoline by gas chromatography/mass spectrometry, principal component analysis and artificial neural networks, *Forensic Science International*, vol. 132, no. 1, pp. 26–39.

Dobrowolski, M., & Iciek, J. 2003, First carbonation alkalinity determination using artificial neural networks, *International Sugar Journal*, vol. 105, no. 1256, pp. 358, 360–361, 363–366.

Dohnal, V. 1998, Artificial neural networks for optimization in the capillary zone electrophoresis, *Chemicke Listy*, vol. 92, no. 8, pp. 669–672.

Dohnal, V., Farkova, M., & Havel, J. 1999, Prediction of chiral separations using a combination of experimental design and artificial neural networks, *Chirality*, vol. 11, no. 8, pp. 616–621.

Dohnal, V., Li, H., Farkova, M., & Havel, J. 2002, Quantitative analysis of chiral compounds from unresolved peaks in capillary electrophoresis using multivariate calibration with experimental design and artificial neural networks, *Chirality*, vol. 14, no. 6, pp. 509–518.

Dolenko, T. A., Churina, I. V., Fadeev, V. V., & Glushkov, S. M. 2000, Valence-band of liquid water Raman-scattering – some peculiarities and applications in the diagnostics of water media, *Journal of Raman Spectroscopy*, vol. 31, no. 8–9, pp. 863–870.

Dolmatova, L., Ruckebusch, C., Dupuy, N., Huvenne, J. P., & Legrand, P. 1997, Quantitative-analysis of paper coatings using artificial neural networks, *Chemometrics and Intelligent Laboratory Systems*, vol. 36, no. 2, pp. 125–140.

Dolmatova, L., Ruckebusch, C., Dupuy, N., Huvenne, J. P., & Legrand, P. 1998, Identification of modified starches using infrared-spectroscopy and artificial neural-network processing, *Applied Spectroscopy*, vol. 52, no. 3, pp. 329–338.

Dolmatova, L., Tchistiakov, V., Ruckebusch, C., Dupuy, N., Huvenne, J. P., & Legrand, P. 1999, Hierarchical neural-network modeling for infrared-spectra interpretation of modified starches, *Journal of Chemical Information and Computer Sciences*, vol. 39, no. 6, pp. 1027–1036.

Du, Y. G., Sreekrishnan, T. R., Tyagi, R. D., & Campbell, P. G. C. 1994, Effect of pH on metal solubilization from sewage-sludge – a neural-net-based approach, *Canadian Journal of Civil Engineering*, vol. 21, no. 5, pp. 728–735.

Dumas, M. E., Debrauwer, L., Beyet, L., Lesage, D., Andre, F., Paris, A., & Tabet, J. C. 2002, Analyzing the physiological signature of anabolic steroids in cattle urine using pyrolysis/metastable atom bombardment mass spectrometry and pattern recognition, *Analytical Chemistry*, vol. 74, no. 20, pp. 5393–5404.

Duponchel, L., Ruckebusch, C., Huvenne, J. P., & Legrand, P. 1999a, Standardisation of near infrared spectrometers using artificial neural networks, *Journal of Near Infrared Spectroscopy*, vol. 7, no. 3, pp. 155–166.

Duponchel, L., Ruckebusch, C., Huvenne, J. P., & Legrand, P. 1999b, Standardisation of near-IR spectrometers using artificial neural networks, *Journal of Molecular Structure*, vol. 481, pp. 551–556.

Dutta, R., Kashwan, K. R., Bhuyan, M., Hines, E. L., & Gardner, J. W. 2003a, Electronic nose-based tea quality standardization, *Neural Networks*, vol. 16, no. 5–6, pp. 847–853.

Dutta, R., Hines, E. L., Gardner, J. W., Kashwan, K. R., & Bhuyan, M. 2003b, Tea quality prediction using a tin-oxide-based electronic nose: an artificial intelligence approach, *Sensors and Actuators B-Chemical*, vol. 94, pp. 228–237.

Edelmann, A., & Lendl, B. 2002, Toward the optical tongue: flow-through sensing of tannin-protein interactions based on FTIR spectroscopy, *Journal of the American Chemical Society*, vol. 124 no. 49, pp. 14741–14747.

Egan, W. J., Lloyd, L., Angel, S. M., & Morgan, S. L. 1995, Evaluation of Raman spectra of environmental contaminants using neural networks, *Proceedings of the PITCON*, paper 715.

Egan, W. J., Angel, S. M., & Morgan, S. L., 2001, Rapid optimization and minimal complexity in computational neural-network multivariate calibration of chlorinated hydrocarbons using Raman-spectroscopy, *Journal of Chemometrics*, vol. 15, no. 1, pp. 29–48.

Eghbaldar, A., Forrest, T. P., Cabrol-Bass, D., Cambon. A., & Guigonis, J. M. 1996, Identification of structural features from mass spectrometry using a neural network approach: application to trimethylsilyl derivatives used for medical diagnosis, *Journal of Chemical Information and Computer Sciences*, vol. 36, no. 4, pp. 637–643.

Eghbaldar, A., Forrest, T. P., Cabrol-Bass, D. 1998, Development of neural networks for the identification of structural features from mass spectral data, *Analytica Chimica Acta*, vol. 359, pp. 283–301.

Ehrentreich, F. 1999, Joined knowledge and signal processing for infrared spectrum interpretation, *Analytica Chimica Acta*, vol. 393, no. 1–3, pp. 193–200.

Ehrentreich, F. 2001, Three-step procedure for infrared spectrum interpretation, *Analytica Chimica Acta*, vol. 427, no. 2, pp. 233–244.

Ehrman, J. M., Clair, T. A., & Bouchard, A. 1996, Using neural networks to predict pH changes in acidified eastern Canadian lakes, *AI Applications*, vol. 10, no. 2, pp. 1–8.

Eiceman, G. A., Nazarov, E. G., Rodriguez, J. E. 2001a, Chemical class information in ion mobility spectra at low and elevated temperatures, *Analytica Chimica Acta*, vol. 433 no. 1, pp. 53–70.

Eiceman, G. A., Nazarov, E. G., Rodriguez, J. E., & Stone, J. A. 2001b, Analysis of a drift tube at ambient pressure: models and precise measurements in ion mobility spectrometry, *Review of Scientific Instruments*, vol. 72, no. 9, pp. 3610–3621.

Eklov, T. & Lundstrom, I. 1999a, Gas-mixture analysis using a distributed chemical sensor system, *Sensors and Actuators B-Chemical*, vol. 57, no. 1–3, pp. 274–282.

Eklov, T. & Lundstrom, I. 1999b, Distributed sensor system for quantification of individual components in a multiple gas-mixture, *Analytical Chemistry*, vol. 71, no. 16, pp. 3544–3550.

Eklov, T., Martensson, P., & Lundstrom, I. 1999c, Selection of variables for interpreting multivariate gas sensor data, *Analytica Chimica Acta*, vol. 381, no. 2–3, pp. 221–232.

Emilio, C. A., Litter, M. I., & Magallanes, J. F. 2002, Semi empirical modeling with application of artificial neural networks for the photocatalytic reaction of ethylenediaminetetraacetic acid (EDTA) over titanium-oxide (TiO2), *Helvetica Chimica Acta*, vol. 85, no. 3, pp. 799–813.

Estienne, F., & Massart, D. L. 2001, Mutivariate calibration with Raman data using fast principal component regression and partial least-squares methods, *Analytica Chimica Acta*, vol. 450, no. 1–2, pp. 123–129.

Facchin, I., Mello, C., Bueno, M. I. M. S., & Poppi, R. J. 1999, Simultaneous determination of lead and sulfur by energy-dispersive x-ray spectrometry. Comparison between artificial neural networks and other multivariate calibration methods, *X-Ray Spectrometry*, vol. 28, no. 3, pp. 173–177.

Farkova, M., Pena-Mendez, E. M., & Havel, J. 1999, Use of artificial neural networks in capillary zone electrophoresis, *Journal of Chromatography A*, vol. 848, no. 1–2, pp. 365–374.

Fatemi, M. H. 2002, Simultaneous modeling of the Kovats retention indices on OV-1 and SE-54 stationary phases using artificial neural networks, *Journal of Chromatography A*, vol. 955, no. 2, pp. 273–280.

Fatemi, M. H. 2003, Quantitative structure-property relationship studies of migration index in microemulsion electrokinetic chromatography using artificial neural networks, *Journal of Chromatography A*, vol. 1002, no. 1–2, pp. 221–229.

Feraud, R. & Clerot, F. 2002, A methodology to explain neural-network classification, *Neural Networks*, vol. 15, no. 2, pp. 237–246.

Fernandez-Caceres, P. L., Martin, M. J., Pablos, F., & Gonzalez, A. G. 2001, Differentiation of tea (Camellia sinensis) varieties and their geographical origin according to their metal content, *Journal of Agricultural and Food Chemistry*, vol. 49, no. 10, pp. 4775–4779.

Ferreira, L. S., De Souza, M. B., & Folly, R. O. M. 2001, Development of an alcohol fermentation control system based on biosensor measurements interpreted by neural networks, *Sensors and Actuators B-Chemical*, vol. 75, no. 3, pp. 166–171.

Ferreira, L. S., Souza, M. B., Trierweiler, J. O., Hitzmann, B., & Folly, R. O. M. 2003, Analysis of experimental biosensor/FIA lactose measurements, *Brazilian Journal of Chemical Engineering*, vol. 20, no. 1, pp. 7–13.

Ferrer, R., Guiteras, J., & Beltran, J. L. 1999, Artificial neural networks (ANNs) in the analysis of polycyclic aromatic hydrocarbons in water samples by synchronous fluorescence, *Analytica Chimica Acta*, vol. 384, no. 3, pp. 261–269.

Fidencio, P. H., Ruisanchez, I., & Poppi, R. J. 2001, Application of artificial neural networks to the classification of soils from Sao Paulo state using near-infrared spectroscopy, *Analyst*, vol. 126, no. 12, pp. 2194–2200.

Forshed, J., Andersson, F. O., & Jacobsson, S. P. 2002, NMR and Bayesian regularized neural-network regression for impurity determination of 4-aminophenol, *Journal of Pharmaceutical & Biomedical Analysis*, vol. 29, 3, pp. 495–505.

Forsstrom, J. J. & Dalton, K. J. 1995, Artificial neural networks for decision-support in clinical medicine, *Annals of Medicine*, vol. 27, no. 5, pp. 509–517.

Frake, P., Gill, I., Luscombe, C. N., Rudd, D. R., Waterhouse, J., & Jayasorriya, U. A. 1998, Near-infrared mass median particle size determination of lactose monohydrate, evaluating several chemometric approaches, *Analyst*, vol. 123, no. 10, pp. 2043–2046.

Francelin, R. A., Gomide, F. A. C., & Lancas, F. M. 1993, Use of artificial neural networks for the classification of vegetable-oils after GC analysis, *Chromatographia*, vol. 35, no. 3–4, pp. 160–166.

Freeman, R., Goodacre, R., Sisson, P. R., Magee, J. G., Ward, A. C., & Lightfoot, N. F. 1994, Rapid identification of species within the mycobacterium-tuberculosis complex by artificial neural-network analysis of pyrolysis mass-spectra, *Journal of Medical Microbiology*, vol. 40, no. 3, pp. 170–173.

Freeman, R., Sisson, P. R., & Ward, A. C. 1995, Resolution of batch variations in pyrolysis mass-spectrometry of bacteria by the use of artificial neural-network analysis, *Antonie Van Leeuwenhoek International Journal of General and Molecular Microbiology*, vol. 68, no. 3, pp. 253–260.

Freeman, R., 2003, Beg, borrow or steal – finding ideas for new NMR experiments, *Concepts in Magnetic Resonance, Part A*, vol. 17A, no. 1, pp. 71–85.

Frenich, A. G., Galera, M. M., Garcia, M. D. G., Vidal, J. L. M., Catasus, M., Marti, L., & Mederos, M. V. 2001, Resolution of HPLC-DAD highly overlapping analytical signals for quantitation of pesticide mixtures in groundwater and soil using multicomponent analysis and neural networks, *Journal of Liquid Chromatography & Related Technologies*, vol. 24, no. 5, pp. 651–668.

Fu, C. Y. 2002, Non-invasive diagnostic and monitoring system based on odor detection, *U.S. Pat. Appl.* 2002-87049.

Fukuoka, Y., Matsuki, H., Miniamitani, H., & Ishida, A. 1998, A modified back-propagation method to avoid false local minima, *Neural Networks*, vol. 11, no. 6, pp. 1059–1072.

Gajdosova, D., Novotna, K., Prosek, P., & Havel, J. 2003, Separation and characterization of humic acids from Antarctica by capillary electrophoresis and matrix-assisted laser desorption ionization time-of-flight mass spectrometry inclusion complexes of humic acids with cyclodextrins, *Journal of Chromatography A*, vol. 1014, no. 1–2, pp. 117–127.

Gallardo, J., Alegret, S., Munoz, R., De-Roman, M., Leija, L., Hernandez, P. R., & Del Valle, M. 2003, An electronic tongue using potentiometric all-solid-state PVC-membrane sensors for the simultaneous quantification of ammonium and potassium ions in water, *Analytical and Bioanalytical Chemistry*, vol. 377, no. 2, pp. 248–256.

Galdikas, A., Mironas, A., Senuliene, D., Strazdiene, V., Setkus, A., & Zelnin, D. 2000, Reponses time based output of metal oxide gas sensors applied to evaluation of meat freshness with neural signal analysis, *Sensors and Actuators B-Chemical*, vol. 69, pp. 258–265.

Galvan, I. M., Zaldivar, J. M., Hernandez, H., & Molga, E. 1996, The use of neural networks for fitting complex kinetic data, *Computers & Chemical Engineering*, vol. 20, no. 12, pp. 1451–1465.

Gang, S., Chen, X.-G., Zhao, Y.-K., Liu, M.-C., & Hu, Z., Determination of Pd(II) by application of an on-line microwave oven and artificial neural networks in flow injection analysis, *Analytica Chimica Acta*, vol. 420, no. 1, pp. 123–131.

Garcia, J. A., Arroyo, V., Sanchez, L., & Pino, J. A. 1999, Practical application of neural networks to predict DO concentration, *Progress in Water Resources, vol. 1 (Water Pollution V)*, pp. 331–340.

Garcia-Alvarez-Coque, M. C., Torres-Lapasio, J. R., & Baeza-Baeza, J. J. 1997, Modelling of retention behaviour of solutes in micellar liquid chromatography, *Journal of Chromatography A*, vol. 780, no. 1–2, pp. 129–148.

Garcia-Baez, P., Suarez-Araujo, C. P., Santana-Rodriguez, J. J., & Garcia, J. H. 1999, Spectrofluorometric identification of polychlorinated biphenyls using an unsupervised modular artificial neural network, *Biomedical Chromatography*, vol. 13, no. 2, pp. 181–183.

Garcia-Gonzalez, D. L. & Aparicio, R. 2003, Virgin olive oil quality classification combining neural network and MOS sensors, *Journal of Agricultural and Food Chemistry*, vol. 51, no. 12, pp. 3515–3519.

Garcia-Parrilla, M. C., Gonzalez, G. A., Heredia, F. J., & Troncoso, A. M. 1997, Differentiation of wine vinegars based on phenolic composition, *Journal of Agricultural and Food Chemistry*, vol. 45, no. 9, pp. 3487–3492.

Garcia-Ruiz, C., Jimenez, O., & Marina, M. L. 2003, Retention modeling and resolution optimization for a group of n-phenylpyrazole derivatives in micellar electrokinetic chromatography using empirical and physicochemical models, *Electrophoresis*, vol. 24, no. 3, pp. 325–335.

Garcia-Valls, R., Hrdlicka, A., Perutka, J., Havel, J., Deorkar, N. V., Tavlarides, L. L., Munoz, M., & Valiente, M. 2001, Separation of rare earth elements by high performance liquid chromatography using a covalent modified silica gel column, *Analytica Chimica Acta*, vol. 439, no. 2, pp. 247–253.

Garcia-Villar, N., Saurina, J., & Hernandez-Cassou, S. 2001, Potentiometric sensor array for the determination of lysine in feed samples using multivariate calibration methods, *Fresenius Journal of Analytical Chemistry*, vol. 371, no. 7, pp. 1001–1008.

Gardner, J. W., Hines, E. L., & Tang, H. C. 1992, Detection of vapors and odors form a multisensor array using pattern-recognition techniques. 2. Artificial neural networks, *Sensors and Actuators B-Chemical*, vol. 9, no. 1, pp. 9–15.

Gardner, J. W., Hines, E. L., Molinier, F., Bartlett, P. N., & Mottram, T. T. 1999, Prediction of health of dairy-cattle from breath samples using neural-network with parametric model of dynamic-response of array of semi-conducting gas sensors, *IEE Proceedings: Science, Measurements and Technologies*, vol. 146, no. 2, pp. 102–106.

Gardner, J. W., Shin, H. W., & Hines, E. L. 2000a, An electronic nose system to diagnose illness, *Sensors and Actuators B-Chemical*, vol. 70, no. 1–3, pp. 19–24.

Gardner, J. W., Shin, H. W., Hines, E. L., & Dow, C. S. 2000b, An electronic nose system for monitoring the quality of potable water, *Sensors and Actuators B-Chemical*, vol. 69, no. 3, pp. 336–341.

Gasteiger, J. & Zupan, J. 1993, Neural networks in chemistry, *Angewandte Chemie – International Edition in English*, vol. 32, no. 4, pp. 503–527.

Gasteiger, J., Li, X., Simon, V., Novic, M., & Zupan, J. 1993, Neural nets for mass and vibrational-spectra, *Journal of Molecular Structure*, vol. 292, pp. 141–159.

Geeraerd, A. H., Herremans, C. H., Cenens, C., & Van Impe, J. F. 1998, Application of artificial neural networks as a non-linear modular modeling technique to describe bacterial growth in chilled food products, *International Journal of Food Microbiology*, vol. 44, no. 1–2, pp. 49–68.

Gemperline, P. J. 1997, Rugged spectroscopic calibration for process control, *Chemometrics and Intelligent Laboratory Systems*, vol. 39, no. 1, pp. 29–40.

Gennaro, M. C., & Marengo, E. 2003 Strategies for optimization of chromatographic methods. Use of multivariate analysis, *Chimica e l'Industria*, vol. 85, no. 5, pp. 49–51.

Geng, L., Luo, A. Q., Fu, R. N., & Li, J. 2000, Identification of Chinese herbal medicine using artificial neural network in pyrolysis-gas chromatography, *Chinese Journal of Analytical Chemistry*, vol. 28, no. 5, pp. 549–553.

Gerbi, V., Zeppa, G., Beltramo, R., Carnacini, A., & Antonelli, A. 1998, Characterisation of white vinegars of different sources with artificial neural networks, *Journal of the Science of Food and Agriculture*, vol. 78, no. 3, pp. 417–422.

Gerdova, I. V., Dolenko, S. A., Dolenko, T. A., Persiantsev, I. G., Fadeev, V. V., & Churina, I. V. 2002, New opportunity solutions to inverse problems in laser spectroscopy involving artificial neural networks, *Izvestiya Akademii Nauk Seriya Fizicheskaya*, vol. 66, no. 8, pp. 1116–1124.

Getino, J., Ares, L., Robla, J. I., Del Carmen Horrillo, M., Sayago, I., Fernandez, M. J., Rodrigo, J., & Gutierrez, F. J. 1998, Gas sensor arrays for contaminated soils, *Proceedings of the 5th International Symposium on Olfaction and the Electronic Nose*, Baltimore, pp. 329–338.

Ghassempour, A., Vaezi, F., Salehpour, P., Nasiri-Aghdam, M., & Adrangui, M. 2002, Monitoring of enzymatic hydrolysis of penicillin G by pyrolysis-negative ion mass spectrometry, *Journal of Pharmaceutical and Biomedical Analysis*, vol. 29, no. 3, pp. 569–578.

Gibson, T. D., Prosser, O., Hulbert, J. N., Marshall, R. W., Corcoran, P., Lowery, P., Ruck-Keene, E. A., & Heron, S. 1997, Detection and simultaneous identification of microorganisms from headspace samples using an electronic nose, *Sensors and Actuators B-Chemical*, vol. 44, no. 1–3, pp. 413–422.

Goicoechea, H. C. & Olivieri, A. C. 2001, Sustained prediction ability of net analyte preprocessing methods using reduced calibration sets – theoretical and experimental study involving the spectrophotometric analysis of multicomponent mixtures, *Analyst*, vol. 126, no. 7, pp. 1105–1112.

Goicoechea, H. C., Collado, M. S., Satuf, M. L., & Olivieri, A. C. 2002a, Complementary use of partial least-squares and artificial neural networks for the nonlinear spectrophotometric analysis of pharmaceutical samples, *Analytical & Bioanalytical Chemistry*, vol. 374, no. 3, pp. 460–465.

Goicoechea, H. C. & Olivieri, A. C. 2002b, Chemometric assisted simultaneous spectrophotometric determination of 4-component nasal solutions with a reduced number of calibration samples, *Analytica Chimica Acta*, vol. 453, no. 2, pp. 289–300.

Gonzalez, G., Mendez, E. M. P., Sanchez, M. J. S., & Havel, J. 2000, Data evaluation for soft drink quality control using principal component analysis and back-propagation neural networks, *Journal of Food Protection*, vol. 63, no. 12, pp. 1719–1724.

Gonzalez Martin, Y., Cerrato Oliveros, M. C., Perez Pavon, J. L., Garcia Pinto, C., & Moreno Cordero, B. 2001, Electronic nose based on metal oxide semiconductor sensors and pattern recognition techniques: characterisation of vegetable oils, *Analytica Chimica Acta*, vol. 449, no. 1–2, pp. 69–80.

Goodacre, R. & Kell, D. B. 1993a, Rapid and quantitative-analysis of bioprocesses using pyrolysis mass-spectrometry and neural networks – application to indole production, *Analytica Chimica Acta*, vol. 279, no. 1, pp. 17–26.

Goodacre, R., Edmonds, A. N., & Kell, D. B. 1993b, Quantitative analysis of the pyrolysis mass-spectra of complex mixtures using artificial neural networks – application to amino-acids in glycogen, *Journal of Analytical and Applied Pyrolysis*, vol. 26, no. 2, pp. 93–114.

Goodacre, R., Kell, D. B., & Bianchi, G. 1993c, Rapid assessment of the adulteration of virgin olive oils by other seed oils using pyrolysis mass-spectrometry and artificial neural networks, *Journal of the Science of Food and Agriculture*, vol. 63, no. 3, pp. 297–307.

Goodacre, R., Neal, M. J., Kell, D. B., Greenham, L. W., Noble, W. C., & Harvey, R. G. 1994a, Rapid identification using pyrolysis mass-spectrometry and artificial neural networks of Propionibacterium acnes isolated from dogs, *Journal of Applied Bacteriology*, vol. 76, no. 2, pp. 124–134.

Goodacre, R., Karim, A., Kaderbhai, M. A., & Kell, D. B. 1994b, Rapid and quantitative-analysis of recombinant protein expression using pyrolysis mass-spectrometry and artificial neural networks – application to mammalian Cytochrome B(5) in Escherichia-Coli, *Journal of Biotechnology*, vol. 34, no. 2, pp. 185–193.

Goodacre, R., Trew, S., Wrigleyjones, C., Neal, M. J., Maddock, J., Ottley, T. W., Porter, N., & Kell, D. B. 1994c, Rapid screening for metabolite overproduction in fermentor broths, using pyrolysis mass-spectrometry with multivariate calibration and artificial neural networks, *Biotechnology and Bioengineering*, vol. 44, no. 10, pp. 1205–1216.

Goodacre, R., Neal, M. J., & Kell, D. B. 1994d, Rapid and quantitative-analysis of the pyrolysis mass-spectra of complex binary and tertiary mixtures using multivariate calibration and artificial neural networks, *Analytical Chemistry*, vol. 66, no. 7, pp. 1070–1085.

Goodacre, R., Trew, S., Wrigleyjones, C., Saunders, G., Neal, M. J., Porter, N., & Kell, D. B. 1995, Rapid and quantitative-analysis of metabolites in fermenter broths using pyrolysis mass-spectrometry with supervised learning – application to the screening of Penicillium-Chrysogenum fermentations for the overproduction of Penicillins, *Analytica Chimica Acta*, vol. 313, no. 1–2, pp. 25–43.

Goodacre, R., Hiom, S. J., Cheeseman, S. L., Murdoch, D., Weightman, A. J., & Wade, W. G. 1996a, Identification and discrimination of oral asaccharolytic Eubacterium spp by pyrolysis mass spectrometry and artificial neural networks, *Current Microbiology*, vol. 32, no. 2, pp. 77–84.

Goodacre, R. & Kell, D. B. 1996b, Correction of mass spectral drift using artificial neural networks, *Analytical Chemistry*, vol. 68, no. 2, pp. 271–280.

Goodacre, R., Pygall, J., & Kell, D. B. 1996c, Plant seed classification using pyrolysis mass spectrometry with unsupervised learning: the application of auto-associative and Kohonen artificial neural networks, *Chemometrics and Intelligent Laboratory Systems*, vol. 34, no. 1, pp. 69–83.

Goodacre, R., Howell, S. A., Noble, W. C., & Neal, M. J. 1996d, Sub-species discrimination, using pyrolysis mass spectrometry and self-organising neural networks, of Propionibacterium acnes isolated from normal human skin, *Zentralblatt fur Bakteriologie-International Journal of Medical Microbiology Virology Parasitology and Infectious Diseases*, vol. 284, no. 4, pp. 501–515.

Goodacre, R., Rischert, D. J., Evans, P. M., & Kell, D. B. 1996e, Rapid authentication of animal cell lines using pyrolysis mass spectrometry and auto-associative artificial neural networks, *Cytotechnology*, vol. 21, no. 3, pp. 231–241.

Goodacre, R., Neal, M. J., & Kell, D. B. 1996f, Quantitative analysis of multivariate data using artificial neural networks: a tutorial review and applications to the deconvolution of pyrolysis mass spectra, *Zentralblatt für Bakteriologie – International Journal of Medical Microbiology Virology Parasitology and Infectious Diseases*, vol. 284, no. 4, pp. 516–539.

Goodacre, R., Hammond, D., & Kell, D. B. 1997a, Quantitative analysis of the adulteration of orange juice with sucrose using pyrolysis mass spectrometry and chemometrics, *Journal of Analytical and Applied Pyrolysis*, vol. 40–1, pp. 135–158.

Goodacre, R. 1997b, Use of pyrolysis mass spectrometry with supervised learning for the assessment of the adulteration of milk of different species, *Applied Spectroscopy*, vol. 51, no. 8, pp. 1144–1153.

Goodacre, R., Timmins, E. M., Jones, A., Kell, D. B., Maddock, J., Heginbothom, M. L., & Magee, J. T. 1997c, On mass spectrometer instrument standardization and interlaboratory calibration transfer using neural networks, *Analytica Chimica Acta*, vol. 348, no. 1–3, pp. 511–532.

Goodacre, R., Harvey, R., Howell, S. A., Greenham, L. W., & Noble, W. C. 1997d, An epidemiological study of Staphylococcus intermedius strains isolated from dogs, their owners and veterinary surgeons, *Journal of Analytical and Applied Pyrolysis*, vol. 44, no. 1, pp. 49–64.

Goodacre, R., Rooney, P. J., & Kell, D. B. 1998a, Discrimination between methicillin-resistant and methicillin-susceptible Staphylococcus aureus using pyrolysis mass spectrometry and artificial neural networks, *Journal of Antimicrobial Chemotherapy*, vol. 41, no. 1, pp. 27–34.

Goodacre, R., Timmins, E. M., Burton, R., Kaderbhai, N., Woodward, A. M., Kell, D. B., & Rooney, P. J. 1998b, Rapid identification of urinary tract infection bacteria using hyperspectral whole-organism fingerprinting and artificial neural networks, *Microbiology-UK*, vol. 144, pp. 1157–1170.

Goodacre, R. & Gilbert, R. J. 1999, The detection of caffeine in a variety of beverages using Curie-point pyrolysis mass spectrometry and genetic programming, *Analyst*, vol. 124, no. 7, pp. 1069–1074.

Goodacre, R., Shann, B., Gilbert, R. J., Timmins, E. M., McGovern, A. C., Alsberg, B. K., Kell, D. B., & Logan, N. A. 2000, Detection of the dipicolinic acid biomarker in Bacillus spores using Curie-point pyrolysis mass spectrometry and Fourier transform infrared spectroscopy, *Analytical Chemistry*, vol. 72, no. 1, pp. 119–127.

Goodacre, R., Vaidyanathan, S., Bianchi, G., & Kell, D. B. 2002a, Metabolic profiling using direct infusion electrospray ionisation mass spectrometry for the characterisation of olive oils, *Analyst*, vol. 127, no. 11, pp. 1457–1462.

Goodacre, R., Radovic, B. S., & Anklam, E. 2002b, Progress toward the rapid nondestructive assessment of the floral origin of European honey using dispersive Raman-spectroscopy, *Applied Spectroscopy*, vol. 56, no. 4, pp. 521–527.

Goodacre, R., York, E. V., Heald, J. K., & Scott, I. M. 2003a, Chemometric discrimination of unfractionated plant extracts analyzed by electrospray mass spectrometry, *Phytochemistry*, vol. 62, no. 6, pp. 859–863.

Goodacre, R. 2003b, Explanatory analysis of spectroscopic data using machine learning of simple, interpretable rules, *Vibrational Spectroscopy*, vol. 32, no. 1, pp. 33–45.

Goodfellow, M., Freeman, R., & Sisson, P. R. 1997, Curie-point pyrolysis mass spectrometry as a tool in clinical microbiology, *Zentralblatt für Bakteriologie – International Journal of Medical Microbiology Virology Parasitology and Infectious Diseases*, vol. 285, no. 2, pp. 133–156.

Gordon, S. H., Wheeler, B. C., Schudy, R. B., Wicklow, D. T., & Greene, R. V. 1998, Neural network pattern recognition of photoacoustic FTIR spectra and knowledge-based techniques for detection of mycotoxigenic fungi in food grains, *J.Food Prot.*, vol. 61, no. 2, pp. 221–230.

Gordon, S., Hammond, R., Roberts, K., Savelli, N., & Wilkinson, D. 2000, In-process particle characterization by spectral extinction, *Transactions of the Institute of Chemical Engineers*, vol. 78, Part A, pp. 1147–1152.

Gorodkin, J., Sogaard, B., Bay, H., Doll, H., Kolster, P., & Brunak, S. 2001, Recognition of environmental and genetic effects on barley phenolic fingerprints by neural networks, *Computers & Chemistry*, vol. 25, no. 3, pp. 301–307.

Groves, W. A., & Zellers, E. T. 2001, Analysis of solvent vapors in breath and ambient air with a surface acoustic wave sensor array, *Annals of Occupational Hygiene*, vol. 45, no. 8, pp. 609–623.

Grus, F. H., Augustin, A. J., & Zimmermann, C. W. 1997, Computer-supported analysis (MegaBlot) of allopurinol-induced changes in the autoantibody repertoires of rats suffering front experimental lens-induced uveitis, *Electrophoresis*, vol. 18, no. 3–4, pp. 516–519.

Grus, F. H., Sabuncuo, P., & Augustin, A. J. 2001, Quantitative analysis of tear film changes in soft contact lens wearers – a clinical study, *Klinische Monatsblatter für Augenheilkunde*, vol. 218, no. 4, pp. 239–242.

Guernion, N., Ratcliffe, N. M., Spencer-Phillips, P. T. N., & Howe, R. A. 2001, Identifying bacteria in human urine: current practice and the potential for rapid, near-patient diagnosis by sensing volatile organic compounds, *Clinical Chemistry and Laboratory Medicine*, vol. 39, no. 10, pp. 893–906.

Guiberteau Cabanillas, A., Galeano Diaz, T., Mora Diez, N. M., Salinas, F., Ortiz Burguillos, J. M., & Vire, J.-C. 2000, Resolution by polarographic techniques of atrazine-simazine and terbutryn-prometryn binary mixtures by using PLS calibration and artificial neural networks *Analyst*, vol. 125, no. 5, pp. 909–914.

Guiberteau, A., Galeano, T., Mora, N., Salinas, F., Ortiz, J. M., & Vire, J. C. 2001, Resolution by polarographic techniques of the ternary mixture of captan, captafol and folpet by using PLS calibration and artificial neuronal networks, *Computers & Chemistry*, vol. 25, no. 5, pp. 459–473.

Guterman, H., 1994, Application of principal component analysis to the design of neural networks, *Neural Parallel & Scientific Computing*, vol. 2, pp. 43–53.

Guo, W., Lu, Y., & Zheng, X. M. 2000a, The predicting study for chromatographic retention index of saturated alcohols by MLR and ANN, *Talanta*, vol. 51, no. 3, pp. 479–488.

Guo, W. Q. & Brodowsky, H. 2000b, Determination of the trace 1,4-dioxane, *Microchemical Journal*, vol. 64, no. 2, pp. 173–179.

Guo, W. Q., Lu, Y., & Zheng, X. M. 2001, The application of artificial neural networks in the study of quantitative structure-retention relationships for saturated alcohols, *Chinese Journal of Analytical Chemistry*, vol. 29, no. 4, pp. 416–420.

Guwy, A. J., Hawkes, F. R., Wilcox, S. J., & Hawkes, D. L. 1997, Neural network and on-off control of bicarbonate alkalinity in a fluidised-bed anaerobic digester, *Water Research*, vol. 31, no. 8, pp. 2019–2025.

Hadjiiski, L., Geladi, P., & Hopke, P. 1999, A comparison of modeling nonlinear systems with artificial neural networks and partial least squares, *Chemometrics and Intelligent Laboratory Systems*, vol. 49, pp. 91–103.

Hajek, M., & Judd, M. R. 1995, Use of neural networks in modeling the interactions between gas analyzers at coal gasifiers, *Fuel*, vol. 74, no. 9, pp. 1347–1351.

Hajmeer, M. N., Basheer, I. A., & Najjar, Y. M. 1997, Computational neural networks for predictive microbiology. 2. Application to microbial growth, *International Journal of Food Microbiology*, vol. 34, no. 1, pp. 51–66.

Hajmeer, M. N., Basheer, I. A., Fung, D. Y. C., & Marsden, J. L. 1998, Nonlinear response surface model based on artificial neural networks for growth of Saccharomyces cerevisiae, *Journal of Rapid Methods and Automation in Microbiology*, vol. 6, no. 2, pp. 103–118.

Hammer, C. L., Small, G. W., Combs, R. J., Knapp, R. B., & Kroutil R. T. 2000, Artificial neural networks for the automated detection of trichloroethylene by passive Fourier transform infrared spectrometry, *Analytical Chemistry*, vol. 72, no. 7, pp. 1680–1689.

Hanaki, S., Nakamoto, T., & Moriizumi, T. 1996, Artificial odor-recognition system using neural-network for estimating sensory quantities of blended fragrance, *Sensors and Actuators A-Physical*, vol. 57, no. 1, pp 65–71.

Hanley, J. A., & McNeil, B. J. 1982, The meaning and the use of the area under a Receiving Operating Characteristics (ROC) curve, *Diagnostic Radiology*, vol. 143, no. 1, pp. 29–36.

Harrington, P. D. 1993, Minimal neural networks – concerted optimization of multiple decision planes, *Chemometrics and Intelligent Laboratory Systems*, vol. 18, no. 2, pp. 157–170.

Harrington, P. D. 1994a, Temperature-constrained backpropagation neural networks, *Analytical Chemistry*, vol. 66, no. 6, pp. 802–807.

Harrington, P. D. 1994b, Quantitative analysis of volatile organic compounds using ion mobility spectra and cascade correlation neural networks, *Proceedings of the 3rd International Workshop on Ion Mobility Spectroscopy*, Galveston, Texas.

Harrington, P. D. 1998, Temperature-constrained cascade correlation networks, *Analytical Chemistry*, vol. 70, no. 7, pp. 1297–1306.

Harrington, P. D., Urbas, A., & Wan, C. H. 2000, Evaluation of neural-network models with generalized sensitivity analysis, *Analytical Chemistry*, vol. 72, no. 20, pp. 5004–5013.

Harrington, P. D., Voorhees, K. J., Basile, F., & Hendricker, A. D. 2002, Validation using sensitivity and target transform factor analyses of neural network models for classifying bacteria from mass spectra, *Journal of the American Society for Mass Spectrometry*, vol. 13, no. 1, pp. 10–21.

Havel, J., Pena, E. M., Rojas-Hernandez, A., Doucet, J. P., & Panaye, A. 1998, Neural networks for optimization of high-performance capillary zone electrophoresis methods – a new method using a combination of experimental design and artificial neural networks, *Journal of Chromatography A*, vol. 793, no. 2, pp. 317–329.

Havel, J., Breadmore, M., Macka, M., & Haddad, P. R. 1999a, Artificial neural networks for computer-aided modelling and optimisation in micellar electrokinetic chromatography, *Journal of Chromatography A*, vol. 850, no. 1–2, pp. 345–353.

Havel, J., Madden, J. E., & Haddad, P. R. 1999b, Prediction of retention times for anions in ion chromatography using artificial neural networks, *Chromatographia*, vol. 49, no. 9–10, pp. 481–488.

Havel, J., Soto-Guerrero, J., & Lubal, P. 2002a, Spectrophotometric study of uranyl-oxalate complexation in solution, *Polyhedron*, vol. 21, no. 14–15, pp. 1411–1420.

Havel, J., Lubal, P., & Farkova, M. 2002b, Evaluation of chemical equilibria with the use of artificial neural networks, *Polyhedron*, vol. 21, no. 14–15, pp. 1375–1384.

Havel, J., Kanicky, V., & Lubal, P. 2003, Spectrophotometric study of uranium(VI) complex equilibria with N,N-dimethylformamide, *Journal of Radioanalytical and Nuclear Chemistry*, vol. 257, no. 2, pp. 391–398.

Havlis, J., Madden, J. E., Revilla, A. L., & Havel, J. 2001, High-performance liquid chromatographic determination of deoxycytidine monophosphate and methyldeoxycytidine monophosphate for DNA demethylation monitoring: experimental design and artificial neural networks optimisation, *Journal of Chromatography B*, vol. 755, no. 1–2, pp. 185–194.

He, H. Q., Xu, G. X., Ye, X. S., & Wang, P. 2003a, A novel chemical image sensor consisting of integrated microsensor array chips and pattern-recognition, *Measurement Science and Technology*, vol. 14, no. 7, pp. 1040–1046.

He, L.-M., Kear-Padilla, L. L., Lieberman, S. H., & Andrews, J. M. 2001, New generation of online oil-in-water monitor, *Abstracts of Papers – American Chemical Society*, no. 221, ENVR-005.

He, L. M., Kear-Padilla, L. L., Lieberman, S. H., & Andrews, J. M. 2003b, Rapid in situ determination of total oil concentration in water using ultraviolet fluorescence and light scattering coupled with artificial neural networks, *Analytica Chimica Acta*, vol. 478, no. 2, pp. 245–258.

Heberger, K. & Borosy, A. P. 1999, Comparison of chemometric methods for prediction of rate constants and activation energies of radical addition reactions, *Journal of Chemometrics*, vol. 13, no. 3–4, pp. 473–489.

Henkel, K. & Schmeisser, D. 2002, Back-propagation-based neural-network with a 2 sensor system for monitoring carbon-dioxide and relative-humidity, *Analytical & Bioanalytical Chemistry*, vol. 374, no. 2, pp. 329–337.

Herrador, M. A., & Gonzalez, A. G. 2001, Pattern recognition procedures for differentiation of Green, Black and Oolong teas according to their metal content from inductively coupled plasma atomic emission spectrometry, *Talanta*, vol. 53, no. 6, pp. 1249–1257.

Hernandez-Caraballo, E. A., Avila-Gomez, R. M., Capote, T., Rivas, F., & Perez, A. G. 2003, Classification of Venezuelan spirituous beverages by means of discriminant analysis and artificial neural networks based on their Zn, Cu and Fe concentrations, *Talanta*, vol. 60, no. 6, pp. 1259–1267.

Hervas, C., Ventura, S., Silva, M., & Perez-Bendito, D. 1998, Computational neural networks for resolving nonlinear multicomponent systems based on chemiluminescence methods, *Journal of Chemical Information and Computer Sciences*, vol. 38 no. 6, pp. 1119–1124.

Hervas, C., Algar, J. A., & Silva, M. 2000, Correction of temperature variations in kinetic-based determinations by use of pruning computational neural networks in conjunction with genetic algorithms, *Journal of Chemical Information and Computer Sciences*, vol. 40, no. 3, pp. 724–731.

Hervas, C., Toledo, R., & Silva, M. 2001, Use of pruned computational neural networks for processing the response of oscillating chemical reactions with a view to analyzing nonlinear multicomponent mixtures, *Journal of Chemical Information and Computer Sciences*, vol. 41, no. 4, pp. 1083–1092.

Herber, S., Grus, F. H., Sabuncuo, P., & Augustin, A. J. 2001, 2-Dimensional analysis of tear protein-patterns of diabetic patients, *Electrophoresis*, vol. 22, no. 9, pp. 1838–1844.

Hilder, E. F., Klampfl, C. W., & Haddad, P. R. 2000, Pressurized-flow anion-exchange capillary electrochromatography using a polymeric ion-exchange stationary phase, *Journal of Chromatography A*, vol. 890, no. 2, pp. 337–345.

Hill, B. D., Jones, S. D. M., Robertson, W. M., & Major, I. T. 2000, Neural network modeling of carcass measurements to predict beef tenderness, *Canadian Journal of Animal Science*, vol. 80, no. 2, pp. 311–318.

Hitzmann, B. & Kullick, T. 1994, Evaluation of pH field-effect transistor measurement signals by neural networks, *Analytica Chimica Acta*, vol. 294, no. 3, pp. 243–249.

Hitzmann, B., Ritzka, A., Ulber, R., Scheper, T., & Schugerl, K. 1997, Computational neural networks for the evaluation of biosensor FIA measurements, *Analytica Chimica Acta*, vol. 348, no. 1–3, pp. 135–141.

Hobson, R. S., Clausi, A., Oh, T., & Guiseppi-Elie, A. 2003, Temperature correction to chemoresistive sensors in an e-NOSE-ANN system, *IEEE Sensors Journal*, vol. 3, no. 4, pp. 484–489.

Hoffheins, B. S., & Lauf, R. J. 1992, Performance of simplified chemical sensor arrays in neural network-based analytical instrument, *Analusis*, vol. 20, pp. 201–207.

Hong, H. K., Shin, H. W., Yun, D. H., Kim, S. R., Kwon, C. H., Lee, K., & Moriizumi, T. 1996, Electronic nose system with micro gas sensor array, *Sensors and Actuators B-Chemical*, vol 36, no. 1–3, pp 338–341.

Hopke, P. K. 2003, The evolution of chemometrics, *Analytica Chimica Acta*, vol. 500, pp. 365–377.

Horimoto, Y., Lee, K., & Nakai, S. 1997, Classification of microbial defects in milk using a dynamic headspace gas chromatograph and computer-aided data processing. 2. Artificial neural networks, partial least-squares regression analysis, and principal component regression analysis, *Journal of Agricultural and Food Chemistry*, vol. 45, no. 3, pp. 743–747.

Hornik, K., Stinchcombe, M. & White, H. 1989, Multilayer feedforward networks are universal approximators, *Neural Networks*, vol. 2, pp. 359–366.

Hsueh, C.-C., Liu, Y., Henry, M., & Freund, M. S. 1999, Chemically diverse modified electrodes: a new approach to the design and implementation of sensor arrays, *Analytica Chimica Acta*, vol. 397, no. 1–3, pp. 135–144.

Hu, Y., Zhou, G., Kang, J., Du, Y., Huang, F., & Ge, J. 1996, Assessment of chromatographic peak purity by means of artificial neural networks, *Journal of Chromatography A*, vol. 734, no. 2, pp. 259–270.

Huang, Y., Cavinato, A. G., Mayes, D. M., Kangas, L. J., Bledsoe, G. E., & Rasco, B. A. 2003, Nondestructive determination of moisture and sodium chloride in cured Atlantic salmon (Salmo salar) (Teijin) using short-wavelength near-infrared spectroscopy (SW-NIR), *Journal of Food Science*, vol. 68, no. 2, pp. 482–486.

Hudon, G., Guy, C., & Hermia, J. 2000, Measurement of odor intensity by an electronic nose, *Journal of the Air & Waste Management Association*, vol. 50, no. 10, pp. 1750–1758

Irudayaraj, J., Yang, H., & Sakhamuri, S. 2002, Differentiation and detection of microorganisms using Fourier-transform infrared photoacoustic-spectroscopy, *Journal of Molecular Structure*, vol. 606, no. 1–3, pp 181–188.

Irudayaraj, J., Xu, F., & Tewari, J. 2003, Rapid determination of invert cane sugar adulteration in honey using FTIR spectroscopy and multivariate analysis, *Journal of Food Science*, vol. 68, no. 6, pp. 2040–2045.

Izquierdo, A., Lopez-Cueto, G., Medina, J. F. R., & Ubide, C. 1998, Simultaneous determination of niobium and tantalum with 4-(2-pyridylazo) resorcinol using partial least squares regression and artificial neural networks, *Quimica Analitica*, vol. 17, no. 2, pp. 67–74.

Jacobsen, S. P., & Hagman, A. 1993, Chemical composition analysis of carrageenans by infrared spectroscopy using partial least square and neural networks, *Analytica Chimica Acta*, vol. 284, pp. 137–147.

Jacobsen, S., Nesic, L., Petersen, M., & Sondergaard, I. 2001, Classification of wheat-varieties – use of 2-dimensional gel-electrophoresis for varieties that cannot be classified by matrix-assisted laser desorption/ionization-time of flight-mass spectrometry and an artificial neural-network, *Electrophoresis*, vol. 22, no. 6, pp. 1242–1245.

Jakus, V. 1992, Artificial intelligence in chemistry, *Collection of Czechoslovak Chemical Communications*, vol. 57, no. 12, pp. 2413–2451.

Jakus, V. 1993, Utilization of artificial intelligence in chemistry, *Chemicke Listy*, vol. 87, no. 4, pp. 262–279.

Jalali-Heravi, M. & Fatemi, M. H. 1998, Prediction of flame ionization detector response factors using an artificial neural network, *Journal of Chromatography A*, vol. 825, no. 2, pp. 161–169.

Jalali-Heravi, M. & Fatemi, M. H. 2000a, Simulation of mass spectra of noncyclic alkanes and alkenes using artificial neural network, *Analytica Chimica Acta*, vol. 415, no. 1–2, pp. 95–103.

Jalali-Heravi, M. & Fatemi, M. H. 2000b, Prediction of thermal-conductivity detection response factors using an artificial neural-network, *Journal of Chromatography A*, vol. 897, no. 1–2, pp. 227–235.

Jalali-Heravi, M. & Parastar, F. 2000c, Development of comprehensive descriptors for multiple linear-regression and artificial neural-network modeling of retention behaviors of a variety of compounds on different stationary phases, *Journal of Chromatography A*, vol. 903, no. 1–2, pp. 145–154.

Jalali-Heravi, M. & Garkani-Nejad, Z. 2001a, Prediction of electrophoretic mobilities of sulfonamides in capillary zone electrophoresis using artificial neural networks, *Journal of Chromatography A*, vol. 927, no. 1–2, pp. 211–218.

Jalali-Heravi, M. & Fatemi, M. H. 2001b, Artificial neural network modeling of Kovats retention indices for noncyclic and monocyclic terpenes, *Journal of Chromatography A*, vol. 915, no. 1–2, pp. 177–183.

Jalali-Heravi, A. & Garkani-Nejad, Z. 2002a, Prediction of electrophoretic mobilities of alkyl- and alkenylpyridines in capillary electrophoresis using artificial neural networks, *Journal of Chromatography A*, vol. 971, no. 1–2, pp. 207–215.

Jalali-Heravi, M. & Garkani-Nejad, Z. 2002b, Prediction of relative response factors for flame ionization and photoionization detection using self-training artificial neural networks, *Journal of Chromatography A*, vol. 950, no. 1–2, pp. 183–194.

Jalali-Heravi, M. & Garkani-Nejad, Z. 2002c, Use of self-training artificial neural networks in modeling of gas-chromatographic relative retention times of a variety of organic-compounds, *Journal of Chromatography A*, vol. 945, no. 1–2, pp. 173–184.

Jarvis, B. W., Desfieux, J., Jiminez, J., & Martinez, D. 2003, Quantification of gas concentrations in mixtures of known gases using an array of different tin-oxide sensors, *IEE Proceedings: Science, Measurements and Technologies*, vol. 150, no. 3, pp. 97–106.

Jensen, K., Tygesen, T. K., Kesmir, C., Skovgaard, I. M., & Sondergaard, I. 1997, Classification of potato varieties using isoelectrophoretic focusing patterns, neural nets, and statistical methods, *Journal of Agricultural and Food Chemistry*, vol. 45, no. 1, pp. 158–161.

Jeyamkondan, S., Jayas, D. S., & Holley, R. A. 2001, Microbial growth modelling with artificial neural networks, *International Journal of Food Microbiology*, vol. 64, no. 3, pp. 343–354.

Jimenez, O. & Marina, M. L. 1997a, Retention modeling in micellar liquid chromatography, *Journal of Chromatography A*, vol. 780, no. 1–2, pp. 149–163.

Jimenez, O., Benito, I., & Marina, M. L. 1997b, Neural networks as a tool for modelling the retention behaviour of dihydropyridines in micellar liquid chromatography, *Analytica Chimica Acta*, vol. 353, no. 2–3, pp. 367–379.

Jimenez-Prieto, R., & Silva, M. 1999, The continuous addition of reagent technique as an effective tool for enhancing kinetic-based multicomponent determinations using computational neural networks, *Analytica Chimica Acta*, vol. 389, pp. 131–139.

Johnson, K. J. & Synovec, R. E. 2002, Pattern recognition of jet fuels: comprehensive GC × GC with ANOVA-based feature selection and principal component analysis, *Chemometrics and Intelligent Laboratory Systems*, vol. 60, no. 1–2, pp. 225–237.

Johnson, S. R., Sutter, J. M., Engelhardt, H. L., Jurs, P. C., White, J., Kauer, J. S., Dickinson, T. A., & Walt, D. R. 1997, Identification of multiple analytes using an optical sensor array and pattern-recognition neural networks, *Analytical Chemistry*, vol. 69, no. 22, pp 4641–4648.

Joo, B.-S., Lee, S.-M., Lee, Y.-S., Choi, N.-J., Jung, J.-K., Yu, J.-B., Lee, M.-H., Byun, H.-G., Huh, J.-S., & Lee, D.-D. 2001, Pattern classification of gas sensor array with online retrainable RBF network, *Chemical Sensors*, vol. 17 (Suppl. B), pp. 197–199.

Kalelkar, S., Dow, E. R., Grimes, J., Clapham, Ma., & Hu, H. 2002, Automated analysis of proton NMR spectra from combinatorial rapid parallel synthesis using self-organizing maps, *Journal of Combinatorial Chemistry*, vol. 4, no. 6, pp. 622–629.

Kaliszan, R., Baczek, T., Bucinski, A., Buszewski, B., & Sztupecka, M. 2003, Prediction of gradient retention from the linear solvent strength (LSS) model, quantitative structure-retention relationships (QSRR), and artificial neural networks (ANN), *Journal of Separation Science*, vol. 26, no. 3–4, pp. 271–282.

Kalny, D., Albrecht-Gary, A. M., & Havel, J. 2001, Highly sensitive method for Palladium(II) determination as a porphyrinato complex by capillary zone electrophoresis, *Analytica Chimica Acta*, vol. 439, no. 1, pp. 101–105.

Kang, S. G., Kenyon, R. G. W., Ward, A. C., & Lee, K. J. 1998a, Analysis of differentiation state in Streptomyces albidoflavus SMF301 by the combination of pyrolysis mass spectrometry and neural networks, *Journal of Biotechnology*, vol. 62, no. 1, pp. 1–10.

Kang, S. G., Lee, D. H., Ward, A. C., & Lee, K. J. 1998b, Rapid and quantitative analysis of clavulanic acid production by the combination of pyrolysis mass spectrometry and artificial neural network, *Journal of Microbiology and Biotechnology*, vol. 8, no. 5, pp. 523–530.

Kapur, G. S., Sastry, M.I.S., Jaiswal, A.K., & Sarpal, A.S. 2004, Establishing structure–property correlations and classification of base oils using statistical techniques and artificial neural networks, *Analytica Chimica Acta*, vol. 506, no. 1, pp. 57–69.

Karpas, Z., & Boger, Z. 1993, Ion mobility spectrometry interpretation by neural networks, *Proceedings of the Workshop on Ion Mobility Spectrometry*, New Mexico State University, pp. 35–39.

Karul, C., Soyupak, S., & Yurteri, C. 1999, Neural network models as a management tool in lakes, *Hydrobiologia*, vol. 409, pp. 139–144.

Kateman, G. & Smits, J. R. M. 1993a, Colored information from a black-box – validation and evaluation of neural networks, *Analytica Chimica Acta*, vol. 277, no. 2, pp. 179–188.

Kateman, G. 1993b, Neural networks in analytical chemistry, *Chemometrics and Intelligent Laboratory Systems*, vol. 19, no. 2, pp. 135–142.

Kato, K., Kato, Y., Takamatsu, K., Udaka, T., Nakahara, T., Matsuura, Y., & Yoshikawa, K. 2000, Towards the realization of an intelligent gas sensing system utilizing a non-linear dynamic response, *Sensors and Actuators B-Chemical*, vol. 71, pp. 192–196.

Keller, P. E., Kouzes, R. T., & Kangas, L. J. 1994, Neural network based chemical sensor systems for environmental monitoring, *Proceedings of the World Congress on Neural Networks,* San Diego, vol. 1, pp. 269–272.

Keller, P. E. 1999, Overview of electronic nose algorithms, *Proceedings of the International Joint Conference on Neural Networks,* Washington, DC, vol. 1, pp. 309–312.

Kenyon, R. G. W., Ferguson, E. V., & Ward, A. C. 1997, Application of neural networks to the analysis of pyrolysis mass spectra, *Zentralblatt für Bakteriologie – International Journal of Medical Microbiology Virology Parasitology and Infectious Diseases*, vol. 285, no. 2, pp. 267–277.

Kim, J., Mowat, A., Poole, P., & Kasabov, N. 2000, Linear and non-linear pattern recognition models for classification of fruit from visible near-infrared spectra, *Chemometrics and Intelligent Laboratory Systems*, vol. 51, pp. 201–216.

Kim, J.-D., Byun, H.-G, Ham, Y.-K., Lee, J.-S., Shon, W.-R., Heo, N.-U., 2002, Neural network algorithm for VOC gas recognition and concentration estimation at field screening using e-nose, *Proceedings of 2001 International Conference on Control, Automation and Systems*, Korea, pp. 1823–1826.

King, D., Lyons, W. B., Flanagan, C., & Lewis, E. 2003, An optical fibre ethanol concentration sensor utilizing Fourier transform signal processing analysis and artificial neural network pattern recognition, *Journal of Optics A: Pure and Applied Optics*, vol. 5, no. 4, pp. S69–S75.

Klawun, C. & Wilkins, C. L. 1994, A novel algorithm for local minimum escape in backpropagation neural networks – application to the interpretation of matrix-isolation infrared-spectra, *Journal of Chemical Information and Computer Sciences*, vol. 34, no. 4, pp. 984–993.

Klawun, C. & Wilkins, C. L. 1995, Neural network assisted rapid screening of large infrared spectral databases, *Analytical Chemistry*, vol. 67, pp. 374–378.

Klawun, C. & Wilkins, C. L. 1996a, Optimization of functional-group prediction from infrared-spectra using neural networks, *Journal of Chemical Information and Computer Sciences*, vol. 36, no. 1, pp. 69–81.

Klawun, C. & Wilkins, C. L. 1996b, Joint neural network interpretation of infrared and mass spectra, *Journal of Chemical Information and Computer Sciences*, vol. 36, no. 2, pp. 249–257.

Kocjancic, R. & Zupan, J. 1997, Application of a feed-forward artificial neural network as a mapping device, *Journal of Chemical Information and Computer Sciences*, vol. 37, no. 6, pp. 985–989.

Koehler, F. W. IV, Small, G. W., Combs, R. J., Knapp, R. B., & Krouti 2000, Calibration transfer algorithm for automated qualitative analysis by passive Fourier transform infrared spectrometry, *Analytical Chemistry*, vol. 72, no. 7, pp. 1690–1698.

Kolehmainen, M., Ronkko, P., & Raatikainen, A. 2003, Monitoring of yeast fermentation by ion mobility spectrometry measurement and data visualization with self-organizing maps, *Analytica Chimica Acta*, vol. 484, no. 1, pp. 93–100.

Kompany-Zareh, M., Tavallali, H., & Sajjadi, M. 2002a, Application of generalized artificial neural networks coupled with an orthogonal design to optimization of a system for the kinetic spectrophotometric determination of Hg(II), *Analytica Chimica Acta*, vol. 469, no. 2, pp. 303–310.

Kompany-Zareh, M., Mansourian, M., & Ravaee, F. 2002b, Simple method for colorimetric spot-test quantitative analysis of Fe(III) using a computer controlled hand-scanner, *Analytica Chimica Acta*, vol. 471, no. 1, pp. 97–104.

Kratochwil, N. A., Huber, W., Muller, F., Kansy, M., & Gerber, P. R. 2002, Predicting plasma protein binding of drugs: a new approach, *Biochemical Pharmacology*, vol. 64, no. 9, pp. 1355–1374.

Kunt, T., Mc Avoy, T. J., Cavicchi, R. E. & Semancik, S. 1997, Dynamic modeling and optimization of micro-hotplate chemical gas sensors, *Proceedings of the IFAC ADCHEM Conference*, Banff, Canada, pp. 91–95.

Kurbatova, S. V., Yashkin, S. N., Moiseev, I. K., & Zemtsova, M. N. 1999, Gas chromatography of alkyladamantanes, *Russian Journal of Physical Chemistry*, vol. 73, no. 9, pp. 1477–1481.

Kushalappa, A. C., Lui, L. H., Chen, C. R., & Lee, B. 2002, Volatile fingerprinting (SPME-GC-FID) to detect and discriminate diseases of potato tubers, *Plant Disease*, vol. 86, no. 2, pp. 131–137.

Kuzmanovski, I., Zografski, Z., Trpkovska, M., Soptrajanov, B., & Stefov, V. 2001, Simultaneous determination of composition of human urinary calculi by use of artificial neural networks, *Fresenius Journal of Analytical Chemistry*, vol. 370, no. 7, pp. 919–923.

Kuzmanovski, I., Trpkovska, M., Soptrajanov, B., & Stefov, V. 2003, Determination of the composition of human urinary calculi composed of whewellite, weddellite and carbonate apatite using artificial neural networks, *Analytica Chimica Acta*, vol. 491, no. 2, pp. 211–218.

Kuzmanovski, I., Ristova, M., Soptrajanov, B., Stefov, V., & Popovski, V. 2004, Determination of the composition of sialoliths composed of carbonate apatite and albumin using artificial neural networks, *Talanta*, vol. 62, no. 4, pp. 813–817.

Kuznetsov, V. V., & D'yakov, A. N. 1999, Estimating parameters of optical membrane pH-sensors using artificial neural networks, *Journal of Analytical Chemistry*, vol. 54, no. 11, pp. 1076–1081.

Kvasnicka, V., Sklenak, S., & Pospichal, J. 1992, Application of recurrent neural networks in chemistry. Prediction and classification of ^{13}C NMR chemical shifts in a series of monosubstituted benzenes, *Journal of Chemical Information and Computer Sciences*, vol. 32, pp. 742–747.

Larsen, J. T., Morgan, W. L. & Goldstein W. H. 1992, Artificial neural networks for plasma x-ray spectroscopic analysis, *Review of Scientific Instruments*, vol. 63, pp. 4775–4777.

Lastres, E., de Armas, G., Catasus, M., Alpizar, J., Garcia, L., & Cerda, V. 1997, Use of neural networks in solving interferences caused by formation of intermetallic compounds in anodic-stripping voltammetry, *Electroanalysis*, vol. 9, no. 3, pp. 251–254.

Latorre, R. M., Hernandez-Cassou, S., & Saurina, J. 2001, Artificial neural networks for quantification in unresolved capillary electrophoresis peaks, *Journal of Separation Science*, vol. 24, no. 6, pp. 427–434.

Lau, K.-T., McAlernon, P., & Slater, J. M. 2000, Discrimination of chemically similar organic vapours and vapour mixtures using the Kohonen network, *Analyst*, vol. 125, no. 1, pp. 65–70.

Lawrence, S., Giles, C. L. & Tsoi, A.C. 1997, Lessons in neural network training: overfitting may be harder than expected, *Proceedings of the 14th National Conference on Artificial Intelligence, AAAI-97*, Menlo Park, pp. 540–545.

Lavigne, J. J. & Anslyn, E. V. 2001, Sensing a paradigm shift in the field of molecular recognition: from selective to differential receptors, *Angewandte Chemie – International Edition*, vol. 40, no. 17, pp. 3119–3130.

Lavine, B. K. & Moores, A. J. 1999a, Genetic algorithms in analytical chemistry, *Analytical Letters*, vol. 32, no. 3, pp. 433–445.

Lavine, B. K., Moores, A., & Helfend, L. K. 1999b, A genetic algorithm for pattern recognition analysis of pyrolysis gas chromatographic data, *Journal of Analytical and Applied Pyrolysis*, vol. 50, no. 1, pp. 47–62.

Leardi, R. 2002, Chemometrics: from classical to genetic algorithms, *Grasas y Aceites*, vol. 53, no. 1, pp. 115–127.

Lee, D. S., Noh, B. S., Bae, S. Y., & Kim, K. 1998, Characterization of fatty acids composition in vegetable oils by gas chromatography and chemometrics, *Analytica Chimica Acta*, vol. 358, no. 2, pp. 163–175.

Lee, D.-S., Lee, M., Huh, J.-S, Byun, H.-G., & Lee, D.-D. 2001, VOC recognition with sensor array and neuro-fuzzy network, *Proceedings of the International Conference on Artificial Chemical Sensing: Olfaction and the Electronic Nose*, pp. 139–143.

Lee, D.-S., Huh, & Lee, D.-D. 2003a, Classifying combustible gases using micro-sensor array, *Sensors and Actuators B-Chemical*, vol. 93, pp. 1–6.

Lee, K.-M., Yu, J.-B., Jun, H.-K., Lim, J.-O., Lee, D.-D., Byun, H.-G., & Huh, J.-S. 2003a, Volatile organic gas recognition using conducting polymer sensor array, *Materials Science Forum*, vol. 439 (Eco-Materials Processing & Design), pp. 344–351.

Legin, A. V., Seleznev, B. L., Rudnitskaya, A. M., Vlasov, Yu. G., Tverdokhlebov, S. V., Mack, B., Abraham, A., Arnold, T., Baraniak, L., & Nitsche, H. 1999, Multisensor system for determination of iron(II), iron(III), uranium(VI) and uranium(IV) in complex solutions, *Czechoslovak Journal of Physics*, vol. 49 (Suppl. 1, Pt. 2, 13th Radiochemical Conference, 1998), pp. 679–685.

Lewis, I. R., Daniel, N. W., Chaffin, N. C., & Griffiths, P. R. 1994, Raman-spectrometry and neural networks for the classification of wood types, *Spectrochimica Acta part A-Molecular Spectroscopy*, vol. 50, no. 11, pp 1943–1958.

Li, H., Zhang, Y. X., Polaskova, P., & Havel, J. 2002, Enhancement of precision in the analysis of medicines by capillary electrophoresis using artificial neural networks, *Acta Chimica Sinica*, vol. 60, no. 7, pp. 1264–1268.

Li, Q.-F., Yao, X.-J., Chen, X.-G., Liu, M.-C., Zhang, R.-H., Zhang, X.-Y., Hu, & Z. 2000a, Application of artificial neural networks for the simultaneous determination of a mixture of fluorescent dyes by synchronous fluorescence, *Analyst*, vol. 125, pp. 2049–2053.

Li, Q. F., Zhou, Y. Y., Wang, H. W., Zhang, H. Y., Liu, S. H., Chen, X. G., & Hu, Z. D. 2000b, Application of artificial neural networks in multifactor optimization of selectivity in capillary electrophoresis, *Analytical Letters*, vol. 33, no. 11, pp. 2333–2347.

Li, Q.-F., Yao, X.-J., Chen, X.-G., Wang, Liu, M.-C., Zhang, R.-H., Zhang, X.-Y., & Hu, Z.-D. 2000c, Application of artificial neural networks for the simultaneous determination of a mixture of fluorescent dyes by synchronous fluorescence, *Analyst*, vol. 125, pp. 2049–2053.

Li, X., Gasteiger, J., & Zupan, J. 1993, On the topology distortion in self-organizing feature maps, *Biological Cybernetics*, vol. 70, no. 2, pp. 189–198.

Li, Y., Wang, J. D., Gu, B. H., & Meng, G. Z. 1999a, Artificial neural network and its application to analytical chemistry, *Spectroscopy and Spectral Analysis*, vol. 19, no. 6, pp. 844–849.

Li, Y., Jiang, J.-H., Chen, Z.-P., Xu, C.-J., & Yu, R.-Q. 1999b, A new method based on counterpropagation network algorithm for chemical pattern recognition, *Analytica Chimica Acta*, vol. 338, pp. 161–170.

Li, Y. W., & Vanespen, P. 1994, Study of the influence of neural-network parameters on the performance-characteristics in pattern-recognition, *Chemometrics and Intelligent Laboratory Systems*, vol. 25, no. 2, pp 241–248.

Li, Y., Yang, S.-L., Wang, J.-D., Gu, B.-H., & Liu, F. 1999c, Simultaneous determination of multi-components in air toxic organic compounds using artificial neural networks in FTIR spectroscopy, *Spectroscopy Letters*, vol. 32, no. 3, pp. 421–429.

Lichan, E. 1994, Developments in the detection of adulteration of olive oil, *Trends in Food Science & Technology*, vol. 5, no. 1, pp. 3–11.

Liden, H., Gorton, L., Bachinger, T., & Mandenius, C.-F. 2000, On-line determination of non-volatile or low-concentration metabolites in a yeast cultivation using an electronic nose, *Analyst*, vol. 125, no. 6, pp. 1123–1128.

Liebman, S. A., Phillips, C., Fitzgerald, W., Pescerodriguez, R. A., Morris, J. B., & Fifer, R. A. 1994, Integrated intelligent instruments for materials science, *Hyphenated Techniques in Polymer Character-ization*, vol. 581, pp. 12–24.

Lin, C. W., Lamanna, J. C., & Takefuji, Y. 1992, Quantitative measurement of 2-component pH-sensi-tive colorimetric spectra using multilayer neural networks, *Biological Cybernetics*, vol. 67, no. 4, pp. 303–308.

Lipkovitz, K. B. & Boyd, D. B., (eds.) 2000, *Reviews in Computational Chemistry*, vol. 16, Wiley VCH, p. 10.

Liu, B. F., Zhang, J. F., & Lu, Y. T. 2002, Predicting and evaluating separation quality of micellar electrokinetic capillary chromatography by artificial neural networks, *Electrophoresis*, vol. 23, no. 9, pp. 1279–1284.

Liu, H. T., Wang, K. T., Xu, H. P., Chen, X. G., & Hu, Z. D. 2002, Application of experimental design and artificial neural networks to separation and determination of active components in traditional Chinese medicinal preparations by capillary electrophoresis, *Chromatographia*, vol. 55, no. 9–10, pp. 579–583.

Lleti, R., Sarabia, L. A., Ortiz, M. C., Todeschini, R., & Colombini, M. P. 2003, Application of the Kohonen artificial neural-network in the identification of proteinaceuos binders in samples of panel painting using gas chromatography-mass spectrometry, *Analyst*, vol. 128, no. 3, pp. 281–286.

Llobet, E., Brezmes, J., Vilanova, X., Sueiras, J. E., & Correig, X. 1997, Qualitative and quantitative analysis of volatile organic compounds using transient and steady-state responses of a thick-film tin oxide gas sensor array, *Sensors and Actuators B-Chemical*, vol. 41, no. 1–3, pp. 13–21.

Lo, S.-C. 1995, Speeding up backpropagation in feedforward neural networks for multivariate calibra-tion problems, *Proceedings of the PITCON*, paper 713.

Lohninger, H. 1991, Classification of mass spectral data using neural networks, *Proceedings of the 5th Software Development Workshop on Computers in Chemistry*, pp. 159–68.

Lohninger, H. & Stancl, F. 1992, Comparing the performance of neural networks to well-established methods of multivariate data-analysis – the classification of mass-spectral data, *Fresenius Journal of Analytical Chemistry*, vol. 344, no. 4–5, pp. 186–189.

Long, J. R., Gregoriou, V. G., & Gemperline, P. J. 1990, Spectroscopic calibration and quantitation using artificial neural networks, *Analytical Chemistry*, vol. 62, pp. 1791–1797.

Lopes, M. F. S., Pereira, C. I., Rodrigues, F. M. S., Martins, M. P., Mimoso, M. C., Barros, T. C., Marques, J. J. Figueiredo, Tenreiro, R. P., Almeida, J. S., & Crespo, M. T. B. 1999, Registered designation of origin areas of fermented food products defined by microbial phenotypes and artificial neural networks, *Applied and Environmental Microbiology*, vol. 65, 10, pp. 4484–4489.

Lopez -Cueto, G., Ostra, M., & Ubide, C. 2000, Linear and non-linear multivariate calibration methods for multicomponent kinetic determinations in cases of severe non-linearity, *Analytica Chimica Acta*, vol. 405, no. 1–2, pp. 285–295.

Lopez-Molinero, A., Pino, J., Castro, A., & Castillo, J. R. 2003, Artificial neural networks applied to the classification of emission lines in inductively coupled plasma-atomic emission spectroscopy, *Analytical Letters*, vol. 36, no. 1, pp. 245–262.

Loukas, Y. L. 2000, Artificial neural networks in liquid chromatography: efficient and improved quanti-tative structure-retention relationship models, *Journal of Chromatography A*, vol. 904, no. 2, pp. 119–129.

Loukas, Y. L. 2001, Radial basis function networks in liquid chromatography: improved structure-reten-tion relationships compared to principal components regression (PCR) and nonlinear partial least squares regression (PLS), *Journal of Liquid Chromatography & Related Technologies*, vol. 24, no. 15, pp. 2239–2256.

Lozano, J., Novic, M., Rius, F. X., & Zupan, J. 1995a, Modeling metabolic energy by neural networks, *Chemometrics and Intelligent Laboratory Systems*, vol. 28, no. 1, pp. 61–72.

Lozano, J., Ruisanchez, I., Larrechi, M. S., Rius, F. X., Novic, M., & Zupan, J. 1995b, Analytical signal diagnosis by neural networks, *Proceedings of the 4th International Conference on Automation, Robotics and Artificial Intelligence applied to Analytical Chemistry and Laboratory Medicine,* Montreux, Switzer-land.

Lu, Y., Bian, L. P., & Yang, P. Y. 2000, Quantitative artificial neural network for electronic noses, *Analytica Chimica Acta*, vol. 417, no. 1, pp. 101–110.

Lubal, P., Fetsch, D., Siroky, D., Lubalova, M., Senkyr, J., & Havel, J. 2000a, Potentiometric and spectroscopic study of uranyl complexation with humic acids, *Talanta*, vol. 51, no. 5, pp. 977–991.

Lubal, P., Perutka, J., & Havel, J. 2000b, Potentiometric study of tungsten(VI) complex formation with tartarate, *Chemical Papers – Chemicke Zvesti*, vol. 54, no. 5, pp. 289–295.

Lubal, P., Godula, T., & Havel, J. 2001a, The complexation of palladium(II) by chloroacetate, *Chemical Papers – Chemicke Zvesti*, vol. 55, no. 3, pp. 157–161.

Lubal, P., Kyvala, M., Hermann, P., Holubova, J., Rohovec, J., Havel, J., & Lukes, I. 2001b, Thermodynamic and kinetic study of copper(II) complexes with N-methylene(phenylphosphinic acid) derivatives of cyclen and cyclam, *Polyhedron*, vol. 20, no. 1–2, pp. 47–55.

Luo, L.-Q., Ma, G.-Z., Liang, G.-L. Guo, C.-L., & Ji, A. 1996, Application of x-ray fluorescence analysis and chemometrics to geology, *Diqiu Xuebao, Spec. Issue*, pp. 254–260.

Luo, L-Q., Guo, C.-L., Ma, G.-Z., & Ji, A. 1997, Choice of optimum model parameters in artificial neural networks and application to x-ray fluorescence analysis, *X-Ray Spectrometry*, vol. 26, no. 1, pp. 15–22.

Luo, L.-Q., Ji, A., Ma, G.-Z., & Guo, C.-L. 1998, Focusing on one component each time – comparison of single and multiple component prediction algorithms in artificial neural networks for x-ray fluorescence analysis, *X-Ray Spectrometry*, vol. 27, no. 1, pp. 17–22.

Luo, L.-Q., 2000a, Predictability comparison of four neural network structures for correcting matrix effects in x-ray fluorescence spectrometry, *Journal of Trace and Microprobe Techniques*, vol. 18, no. 3, pp 349–360.

Luo, L.-Q., Wu, X.-J., Gan, L., Liang, G.-L., & Ma, G.-Z. 2000b, X-ray fluorescence analysis based on knowledge system, *Advances in X-Ray Analysis*, vol. 44, pp. 374–379.

Luo, M. L. & Li, M. L. 2000c, Using modified BP neural-network for nonlinear modeling in chemistry, *Acta Chim.Sin.*, vol. 58, no. 11, pp. 1409–1412.

Lyons, W. B., Ewald, H., Flanagan, C., Lochmann, S., & Lewis, E. 2001, A neural networks based approach for determining fouling of multi-point optical fibre sensors in water systems, *Measurement Science and Technology*, vol. 12 no. 7, pp. 958–965.

Lyons, W. B., Ewald, H., & Lewis, E. 2002, An optical fibre distributed sensor based on pattern recognition, *Journal of Materials Processing Technology*, vol. 127, no. 1, pp. 23–30.

Ma, Q.-L., Yan, A.-X., Hu, Z., Li, Z.-X., & Fan, B.-T. 2000, Principal component analysis and artificial neural networks applied to the classification of Chinese pottery of Neolithic age *Analytica Chimica Acta*, vol. 406, no. 2, pp. 247–256.

MacDonald, D. D. 2002, A brief history of electrochemical impedance spectroscopy, *Proceedings – Electrochemical Society*, vol. 2002–13 (Corrosion Science), pp. 72–88.

Madden, J. E., Avdalovic, N., Haddad, P. R., & Havel, J. 2001, Prediction of retention times for anions in linear gradient elution ion chromatography with hydroxide eluents using artificial neural networks, *Journal of Chromatography A*, vol. 910, no. 1, pp. 173–179.

Magallanes, J. F., Vazquez, C. 1998, Automatic classification of steels by processing energy-dispersive x-ray spectra with artificial neural networks, *Journal of Chemical Information and Computer Sciences*, vol. 38, no. 4, pp. 605–609.

Magallanes, J. F., Smichowski, P., & Marrero, J. 2001, Optimization and empirical modeling of HG-ICP-AES analytical technique through artificial neural networks, *Journal of Chemical Information and Computer Sciences*, vol. 41, no. 3, p. 824.

Magallanes, J. F., Zupan, J., Gomez, D., Reich, S., Dawidowski, L., & Groselj, N. 2003, The mean angular distance among objects and its relationships with Kohonen artificial neural networks, *Journal of Chemical Information and Computer Sciences*, vol. 43, no. 5, pp. 1403–1411.

Magelssen, G. R. & Elling, J. W. 1997, Chromatography pattern recognition of Aroclors using iterative probabilistic neural networks, *Journal of Chromatography A*, vol. 775, no. 1–2, pp. 231–242.

Majcen, N., Rajerkanduc, K., Novic, M., & Zupan, J. 1995, Modeling of property prediction from multicomponent analytical data using different neural networks, *Analytical Chemistry*, vol. 67, no. 13, pp. 2154–2161.

Majcen, N., Rius, F. X., & Zupan, J. 1997, Linear and non-linear multivariate analysis in the quality control of industrial titanium dioxide white pigment, *Analytica Chimica Acta*, vol. 348, no. 1–3, pp. 87–100.

Malovana, S., Gajdosova, D., Benedik, J., & Havel, J. 2001, Determination of esmolol in serum by capillary zone electrophoresis and its monitoring in course of heart surgery, *Journal of Chromatography B – Analytical Technologies in the Biomedical and Life Sciences*, vol. 760, no. 1, pp. 37–43.

Malovana, S., Frias-Garcia, S., & Havel, J. 2002, Artificial neural networks for modeling electrophoretic mobilities of inorganic cations and organic cationic oximes used as antidote contra nerve paralytic chemical weapons, *Electrophoresis*, vol. 23, no. 12, pp. 1815–1821.

Mandenius, C.-F. 2000, Electronic noses for bioreactor monitoring, *Advances in Biochemical Engineering/ Biotechnology*, vol. 66 (Bioanalysis and Biosensors for Bioprocess Monitoring), pp. 65–82.

Maquelin, K., Choo-Smith, L. P., van Vreeswijk, T., Endtz, H. P., Smith, B., Bennett, R., Bruining, H. A., & Puppels, G. J. 2000, Raman spectroscopic method for identification of clinically relevant microorganisms growing on solid culture medium, *Analytical Chemistry*, vol. 72, no. 1, pp. 12– 19.

Maquelin, K., Kirschner, C., Choo-Smith, L. P., van den Braak, N., Endtz, H. P., Naumann, D., & Puppels, G. J. 2002a, Identification of medically relevant microorganisms by vibrational spectroscopy, *Journal of Microbiological Methods*, vol. 51, no. 3, pp. 255–271.

Maquelin, K., Choo-Smith, L. P., Endtz, H. P., Bruining, H. A., & Puppels, G. J. 2002b, Rapid identification of Candida species by confocal Raman micro spectroscopy, *Journal of Clinical Microbiology*, vol. 40, no. 2, pp. 594–600.

Maquelin, K., Kirschner, C., Choosmith, L. P., Ngothi, N. A., Vanvreeswijk, T., Stammler, M., Endtz, H. P., Bruining, H. A., Naumann, D., & Puppels, G. J. 2003, Prospective study of the performance of vibrational spectroscopies for rapid identification of bacterial and fungal pathogens recovered from blood cultures, *Journal of Clinical Microbiology*, vol. 41, no. 1, pp. 324–329.

Marengo, E., Gennaro, M. C., & Angelino, S. 1998, Neural network and experimental design to investigate the effect of five factors in ion-interaction high-performance liquid chromatography, *Journal of Chromatography A*, vol. 799, no. 1–2, pp. 47–55.

Marengo, E., Aceto, M., & Maurino, V. 2002, Classification of Nebbiolo-based wines from Piedmont (Italy) by means of solid-phase microextraction-gas chromatography-mass spectrometry of volatile compounds, *Journal of Chromatography A*, vol. 943, no. 1, pp. 123–137.

Marigheto, N. A., Kemsley, E. K., Defernez, M., & Wilson, R. H. 1998, A comparison of midinfrared and Raman spectroscopies for the authentication of edible oils, *Journal of the American Oil Chemists Society*, vol. 75, no. 8, pp 987–992.

Marini, F., Balestrieri, F., Bucci, R., Magri, A. L., & Marini, D. 2003, Supervised pattern recognition to discriminate the geographical origin of rice bran oils: a first study, *Microchemical Journal*, vol. 74, no. 3, pp. 239–248.

Marjoniemi, M. 1994, Nonlinear modeling of chromium tanning solution using artificial neural networks, *Applied Spectroscopy*, vol. 48, no. 1, pp. 21–26.

Marsalek, P., Farkova, M., & Havel, J. 2002, Quantitative MALDI-TOFMS analysis of amino acids applying soft modeling methods, *Chemical Papers – Chemicke Zvesti*, vol. 56, no. 3, pp. 188–193.

Marshall, J., Chenery, S., Evans, E. H., & Fisher, A. 1998, Atomic spectrometry update – atomic emission spectrometry, *Journal of Analytical Atomic Spectrometry*, vol. 13, no. 6, pp. 107R– 128R.

Marti, V., Aguilar, M., & Farran, A. 1999, Experimental designs and response surface modeling applied for the optimization of metal-cyanide complexes analysis by capillary electrophoresis, *Electrophoresis*, vol. 20, no. 17, pp. 3381–3387.

Martin, M. A., Santos, J. P., & Agapito, J. A. 2001, Application of artificial neural networks to calculate the partial gas concentration in a mixture, *Sensors and Actuators B-Chemical*, vol. 77, pp. 468–471.

Matias, V., Ohl, G., Soares, J. C., Barradas, N. P., Vieira, A., Cardoso, S., & Freitas, P. P. 2001, Determination of the composition of light thin films with artificial neural network analysis of Rutherford backscattering experiments, *Physical Review E: Statistical, Nonlinear, and Soft Matter Physics*, vol. 67, no. 4–2.

Mayfield, H. T., Eastwood, D., & Burggraf, L. W. 1999, Classification of infrared spectra of organophosphorus compounds with artificial neural networks, *Proceedings of SPIE – The International Society for Optical Engineering*, vol. 3854 (Pattern Recognition, Chemometrics, and Imaging for Optical Environmental Monitoring), pp. 56–64.

Mazzatorta, P., Benfenati, E., Neagu, D., & Gini, G., 2002, The importance of scaling in data mining for toxicity prediction, *Journal of Chemical Information and Computer Sciences*, vol. 42, no. 5, pp. 1250– 1255.

McAlernon, P., Slater, J. M., & Lau, K.-T. 1999, Mapping of chemical functionality using an array of quartz crystal microbalances in conjunction with Kohonen self-organizing maps, *Analyst*, vol. 124, pp. 851–857.

Mc Avoy, T. J., Wang, N. S., Naidu, S., Bhat, N., Gunter, J. and Simmons, M. 1989, Interpreting biosensor data via backpropagation, *Proceedings of the International Joint Conference on Neural Networks*, Washington, DC, vol. 1, pp. 227–233.

Mc Avoy, T. J., Su, H. T., Wang, N. S., & He, M. 1992, A comparison of neural networks and partial least squares for deconvoluting fluorescence spectra, *Biotechnology and Bioengineering*, vol. 40, pp. 53–62.

McCarrick, C. W., Ohmer, D.T., Gilliland, L. A., & Edwards, P. A. 1996, Fuel identification by neural network analysis of the response of vapor-sensitive sensor arrays, *Analytical Chemistry*, vol. 68, pp. 4264–4269.

McGovern, A. C., Ernill, R., Kara, B. V., Kell, D. B., & Goodacre, R. 1999, Rapid analysis of the expression of heterologous proteins in Escherichia coli using pyrolysis mass spectrometry and Fourier transform infrared spectroscopy with chemometrics: application to alpha 2-interferon production, *Journal of Biotechnology*, vol. 72, no. 3, pp. 157–167.

McGovern, A. C., Broadhurst, D., Taylor, J., Kaderbhai, N., Winson, M. K., Small, D. A., Rowland, J. J., Kell, D. B., & Goodacre, R. 2002, Monitoring of complex industrial bioprocesses for metabolite concentrations using modern spectroscopies and machine learning: application to gibberellic acid production, *Biotechnology and Bioengineering*, vol. 78, no. 5, pp. 527–538.

Meisel, H., Lorenzen, P. C., Martin, D., & Schlimme, E. 1997, Chemometric identification of butter types by analysis of compositional parameters with neural networks, *Nahrung-Food*, vol. 41, no. 2, pp. 75–80.

Mello, C., Poppi, R. J., de Andrade, J. C., & Cantarella, H. 1999, Pruning neural network for architecture optimization applied to near-infrared reflectance spectroscopic measurements. Determination of the nitrogen content in wheat leaves, *Analyst*, vol. 124, pp. 1669–1674.

Melnyk, S., Pogribna, M., Pogribny, I. P., Yi, P., & James, S. J. 2000, Measurement of plasma and intracellular S-adenosylmethionine and S-adenosylhomocysteine utilizing coulometric electrochemical detection: alterations with plasma homocysteine and pyridoxal 5-phosphate concentrations, *Clinical Chemistry*, vol. 46, no. 2, pp. 265–272.

Metting, H. J. & Coenegracht, P. M. J. 1996, Neural networks in high-performance liquid chromatography optimization: response surface modeling, *Journal of Chromatography A*, vol. 728, no. 1–2, pp. 47–53.

Meyer, M., & Wiegelt, T. 1992, Interpretation of infrared spectra by artificial neural networks, *Analytica Chimica Acta*, vol. 265, pp. 183–190.

Meyer, M., Meyer, K., & Hobert, H. 1993, Neural networks for the interpretation of infrared spectra using extremely reduced spectral data, *Analytica Chimica Acta*, vol. 282, pp. 407–415.

Micic, M., Jovancicevic, B., Polic, P., Susic, N., & Markovic, D. 1998, Classification tools based on artificial neural networks for the purpose of identification of origin of organic matter and oil pollution in recent sediments, *Fresenius Environmental Bulletin*, vol. 7, no. 11–12, pp. 648–653.

Milne, G. W. A. 1997, Mathematics as a basis for chemistry, *Journal of Chemical Information and Computer Sciences,* vol. 37, no. 4, pp. 639–644.

Mittermayr, C. R., Drouen, A. C. J. H., Otto, M., & Grasserbauer, M. 1994, Neural networks for library search of ultraviolet-spectra, *Analytica Chimica Acta*, vol. 294, no. 3, pp. 227–242.

Mo, H. & Deng, B. 1996, Simulation calculation for chemical speciation of lead by feedback neural network, *Chinese Chemical Letters*, vol. 7, no. 3, pp. 265–266.

Moberg, M., Bylund, D., Danielsson, R., & Markides, K. 2000, Optimization strategy for liquid chromatography-electrospray ionization mass spectrometry methods, *Analyst*, vol. 125, no. 11, pp. 1970–1976.

Mohamed, E. I., Linder R., Perriello G., Di Daniele N., Poppl S. J., & De Lorenzo A. 2002, Predicting type 2 diabetes using an electronic nose-based artificial neural network analysis, *Diabetes, Nutrition & Metabolism*, vol. 15, no. 4, pp. 215–21.

Mohamed, I., Bruno E., Linder R., Alessandrini M., Di Girolamo A., Poppl S. J., Puija A., De Lorenzo A. 2003, A novel method for diagnosing chronic rhinosinusitis based on an electronic nose, *Anales Otorrinolaringologicos Ibero-Americanos*, vol. 30, no. 5, pp. 447–57.

Mongelli, M., Chang, A., & Sahota, D. 1997, The development of a hybrid expert system for the interpretation of fetal acid-base status, *International Journal of Medical Informatics*, vol. 44, no. 2, pp. 135–144.

Montanarella, L., Bassani, M. R., & Breas, O. 1995, Chemometric classification of some European wines using pyrolysis mass spectrometry, *Rapid Communications in Mass Spectrometry*, vol. 9, no. 15, pp. 1589–1593.

Moon, W. J., Yu, J. H., & Choi, G. M. 2002, The CO and H2 gas selectivity of CuO-doped SnO2-ZnO composite gas sensor, *Sensors and Actuators, B-Chemical*, vol. B87, no. 3, pp. 464–470.

Morales, M. L., Gonzalez, G. A., Casas, J. A., & Troncoso, A. M. 2001, Multivariate analysis of commercial and laboratory produced sherry wine vinegars: influence of acetification and aging, *European Food Research and Technology*, vol. 212, no. 6, pp. 676–682.

Morgan, W. L., Larsen, J. T., & Goldstein, W. H. 1994, The use of artificial neural networks in plasma spectroscopy, *Journal of Quantum Spectroscopy & Radiation Transfer*, vol. 51, no. 1/2, pp. 247–253.

Morris, J. B., Pesce-Rodriguez, R. A., Fifer, R. A., Liebman, S. A., Lurcott, S. M., Levy, E. J., Skeffington, B., & Sanders, A. 1991, Development of expert systems and neural networks in analytical chemistry, *Intelligent Instruments & Computers*, pp. 167–175.

Moya-Hernandez, R., Rueda-Jackson, J. C., Ramirez, M. T., Vazquez, G. A., Havel, J., & Rojas-Hernandez, A. 2002, Statistical study of distribution diagrams for two-component systems: relationships of means and variances of the discrete variable distributions with average ligand number and intrinsic buffer capacity, *Journal of Chemical Education*, vol. 79, no. 3, pp. 389–392.

Mukesh, D. 1996, Applications of neural computing for process chemists. 1. Introduction to neural network, *Journal of Chemical Education*, vol. 73, no. 5, pp 431–433.

Mulholland, M., Hibbert, D. B., Haddad, P. R., & Parslov, P. 1995, A comparison of classification in artificial intelligence, induction versus a self-organising neural networks, *Chemometrics and Intelligent Laboratory Systems*, vol. 30, no. 1, pp. 117–128.

Mulholland, M., Preston, P., Hibbert, D. B., Haddad, P. R., & Compton, P. 1996, Teaching a computer ion chromatography from a database of published methods, *Journal of Chromatography A*, vol. 739, no. 1–2, pp. 15–24.

Muller, C., Hitzmann, B., Schubert, F., & Scheper, T. 1997, Optical chemo- and biosensors for use in clinical applications, *Sensors and Actuators B-Chemical*, vol. 40, no. 1, pp. 71–77.

Murphy, D. M., Middlebrook, A. M., & Warshawsky, M. 2003, Cluster analysis of data from the Particle Analysis by Laser Mass Spectrometry (PALMS) instrument, *Aerosol Science and Technology*, vol. 37, no. 4, pp. 382–391.

Muzikar, M., Havel, J., & Macka, M. 2002, Capillary electrophoretic study of interactions of metal ions with crown ethers, a sulfated beta-cyclodextrin, and zwitterionic buffers present as additives in the background electrolyte, *Electrophoresis*, vol. 23, no. 12, pp. 1796–1802.

Muzikar, M., Havel, J., & Macka, M. 2003, Capillary electrophoresis determinations of trace concentrations of inorganic ions in large excess of chloride: soft modelling using artificial neural networks for optimisation of electrolyte composition, *Electrophoresis*, vol. 24, no. 12–13, pp. 2252–2258.

Nascimento, C. A. O., Guardani, R., & Giulietti, M. 1997 Use of neural networks in the analysis of particle size distributions by laser diffraction, *Powder Technology*, vol. 90, pp. 89–94.

Nakamoto, T., Fukuda, A., & Moriizumi, T. 1991, Improvement of identification capability in odor-sensing system, *Sensors and Actuators B-Chemical*, vol. 3, pp. 221–226.

Nakamoto, T., Takagi, H., Utsumi, S., & Morriizumi, T. 1992, Gas/odor identification by semi-conductor gas-sensor array and an analog artificial-network circuit, *Sensors and Actuators B-Chemical*, vol. 8, pp. 181–186.

Nanto, H., Kondo, K., Habara, M., Douguchi. Y., Waite, R. I., & Nakazumi, H, 1996a, Identification of aromas from alcohols using a Japanese lacquer film-coated quartz resonator gas sensor in conjunction with pattern-recognition analysis, *Sensors and Actuators B-Chemical*, vol. 35, no. 1–3, pp 183–186.

Nanto, H., Tsubakino, S., Habara, M., Kondo, K., Morita, T., Douguchi, Y., Nakazumi, H., & Waite, R. I. 1996b, A novel chemical sensor using Ch3Si(och3)(3) sol-gel thin-film coated quartz-resonator microbalance, *Sensors and Actuators B-Chemical*, vol. 34, no. 1–3:, pp 312–316.

Nazarov, E., Eiceman, G. A., & Bell, S. E. 1999, Quantitative assessment for the training of neural networks with large libraries of ion mobility spectra, *International Journal for Ion Mobility Spectrometry*, vol. 2, no. 1, pp. 45–60.

Nazarov, P. V., Apanasovich, V. V., Lutkovskaya, K. U., Lutkovski, V. M., & Misakov, P. Y. 2003, Neural network data analysis for intracavity laser spectroscopy, *Proceedings of SPIE – The International Society for Optical Engineering,* vol. 5135 (Optical Information, Data Processing and Storage, and Laser Communication Technologies), pp. 61–69.

Ni, H. F., & Hunkeler, D. 1997, Prediction of copolymer composition drift using artificial neural networks – copolymerization of acrylamide with quaternary ammonium cationic monomers, *Polymer*, vol. 38, no. 3, pp 667–675.

Ni, Y.-N., & Liu, C. 1999, Artificial neural networks and multivariate calibration for spectrophotometric differential kinetic determinations of food antioxidants, *Analytica Chimica Acta*, vol. 396, no. 2–3, pp. 221–230.

Ni, Y., Liu, C., & Kobot, S. 2000, Simultaneous kinetic spectrophotometric determination of acetaminophen and phenobarbital by artificial neural networks and partial least squares, *Analytica Chimica Acta*, vol. 419, pp. 185–196.

Ni, Y. N., Chen, S. H., & Kokot, S. 2002, Spectrophotometric determination of metal ions in electroplating solutions in the presence of EDTA with the aid of multivariate calibration and artificial neural networks, *Analytica Chimica Acta*, vol. 463, no. 2, pp. 305–316.

Niebling, G. & Schlachter, A. 1995, Qualitative and quantitative gas-analysis with nonlinear interdigital sensor arrays and artificial neural networks, *Sensors and Actuators B-Chemical*, vol. 27, no. 1–3, pp. 289–292.

Nilsson, T., Bassani, M. R., Larsen, T. O., & Montanarella, L. 1996, Classification of species in the genus Penicillium by Curie point pyrolysis mass spectrometry followed by multivariate analysis and artificial neural networks, *Journal of Mass Spectrometry*, vol. 31, no. 12, pp. 1422–1428.

Noble, P. A., Bidle, K. D., & Fletcher, M. 1997, Natural microbial community compositions compared by a back-propagating neural network and cluster analysis of 5S rRNA, *Applied and Environmental Microbiology*, vol. 63, no. 5, pp. 1762–1770.

Noble, P. A., Almeida, J. S., & Lovell, C. R. 2000, Application of neural computing methods for interpreting phospholipid fatty acid profiles of natural microbial communities, *Applied and Environmental Microbiology*, vol. 66, no. 2, pp. 694–699.

Nord, L. I., & Jacobsson, S. P. 1998, A novel method for examination of the variable contribution to computational neural-network models, *Chemometrics and Intelligent Laboratory Systems*, vol. 44, no. 1–2, pp 153–160.

Nott, P. J. K., Karim, M. N., & Morris, A. J. 1996, Fault and contamination detection in a continuous bakers yeast fermentation, *Computers & Chemical Engineering*, vol. 20, pp. S611–S616.

Olden, J. D. & Jackson, D. A. 2002, Illuminating the black-box – a randomization approach for understanding variable contributions in artificial neural networks, *Ecological Modeling*, vol. 154, no. 1–2, pp. 135–150.

Oliveros, M. C. C., Pavon, J. L. P., Pinto, C. G., Laespada, M. E. F., Cordero, B. M., & Forina, M. 2002, Electronic nose based on metal oxide semiconductor sensors as a fast alternative for the detection of adulteration of virgin olive oils, *Analytica Chimica Acta*, vol. 459, no. 2, pp. 219–228.

Olmos, P., Diaz, J. C., Perez, J. M., Aguayo, P., Gomez, P., & Rodellar, V. 1994, Drift problems in the automatic-analysis of gamma-ray spectra using associative memory algorithms, *IEEE Transactions on Nuclear Science*, vol. 41, no. 3, pp 637–641.

Onjia, A., Vasiljevic, T., Cokesa, D., & Lausevic, M. 2002, Factorial design in isocratic high-performance liquid chromatography of phenolic compounds, *Journal of the Serbian Chemical Society*, vol. 67, no. 11, pp. 745–751.

Otto, M., George, T., Schierle, C., & Wegscheider, W. 1992, Fuzzy-logic and neural networks – applications to analytical-chemistry, *Pure and Applied Chemistry*, vol. 64, no. 4, pp. 497–502.

Pacheco, M. L. & Havel, J. 2001, Capillary zone electrophoretic (CZE) study of Uranium (VI) complexation with humic acids, *Journal of Radioanalytical and Nuclear Chemistry*, vol. 248, no. 3, pp. 565–570.

Pacheco, M. D. & Havel, J. 2002, Capillary zone electrophoresis of humic acids from the American continent, *Electrophoresis*, vol. 23, no. 2, pp. 268–277.

Pacheco, M. L., Pena-Mendez, E. M., & Havel, J. 2003, Supramolecular interactions of humic acids with organic and inorganic xenobiotics studied by capillary electrophoresis, *Chemosphere*, vol. 51, no. 2, pp. 95–108.

Padin, P. M., Pena, R. M., Garcia, S., Iglesias, R., Barro, S., & Herrero, C. 2001, Characterization of Galician (NW Spain) quality brand potatoes – a comparison study of several pattern-recognition techniques, *Analyst*, vol. 126, no. 1, pp. 97–103.

Palacios-Santander, J. M., Jimenez-Jimenez, A., Cubillana-Aguilera, L. M., Naranjo-Rodriguez, I., & Hidalgo-Hidalgo-de-Cisneros, J. L. 2003, Use of artificial neural networks, aided by methods to reduce dimensions, to resolve overlapped electrochemical signals. A comparative study including other statistical methods, *Microchimica Acta*, vol. 142, no. 1–2, pp. 27–36.

Parrilla, M. C. G., Heredia, F. J., & Troncoso, A. M. 1999, Sherry wine vinegars: phenolic composition changes during aging, *Food Research International*, vol. 32, no. 6, pp. 433–440.

Patnaik, P. R. 1996, Neural network applications in the selective separation of biological products, *Indian Journal of Chemical Technology*, vol. 3, no. 1, pp. 11–16.

Patnaik, P. R. 1999, Applications of neural networks to recovery of biological products, *Biotechnology Advances*, vol. 17, no. 6, pp. 477–488.

Paulsson, N., & Winquist, F. 1997, Breath alcohol, multi sensor arrays and electronic noses, *Proceedings of SPIE – The International Society for Optical Engineering,* vol. 2932 (Human Detection and Positive Identification: Methods and Technologies), pp. 84–90.

Paulsson, N. J. P. & Winquist, F. 1999, Analysis of breath alcohol with a multisensor array: instrumental setup, characterization and evaluation, *Forensic Science International*, vol. 105, no. 2, pp. 95–114.

Paulsson, N., Larsson, E., & Winquist, F. 2001, Extraction and selection of parameters for evaluation of breath alcohol measurement with an electronic nose, *Sensors and Actuators, A-Physical,* vol. A84, no. 3, pp. 187–197.

Pavlou, A. K., Magan, N., Sharp, D., Brown, J., Barr, H., & Turner, A. P. F. 2000, An intelligent rapid odour recognition model in discrimination of Helicobacter pylori and other gastroesophageal isolates in vitro, *Biosensors & Bioelectronics*, vol. 15, no. 7–8, pp. 333–342.

Pazourek, J., Gonzalez, G., Revilla, A. L., & Havel, J. 2000, Separation of polyphenols in Canary Islands wine by capillary zone electrophoresis without preconcentration, *Journal of Chromatography A*, vol. 874, no. 1, pp. 111–119.

Penchev, P. N., Andreev, G. N., & Varmuza, K. 1999, Automatic classification of infrared spectra using a set of improved expert-based features, *Analytica Chimica Acta*, vol. 388, no. 1–2, pp. 145–159.

Penchev, P. N. & Varmuza, K. 2001, Characteristic substructures in sets of organic compounds with similar infrared spectra, *Computers & Chemistry*, vol. 25, no. 3, pp. 231–237.

Penza, M., Cassano, G., & Tortorella, F. 2002, Identification and quantification of individual volatile organic-compounds in a binary mixture by saw multisensor array and pattern-recognition analysis, *Measurement Science Technology*, vol. 13, no. 6, pp. 846–858.

Penza, M. & Cassano, G. 2003, Application of principal component analysis and artificial neural-networks to recognize the individual VOCs of methanol/2-propanol in a binary-mixture by saw multisensor array, *Sensors and Actuators B-Chemical,* vol. 89, no. 3, pp. 269–284.

Pereira, E. R., Mello, C., Costa, P. A., Arruda, M. A. Z., & Poppi, R. J. 2001, Neuro-genetic approach for optimisation of the spectrophotometric catalytic determination of cobalt, *Analytica Chimica Acta*, vol. 433, no. 1, pp. 111–117.

Peres, C., Viallon, C., & Berdague, J. L. 2002, Curie point pyrolysis-mass spectrometry applied to rapid characterisation of cheeses, *Journal of Analytical and Applied Pyrolysis*, vol. 65, no. 2, pp. 161–171.

Pérez-Magariño, S., Ortega-Heras, M., González-San José, M.L., & Boger Z. 2004, Comparative study of artificial neural network and multivariate methods to classify Spanish DO rosé wines *Talanta*, vol. 61 no. 5, pp. 983–990.

Perez-Trujillo, J. P., Barbaste, M., & Medina, B. 2003, Chemometric study of bottled wines with denomination of origin from the Canary Islands (Spain) based on ultra-trace elemental content determined by ICP-MS, *Analytical Letters*, vol. 36, no. 3, pp. 679–697.

Peris, M., Chirivella, V., Martinez, S., Bonastre, A., Ors, R., & Serrano, J. 1994, Rule nets: application to the advanced automation of a flow-injection analysis system, *Chemometrics and Intelligent Laboratory Systems: Laboratory Information,* vol. 26, pp. 123.

Peris, M., Ors, R., Bonastre, A., & Gil, P. 1997 Application of rule nets to temporal reasoning in the monitoring of a chemical analysis process, *Laboratory Automation and Information Management*, vol. 33, pp. 49.

Peris, M., Bonastre, A., & Ors, R. 1998, Distributed expert system for the monitoring and control of chemical processes, *Laboratory Robotics and Automation,* vol. 10, pp. 163.

Persaud, K.C., & Byun, H.-G. 1998a, An artificial neural network based encoding of an invariant Sammon map for real-time projection of patterns from odour sensor arrays, *Proceedings of the International Conference on Advances in Pattern Recognition*, pp. 187–194.

Persaud, K. C., Bailey, R. A.; Pisanelli, A. M., Byun, H.-G., Lee, D.-H., & Payne, J. S. 1998b, Conducting polymer sensor arrays, *Proceedings of the 5th International Symposium on Olfaction and the Electronic Nose*, pp. 318–328.

Peterson, K. L. 2000, Artificial neural networks and their use in chemistry, in Lipkovitz, K. B. & Boyd, D. B. (eds.) *Reviews in Computational Chemistry*, vol. 16, Wiley VCH, pp. 53–140.

Petritis, K., Kangas, L. J., Ferguson, P. L., Anderson, G. A., Pasa-Tolic, L., Lipton, M. S., Auberry, K. J., Strittmatter, E. F., Shen, Y. F., Zhao, R., & Smith, R. D. 2003, Use of artificial neural networks for the accurate prediction of peptide liquid chromatography elution times in proteome analyses, *Analytical Chemistry*, vol. 75, no. 5, pp. 1039–1048.

Phares, D. J., Rhoads, K. P., Johnston, M. V., & Wexler, A. S. 2003, Size-resolved ultrafine particle composition analysis – 2. Houston, *Journal of Geophysical Research-Atmospheres*, vol. 108, no. D7.

Ping, W., & Jun, X. 1996, A novel recognition method for electronic nose using artificial neural-network and fuzzy recognition, *Sensors and Actuators B-Chemical*, vol. 37, no. 3, pp 169–174.

Pinto, C. G., Laespada, M. E. F., Pavon, J. L. P., & Cordero, B. M. 2001, Electronic olfactometry. A new tool in analytical chemistry, *Quimica Analitica*, vol. 20, no. 1, pp. 3–11.

Pishvaie, M. R., & Shahrokhi, M. 2000, Approximation of titration curves by artificial neural networks and its application to pH control, *Scientia Iranica*, vol. 7, no. 2, pp. 82–91.

Place, J. F., Truchaud, A., Ozawa, K., Pardue, H., & Schnipelsky, P. 1994, International federation of clinical chemistry. Use of artificial intelligence in analytical systems for the clinical laboratory. IFCC committee on analytical systems, *Clinical Chimica Acta*, vol. 231, no. 2, pp. S5–S34.

Place, J. F., Truchaud, A., Ozawa, K., Pardue, H., & Schnipelsky, P. 1995, Use of artificial intelligence in analytical systems for the clinical laboratory, *Clinical Biochemistry*, vol. 28, no. 4, pp. 373–389.

Pletnev, I. V., & Zernov, V. V. 2002, Classification of metal ions according to their complexing properties: a data-driven approach, *Analytica Chimica Acta*, vol. 455, no. 1, pp. 131–142.

Plumb, A. P., Rowe, R. C., York, P., & Doherty, C. 2002, The effect of experimental design on the modeling of a tablet coating formulation using artificial neural networks, *European Journal of Pharmaceutical Science*, vol. 16, no. 4–5, pp. 281–288.

Pokorna, L., Revilia, A., Havel, J., & Patocka, J. 1999, Capillary zone electrophoresis determination of galanthamine in biological fluids and pharmaceutical preparatives: experimental design and artificial neural network optimization, *Electrophoresis*, vol. 20, no. 10, pp. 1993–1997.

Pokorna, L., Pacheco, M. L., & Havel, J. 2000, Highly reproducible capillary zone electrophoresis of humic acids in cyclodextrin- or oligosaccharide-modified background electrolytes, *Journal of Chromatography A*, vol. 895, no. 1–2, pp. 345–350.

Polaskova, P., Bocaz, G., Li, H., & Havel, J. 2002, Evaluation of calibration data in capillary electrophoresis using artificial neural networks to increase precision of analysis, *Journal of Chromatography A*, vol. 979, no. 1–2, pp. 59–67.

Polster, J., Prestel, G., Wollenweber, M., Kraus, G., & Gauglitz, G. 1995, Simultaneous determination of penicillin and ampicillin by spectral fibre-optical enzyme optodes and multivariate data analysis based on transient signals obtained by flow injection analysis, *Talanta*, vol. 42, no. 12, pp. 2065–2072.

Pompe, M., Razinger, M., Novic, M., & Veber, M. 1997, Modelling of gas chromatographic retention indices using counterpropagation neural networks, *Analytica Chimica Acta*, vol. 348, pp. 215–221.

Poppi, R. J. & Massart, D. L. 1998, The optimal brain surgeon for pruning neural network architecture applied to multivariate calibration, *Analytica Chimica Acta*, vol. 375, pp. 187–195.

Powell, D. A., Turula, V., de Haseth, J. A., van Halbeek, H., & Meyer, B. 1994, Sulphate detection in Glycoprotein-derived oligosaccharides by artificial neural network analysis of Fourier-transform infrared spectra, *Analytical Biochemistry*, vol. 220, pp. 20–27.

Prado, M., Comini, G., Faglia, G., & Sberveglieri, G. 1999a, A systematic way of extracting features from the dynamic response of a sensor array, *Proceedings of the 4th Italian Conference on Sensors and Microsystems*, pp. 271–275.

Prado, M., Niederjaufner, G., Comini, G., Faglia, G., & Sberveglieri, G. 1999b, Testing the resolving power of an electronic nose by discriminating various types of cheeses, *Proceedings of the 4th Italian Conference on Sensors and Microsystems*, pp. 276–281.

Pardo, M., Faglia, G., Sberveglieri, G., Corte, M., Masulli, F., & Riani, M. 2000, Monitoring reliability of sensors in an array by neural networks, *Sensors and Actuators, B-Chemical*, vol. 67, no. 1–2, pp. 128–133.

Pravdova, V., Pravda, M., & Guilbault, G. G. 2002, Role of chemometrics for electrochemical sensors, *Analytical Letters*, vol. 35, no. 15, pp. 2389–2419.

Prieto, M. M. 1993, Informatics in food microbiology: applications for the identification of microorganisms, *Microbiologia*, vol. 9, spec. no., pp. 96–103.

Qi, J. H., Zhang, X. Y., Zhang, R. S., Liu, M. C., Hu, Z. D., Xue, H. F., & Fan, B. T. 2000a, Prediction of programmed-temperature retention values of naphthas by artificial neural networks, *SAR and QSAR in Environmental Research*, vol. 11, no. 2, pp. 117–131.

Qi, Y. H., Yang, J. A., & Xu, L. 2000b, Correlation analysis of the structures and gas-liquid chromatographic retention indices of amines, *Chinese Journal of Analytical Chemistry*, vol. 28, no. 2, pp. 223–227.

Qin, S. J., Yue, H. Y., & Dunia, R. 1997, Self-validating inferential sensors with application to air emission monitoring, *Industrial & Engineering Chemical Research*, vol. 36, no. 5, pp. 1675–1685.

Qin, S. J. & Wu, Z. J. 2001, A new approach to analyzing gas-mixtures, *Sensors and Actuators B-Chemical*, vol. 80, no. 2, pp. 85–88.

Questier, F., Guo, Q., Walczak, B., Massart, D. L., Boucon, C., & Dejong, S. 2002, The neural-gas network for classifying analytical data, *Chemometrics and Intelligent Laboratory Systems*, vol. 61, no. 1–2, pp. 105–121.

Racca, J. M. J., Philibert, A., Racca, R., & Prairie, Y. T. 2001, A comparison between diatom-based pH inference models using artificial neural networks (ANN), weighted averaging (WA) and weighted averaging partial least squares (WA-PLS) regressions, *Journal of Paleolimnology*, vol. 26, no. 4, pp. 411–422.

Raimundo, I. M., & Narayanaswamy, R. 2003, Simultaneous determination of Zn(II), Cd(II) and Hg(II) in water, *Sensors and Actuators, B-Chemical*, vol. 90, no. 1–3, pp. 189–197.

Raj, A. S., Ravi, R., Parthiban, T., & Radhakrishnan, G. 1999, Artificial neural network applications in electrochemistry – a review, *Bulletin of Electrochemistry*, vol. 15, no. 12, pp. 552–555.

Rantanen, J., Rasanen, E., Antikainen, O., Mannermaa, J. P., & Yliruusi, J. 2001, In-line moisture measurement during granulation with a four-wavelength near-infrared sensor: an evaluation of process-related variables and a development of non-linear calibration model, *Chemometrics and Intelligent Laboratory Systems*, vol. 56, no. 1, pp. 51–58.

Rao, R. N. 2000, Role of modern analytical techniques in quality assurance of fuels and lubricants, *Proceedings of the 2nd International Symposium on Fuels and Lubricants*, New Delhi, vol. 1 pp. 185–190.

Regeniter, A., Steiger, J. U., Scholer, A., Huber, P. R., & Siede, W. H. 2003, Windows to the ward – graphically oriented report forms – presentation of complex, interrelated laboratory data for electrophoresis/immunofixation, cerebrospinal-fluid, and urinary protein profiles, *Clinical Chemistry*, vol. 49, no. 1, pp. 41–50.

Reibel, J., Stahl, U., Wessa, T., & Rapp, M. 2000, Gas analysis with SAW sensor systems, *Sensors and Actuators, B-Chemical*, vol. 65, no. 1–3, pp. 173–175.

Reichova, N., Pazourek, J., Polaskova, P., & Havel, J. 2002, Electrophoretic behavior of adamantane derivatives possessing antiviral activity and their determination by capillary zone electrophoresis with indirect detection, *Electrophoresis*, vol. 23, no. 2, pp. 259–262.

Remola, J. A., Lozano, J., Ruisanchez, I., Larrechi, M. S., Rius, F. X., & Zupan, J. 1996, New chemometric tools to study the origin of amphorae produced in the Roman Empire, *TRAC-Trends in Analytical Chemistry*, vol. 15, no. 3, pp. 137–151.

Ren, Y.-L., Gou, Y.-H., Ren, R.-X., Liu, P.-Y., & Guo, Y. 1999, Application of artificial neural network multivariate calibration to near-infrared spectrophotometric determination of powdered pharmaceutical Metronidazole, *Spectroscopy Letters*, vol. 32, no.3, pp. 431–442.

Ren, S.-X., & Gao, L. 2004, Wavelet packet transform and artificial neural network applied to simultaneous kinetic multicomponent determination, *Analytical and Bioanalytical Chemistry*, vol. 378, no. 5, pp. 1392–1398.

Richards, E., Bessant, C., & Saini, S. 2002a, Optimization of a neural-network model for calibration of voltammetric data, *Chemometrics and Intelligent Laboratory Systems*, vol. 61, no. 1–2, pp. 35–49.

Richards, E., Bessant, C., & Saini, S. 2002b, Multivariate data analysis in electroanalytical chemistry, *Electroanalysis*, vol. 14, no. 22, pp. 1533–1542.

Richards, E., Bessant, C., & Saini, S. 2003, A liquid handling-system for the automated acquisition of data for training, validating and testing calibration models, *Sensors and Actuators B-Chemical*, vol. 88, no. 2, pp. 149–154.

Rius, A., Ruisanchez, I., Callao, F. X., & Rius, F. X. 1998, Reliability of analytical systems: use of control chart, time series models and recurrent neural networks (RNN), *Chemometrics and Intelligent Laboratory Systems*, vol. 40, no. 1, pp. 1–18.

Riesenberg, D. & Guthke, R. 1999, High-cell-density cultivation of microorganisms, *Applied Microbiology and Biotechnology*, vol. 51, no. 4, pp. 422–430.

Rodriguez-Mendez, M. L. 2000, Langmuir-Blodgett films of rare-earth lanthanide bisphthalocyanines. Applications as sensors of gases and volatile organic compounds, *Comments on Inorganic Chemistry*, vol. 22, no. 3–4, pp. 227–239.

Roggo, Y., Duponchel, L., & Huvenne, J. P. 2003, Comparison of supervised pattern-recognition methods with McNamara statistical test – application to qualitative-analysis of sugar-beet by near-infrared spectroscopy, *Analytica Chimica Acta*, vol. 477, no. 2, pp. 187–200.

Roppel, T. A., & Wilson, D. M. 2001, Improved chemical identification from sensor arrays using intelligent algorithms, *Proceedings of SPIE - The International Society for Optical Engineering*, 4205 (Advanced Environmental and Chemical Sensing Technology), pp. 260–266.

Rowe, R. C., Mulley, V. J., Hughes, J. C., Nabney, I. T., & Debenham, R. M. 1994, Neural networks for chromatographic peak classification – a preliminary study, *LC GC – Magazine of Separation Science*, vol. 12, no. 9, p. 690.

Ruan, R., Li, Y., Lin, X., & Chen, P. 2002, Non-destructive determination of deoxynivalenol levels in barley using near-infrared spectroscopy, *Applied Engineering in Agriculture*, vol. 18, no. 5, pp. 549–553.

Ruanet, V. V., Kudryavtsev, A. M., & Dadashev, S. Y. 2001, The use of artificial neural networks for automatic analysis and genetic identification of gliadin electrophoretic spectra in durum wheat, *Russian Journal of Genetics*, vol. 37, no. 10, pp. 1207–1209.

Ruckebusch, C., Duponchel, L., Huvenne, J.-P., Legrand, P., Nedjar-Arroume, N., Lignot, B., Dhulster, P., & Guillochon, D. 1999, Hydrolysis of hemoglobin surveyed by infrared spectroscopy II. Progress predicted by chemometrics, *Analytica Chimica Acta*, vol. 396, no. 2–3, pp. 241–251.

Ruckebusch, C., Duponchel, L., & Huvenne, J. P. 2001a, Degree of hydrolysis from mid-infrared spectra, *Analytica Chimica Acta*, vol. 446, no. 1–2, pp. 257–268.

Ruckebusch, C., Sombret, B., Froidevaux, R., & Huvenne, J. P. 2001b, Online midinfrared spectroscopic data and chemometrics for the monitoring of an enzymatic-hydrolysis, *Applied Spectroscopy*, vol. 55, no. 12, pp. 1610–1617.

Rudnitskaya, A., Ehlert, A., Legin, A., Vlasov, Y., & Buttgenbach, S. 2001, Multisensor system on the basis of an array of non-specific chemical sensors and artificial neural networks for determination of inorganic pollutants in a model groundwater, *Talanta*, vol. 55, no. 2, pp. 425–431.

Ruisanchez, I., Potokar, P., & Zupan, J. 1996, Classification of energy dispersion x-ray spectra of mineralogical samples by artificial neural networks, *Journal of Chemical Information and Computer Sciences*, vol. 36, no. 2, pp. 214–220.

Ruisanchez, I., Lozano, J., Larrechi, M. S., Rius, F. X., & Zupan, J. 1997, On-line automated analytical signal diagnosis in sequential injection analysis systems using artificial neural networks, *Analytica Chimica Acta*, vol. 348, pp. 113–127.

Rumelhart, D. E., Hinton, G. E., & Williams, R. J. 1986, Learning internal representations by error propagation, in *Parallel Distributed Processing*, MIT Press, chap. 8, pp. 318–362.

Rustoiu-Csavdari, A., Baldea, S., & Mihai, D. 2002, Catalytic determination of Pb(II) in the presence of Cu(II), *Analytical & Bioanalytical Chemistry*, vol. 374, no. 1, pp. 17–24.

Ryman-Tubb, N. 1992, Designing an electronic wine-taster using neural networks, *Electronic Engineering*, March, pp. 37–42.

Sackett, R. E., Rogers, S. K., Desimio, M. S., Raymer, J. H., Ruck, Dennis W., Kabrisiky, M., & Bleckmann, C. A. 1996, Neural network analysis of chemical compounds in nonrebreathing Fisher-344 rat breath, *Proceedings of SPIE – The International Society for Optical Engineering*, vol. 2760 (Applications and Science of Artificial Neural Networks II), pp. 386–397.

Safavi, A., Absalan, G., & Maesum, S. 2001a, Simultaneous determination of Se(IV) and Te(IV) as catalysts using "neural networks" through a single catalytic kinetic run, *Canadian Journal of Analytical Sciences and Spectroscopy*, vol. 46, no. 1, pp. 23–27.

Safavi, A., Absalan, G., & Maesum, S. 2001b, Simultaneous determination of V(IV) and Fe(II) as catalyst using "neural networks" through a single catalytic kinetic run, *Analytica Chimica Acta*, vol. 432, no. 2, pp. 229–233.

Safavi, A., & Bagheri, M. 2003, Novel optical pH sensor for high and low pH values, *Sensors and Actuators, B-Chemical*, vol. 90, no. 1–3, pp. 143–150.

Saini, S., & Bessant, C. 2000, Analysis of mixtures of organic compounds by voltammetry, *PCT International Application*, WO 99-GB3263, pp. 50.

Saldanha, T. C. B., de Araujo, M. C. U., & Neto, B. D. 1999, Simultaneous multicomponent analysis by UV-VIS spectrophotometry, *Quimica Nova*, vol. 22, no. 6, pp. 847–853.

Salles, E. O. T., de Souza, P. A. Jr., & Garg, V. K. 1994, Artificial neural network for identification of a substance from a Mossbauer data bank, *Nuclear Instruments & Methods in Physics Research, Section B: Beam Interactions with Materials and Atoms*, vol. 94, no. 4, pp. 499–502.

Salles, E. O. T., de Souza, P. A. Jr., & Garg, V. K. 1995, Identification of crystalline structures using Mossbauer parameters and artificial neural network, *Journal of Radioanalytical and Nuclear Chemistry*, vol. 190, no. 2, pp. 439–447.

Salter, G. J., Lazzari, M., Giansante, L., Goodacre, R., Jones, A., Surricchio, G., Kell, D. B., & Bianchi, G. 1997, Determination of the geographical origin of Italian extra virgin olive oil using pyrolysis mass spectrometry and artificial neural networks, *Journal of Analytical and Applied Pyrolysis*, vol. 40–1, pp. 159–170.

Sanchez, J. M., Hidalgo, M., Havel, J., & Salvado, V. 2002, The speciation of rhodium(III) in hydrochloric acid media by capillary zone electrophoresis, *Talanta*, vol. 56, no. 6, pp. 1061–1071.

Sanders, J. C., Breadmore, M. C., Kwok, Y. C., Horsman, K. M., & Landers, J. P. 2003, Hydroxypropyl cellulose as an adsorptive coating sieving matrix for DNA separations – artificial neural-network optimization for microchip analysis, *Analytical Chemistry*, vol. 75, no. 4, pp. 986–994.

Sanni, O. D., Wagner, M. S., Briggs, D., Castner, D. G., & Vickerman, J. C. 2002, Classification of adsorbed protein static ToF-SIMS spectra by principal component analysis and neural networks, *Surface and Interface Analysis*, vol. 33, no. 9, pp. 715–728.

Sante, V. S., Lebert, A., LePottier, G., & Ouali, A. 1996, Comparison between two statistical models for prediction of turkey breast meat colour, *Meat Science*, vol. 43, no. 3–4, pp. 283–290.

Sarabia, L. A., Ortiz, M. C., Arcos, M. J., Sanchez, M. S., Herrero, A., & Sanllorente, S. 2002, Multivariate detection capability using a neural classifier for nonselective signals, *Chemometrics and Intelligent Laboratory Systems*, vol. 61, no. 1–2, pp. 89–104.

Sarle, W. S. 2004, Neural Networks FAQ – weekly reminder to the Usenet newsgroup comp.ai.neural-nets available at ftp://ftp.sas.com/pub/neural/FAQ.html.

Saurina, J., Hernandez-Cassou, S., & Tauler, R. 1997, Multivariate curve resolution and trilinear decomposition methods in the analysis of stopped-flow kinetic data for binary amino acid mixtures, *Analytical Chemistry*, vol. 69, no. 13, pp. 2329–2336.

Saurina, J., Hernandez-Cassou, S., Tauler, R., & Izquierdo-Ridorsa, A. 1998, Multivariate resolution of rank-deficient spectrophotometric data from first-order kinetic decomposition reactions, *Journal of Chemometrics*, vol. 12, no. 3, pp. 183–203.

Saurina, J. & Hernandez-Cassou, S. 1999, A comparison of chemometric methods for the flow injection simultaneous spectrophotometric determination of aniline and cyclohexylamine, *Analyst*, vol. 124, no. 5, pp. 745–749.

Sberveglieri, G., Comini, E., Faglia, G., Niederjaufner, G., Pardo, M., Benussi, G. P., Contarini, G., & Povolo, M. 1998, Distinguishing different heat treatments of milk by an electronic nose based on ANN, *Proceedings of the 3rd Italian Conference on Sensors and Microsystems*, Genoa, pp. 205–210.

Schepers, A. W., Thibault, J., & Lacroix, C. 2000, Comparison of simple neural networks and nonlinear regression models for descriptive modeling of Lactobacillus helveticus growth in pH-controlled batch cultures, *Enzyme and Microbial Technology*, vol. 26, no. 5–6, pp. 431–445.

Schirm, B., Benend, H., & Watzig, H. 2001, Improvements in pentosan polysulfate sodium quality assurance using fingerprint electropherograms, *Electrophoresis*, vol. 22, no. 6, pp. 1150–1162.

Schleiter, I. M., Borchardt, D., Wagner, R., Dapper, T., Schmidt, K. D., Schmidt, H. H., & Werner, H. 1999, Modelling water quality, bioindication and population dynamics in lotic ecosystems using neural networks, *Ecological Modelling*, vol. 120, no. 2–3, pp. 271–286.

Schlimme, E., Lorenzen, P. C., Martin, D., Meisel, H., & Thormahlen, K. 1997, Identification of butter types, *Kieler Milchwirtschaftliche Forschungsberichte*, vol. 49, no. 2, pp. 135–146.

Schnurer, J., Olsson, J., & Borjesson, T. 1999, Fungal volatiles as indicators of food and feeds spoilage, *Fungal Genetics and Biology*, vol. 27, no. 2–3, pp. 209–217.

Schulze, H. G., Blades, M. W., Bree, A. V., Gorzalka, B. B., Greek, S. G., & Turner, R. F. B. 1994, Characteristics of backpropagation neural networks employed in the identification of neurotransmitter Raman spectra, *Applied Spectroscopy*, vol. 48, no. 1, pp. 50–57.

Schulze, H. G., Greek, L. S., Gorzalka, B. B., Bree, A. V., Blades, M. W., & Turner, R. F. B. 1995, Artificial neural-network and classical least-squares methods for neurotransmitter mixture analysis, *Journal of Neuroscience Methods* vol. 56, no. 2, pp 155–167.

Sebastian, I., Viallon, C., Tournayre, P., & Berdague, J. L. 2000, Pyrolysis-mass spectrometry characterisation of ovine fat tissues according to diet, *Analusis*, vol. 28, no. 2, pp. 141–147.

Sedo, O. & Havel, J. 2003, Analysis of peptides by MALDI-TOF mass spectrometry, *Chemicke Listy*, vol. 97, no. 2, pp. 109–113.

Seker, S. & Becerik, I. 2003, Neural-network modeling of an electrochemical process, *Ann.Chim.Rome*, vol. 93, no. 5–6, pp. 551–560.

Selzer, P., Gasteiger, J., Thomas, H., & Salzer, R. 2000, Rapid access to infrared reference spectra of arbitrary organic compounds: scope and limitations of an approach to the simulation of infrared spectra by neural networks, *Chemistry – A European Journal*, vol. 6, no. 5, pp. 920–927.

Sentellas, S., Saurina, J., Hernandez-Cassou, S., Galceran, M. T., & Puignou, L. 2003, Quantitation in multianalyte overlapping peaks from capillary electrophoresis runs using artificial neural networks, *Journal of Chromatographic Science*, vol. 41, no. 3, pp. 145–150.

Shaffer, R. E., & Rose-Pehrsson, S. L. 1999, Improved probabilistic neural network algorithm to chemical sensor array pattern recognition, *Analytical Chemistry*, vol. 71, no. 19, pp. 4263–4271.

Shamsipur, M., Hemmatee-Nejad, B., & Akhond, M. 2002, Simultaneous determination of promethazine, chlorpromazine, and perphenazine by multivariate calibration methods and derivative spectrophotometry, *Journal of AOAC International*, vol. 85, no. 3, pp. 555–562.

Shan, Y., Zhao, R. H., Tian, Y., Liang, Z., & Zhang, Y. K. 2002, Retention modeling and optimization of pH value and solvent composition in HPLC using back-propagation neural networks and uniform design, *Journal of Liquid Chromatography & Related Technologies*, vol. 25, no. 7, pp. 1033–1047.

Shao, X. G., Leung, A. K. M., & Cau, F. T. 2003, Wavelet: a new trend in chemistry, *Accounts of Chemical Research*, vol. 36, no. 4, pp. 276–283.

Sharma, A. K., Sheikh, S., Pelczer, I., & Levy, G. C. 1994, Classification and clustering: using neural networks, *Journal of Chemical Information and Computer Sciences*, vol. 34, pp. 1130–1139.

Shaw, A. D., Kaderbhai, N. Jones, A., Woodward, A. M., Goodacre, R., Rowland, J. J., & Kell, D. B. 1999, Noninvasive, online monitoring of the biotransformation by yeast of glucose to ethanol using dispersive Raman-spectroscopy and chemometrics, *Applied Spectroscopy*, vol. 53, no. 11, pp 1419–1428.

Shen, H., Carter, J. F., Brereton, R. G., & Eckers, C. 2003, Discrimination between tablet production methods using pyrolysis-gas chromatography-mass spectrometry and pattern recognition, *Analyst*, vol. 128, no. 3, pp. 287–292.

Shih, J. S. 2000, Piezoelectric crystal membrane chemical sensors based on fullerene and macrocyclic polyethers, *Journal of the Chinese Chemical Society*, vol. 47, no. 1, pp. 21–32.

Shin, H. W., Llobet, E., Gardner, J. W., Hines, E. L., & Dow, C. S. 2000, Classification of the strain and growth phase of cyanbacteria in potable water using an electronic nose system, *IEE Proceedings: Science, Measurements and Technologies*, vol. 147, no. 4, pp. 158–164.

Shu, Y., Chen, Y., & Lu, W. 1994, A taste-sensing system with BP and SOM neural network, *Proceedings of the 5th International Meeting on Chemical Sensors*, Rome, Italy, pp.1041–1044.

Shukla, K. K., & Raghunath 1999, An efficient global algorithm for supervised training of neural networks, *Computers and Electrical Engineering*, vol. 25, pp. 195–218.

Sigman, M. E., & Rives, S. S. 1994, Prediction of atomic ionization-potentials I-III using an artificial neural-network, *Journal of Chemical Information and Computer Sciences*, vol. 34, no. 3, pp 617–620.

Simhi, R., Gotshal, Y., Bunimovich, D., Sela, B. A., & Katzir, A. 1996, Fiberoptic evanescent-wave spectroscopy for fast multicomponent analysis of human blood, *Applied Optics*, vol. 35, no. 19, pp. 3421–3425.

Simpson, M., Anderson, D. R., Mcleod, C. W., & Cooke, M. 1993, Use of pattern recognition for laser desorption – ion mobility spectrometery of polymeric material, *Analyst*, vol. 118, pp. 1293–1298.

Sinesio, F., Di Natale, C., Quaglia, G. B., Bucarelli, F. M., Moneta, E., Macagnano, A., Paolesse, R., D'Amico, A. 2000, Use of electronic nose and trained sensory panel in the evaluation of tomato quality, *Journal of the Science of Food and Agriculture*, vol. 80, no. 1, pp. 63–71.

Siouffi, A. M. & Phan-Tan-Luu, R. 2000, Optimization methods in chromatography and capillary electrophoresis, *Journal of Chromatography A*, vol. 892, no. 1–2, pp. 75–106.

Sisson, P. R., Freeman, R., Law, D., Ward, A. C., & Lightfoot, N. F. 1995, Rapid detection of verocytotoxin production status in Escherichia-Coli by artificial neural-network analysis of pyrolysis mass-spectra, *Journal of Analytical and Applied Pyrolysis*, vol. 32, pp. 179–185.

Skarpeid, H. J., Kvaal, K., & Hildrum, K. I. 1998, Identification of animal species in ground meat mixtures by multivariate analysis of isoelectric focusing protein profiles, *Electrophoresis*, vol. 19, no. 18, pp. 3103–3109.

Slot, M. 1997, Conditions of the Maillard reaction. Modelling of the process, *Inzynieria Chemiczna I Procesowa*, vol. 18, no. 1, pp. 71–82.

Smits, J. R. M., Melssen, W. J., Derksen, M. W. J., & Kateman, G. 1993, Drift correction for pattern classification with neural networks, *Analytica Chimica Acta*, vol. 248, pp. 91–105.

Song, M. H., Breneman, C. M., Bi, J. B., Sukumar, N., Bennett, K. P., Cramer, S., & Tugcu, N. 2002, Prediction of protein retention times in anion-exchange chromatography systems using support vector regression, *Journal of Chemical Information and Computer Sciences*, vol. 42, no. 6, pp. 1347–1357.

Song, X. H., Hopke, P. K., Fergenson, D. P., & Prather, K. A. 1999, Classification of single particles analyzed by ATOFMS using an artificial neural-network, ART-2A, *Analytical Chemistry*, vol. 71, no. 4, pp. 860–865.

Song, X. H., Faber, N. M., Hopke, P. K., Suess, D. T., Prather, K. A., Schauer, J. J., & Cass, G. R. 2001, Source apportionment of gasoline and diesel by multivariate calibration based on single particle mass spectral data, *Analytica Chimica Acta*, vol. 446, no. 1–2, pp. 329–343.

Sorensen, H. A., Sperotto, M. M., Petersen, M., Kesmir, C., Radzikowski, L., Jacobsen, S., & Sondergaard, I. 2002, Variety identification of wheat using mass spectrometry with neural networks and the influence of mass spectra processing prior to neural network analysis, *Rapid Communications in Mass Spectrometry*, vol. 16, no. 12, pp. 1232–1237.

Spining, M. T., Darsey, J. A., Sumpter, B. G., & Nold, D. W. 1994, Opening up the black box of artificial neural networks, *Journal of Chemical Education*, vol. 71, no. 5, pp. 406–411.

Srecnik, G., Debeljak, Z., Cerjan-Stefanovic, S., Bolanca, T., Novic, M., Lazaric, K., & Gumhalter-Lulic, Z. 2002a, Use of artificial neural networks for retention modelling in ion chromatography, *Croatica Chemica Acta*, vol. 75, no. 3, pp. 713–725.

Srecnik, G., Debeljak, Z., Cerjan-Stefanovic, S., Novic, M., & Bolanca, T. 2002b, Optimization of artificial neural networks used for retention modelling in ion chromatography, *Journal of Chromatography A*, vol. 973, no. 1–2, pp. 47–59.

Sreekrishnan, T. R. & Tyagi, R. D. 1995, Sensitivity of metal-bioleaching operation to process variables, *Process Biochemistry*, vol. 30, no. 1, pp. 69–80.

Stefan, R. I., van Staden, J. F., & Aboul-Enein, H. Y. 1999, Electrochemical sensor arrays, *Critical Reviews in Analytical Chemistry*, vol. 29, no. 2, pp. 133–153.

Stegmann, J. A., & Buenfeld, N. R. 1999, A glossary of basic neural network terminology for regression problems, *Neural Computing & Applications*, vol. 8, pp. 290–296.

Steinhart, H., Stephan, A., & Bucking, M. 2000, Advances in flavor research, *Journal of High Resolution Chromatography*, vol. 23, no. 7–8, pp. 489–496.

Stetter, J. R., Findlay, Jr., M. W., Schroeder, K. M., Yue, C., & Penrose, W. R. 1992, Quality classification of grain using a sensor array and pattern recognition, *Analytica Chimica Acta*, vol. 284, pp. 1–11.

Steyer, J. P., Rolland, D., Bouvier, J. C., & Moletta, R. 1997, Hybrid fuzzy neural network for diagnosis – application to the anaerobic treatment of wine distillery wastewater in a fluidized bed reactor, *Water Science and Technology*, vol. 36, no. 6–7, pp. 209–217.

Suah, F. B. M., Ahmad, M., & Taib, M. N. 2003a, Optimization of the range of an optical fiber pH sensor using feed-forward artificial neural network, *Sensors and Actuators, B-Chemical*, vol. 90, no. 1–3, pp. 175–181.

Suah, F. B. M., Ahmad, M., & Taib, M. N. 2003b, Applications of artificial neural network on signal processing of optical fibre pH sensor based on bromophenol blue doped with sol-gel film *Sensors and Actuators, B-Chemical*, vol. 90, no. 1–3, pp. 182–188.

Sumpter, B. G., Getino, C., & Noid, D. W. 1994, Theory and applications of neural computing in chemical science, *Annual Review of Physical Chemistry*, vol. 45, pp. 439–481.

Sun, G., Chen, X. G., Zhao, Y. K., Liu, M. C., & Hu, Z. D. 2000a, Determination of Pd(II) by application of an on-line microwave oven and artificial neural networks in flow injection analysis, *Analytica Chimica Acta*, vol. 420, no. 1, pp. 123–131.

Sun, G., Chen, X-G., Li, Q.-F., Wang, H.-W., Zhou, Y.-Y., & Hu, Z. 2000b, Evaluation of nonlinear modeling based on artificial neural networks for the spectrophotometric determination of Pd(II) with CPA-mK, *Fresenius Journal of Analytical Chemistry*, vol. 367, no. 3, pp. 215–219.

Sun, G., Zhou, Y.-Y., Wang, H.-W., Chen, H.-L., Chen, X.-G., & Hu, Z. 2000c, Application of artificial neural networks coupled with an orthogonal design and optimization algorithms to multifactor optimization of a new FIA system for the determination of uranium(vi) in ore samples, *Analyst*, vol. 125, no. 5, pp. 921–925.

Sun, L. X., Danzer, K., & Thiel, G. 1997, Classification of wine samples by means of artificial neural networks and discrimination analytical methods, *Fresenius Journal of Analytical Chemistry*, vol. 359, no. 2, pp. 143–149.

Sun, L,-X., Reddy, A. M., Matsuda, N., Takatsu, A., Kato, K., & Okada, T. 2003, Simultaneous determination of methylene blue and new methylene blue by slab optical waveguide spectroscopy and artificial neural networks, *Analytica Chimica Acta*, vol. 487, no. 1, pp. 109–116.

Sutter, J. M., Peterson, T. A., & Jurs, P. C. 1997, Prediction of gas chromatographic retention indices of alkylbenzenes, *Analytica Chimica Acta*, vol. 342, no. 2–3, pp. 113–122.

Suzuki, T., Timofei, S., Iuoras, B. E., Uray, G., Verdino, P., & Fabian, W. M. F. 2001, Quantitative structure-enantioselective retention relationships for chromatographic separation of arylalkylcarbinols on Pirkle type chiral stationary phases, *Journal of Chromatography A*, vol. 922, no. 1–2, pp. 13–23.

Svozil, D., Pospichal, J., & Kvasnicka, V. 1995, Neural-network prediction of C-13 NMR chemical-shifts of alkanes, *Journal of Chemical Information and Computer Sciences*, vol. 35, no. 5, pp. 924–928.

Svozil, D., Kvasnicka, V., & Pospichal, J. 1997, Introduction to multilayer feedforward neural networks, *Chemometrics and Intelligent Laboratory Systems*, vol. 39, no. 1, pp. 43–62.

Syu, M. J. & Chen, B. C. 1998a, Back-propagation neural network adaptive control of a continuous wastewater treatment process, *Industrial & Engineering Chemistry Research*, vol. 37, no. 9, pp. 3625–3630.

Syu, M. J. & Liu, J. Y. 1998b, Neural network signal detection of an SO2 electrode, *Sensors and Actuators B-Chemical*, vol. 49, no. 3, pp. 186–194.

Tafeit, E. & Reibnegger, G. 1999, Artificial neural networks in laboratory medicine and medical outcome prediction, *Clin.Chem.Lab Med.*, vol. 37, no. 9, pp. 845–853.

Taib, M. N., Andres, R., & Narayanaswamy, R. 1996, Extending the response range of an optical fibre pH sensor using an artificial neural network, *Analytica Chimica Acta*, vol. 330, no. 1, pp. 31–40.

Talaie, A. & Romagnoli, J. A. 1996a, An integrated artificial neural network polymer-based pH sensor: a new engineering perspective to conducting polymer technology, *Synthetic Metals*, vol. 82, no. 3, pp. 231–235.

Talaie, A., Boger, Z., Romagnoli, J. A., Adeloju, S. B., & Yuan, Y. J. 1996b, Data acquisition, signal processing and modelling: a study of a conducting polypyrrole formate biosensor. 1. Batch experiment, *Synthetic Metals*, vol. 83, no. 1, pp. 21–26.

Talaie, A. & Romagnoli, J. A. 1996c, Application of artificial neural networks to the real-time operation of conducting polymer sensors: a pattern recognition approach, *Synthetic Metals*, vol. 82, no. 1, pp. 27–33.

Talaie, A., Romagnoli, J. A. 1997, Pattern recognition application in conducting polymer sensors, *Proceedings of the IFAC ADCHEM Conference*, Banff, Canada, pp. 329–333.

Talaie, A., Lee, J. Y., Adachi, K., Taguchi, T., & Romagnoli, J. 1999, Dynamic polymeric electrodes, dynamic computer modeling and dynamic electrochemical sensing, *Journal of Electroanalytical Chemistry*, vol. 468, no. 1, pp. 19–25.

Talaie, A., Lee, J. Y., Lee, Y. K., Adachi, K., Taguchi, T., Maeder, E., Jang, J., & Romagnoli, J. A. 2001, Dynamic modeling of a polymeric composite interface: an introduction to in situ neurocomputing in composite-based pH sensors, *Composite Interfaces*, vol. 8, no. 2, pp. 127–134.

Tanabe, K., Tamura, T., & Uesaka, H. 1992, Neural network systems for the identification of infrared spectra, *Applied Spectroscopy*, 46, no. 5, pp. 807–810.

Tanabe, K., Matsumoto, T., Tamura, T., Hiraishi, J., Saeki, S., Arima, M., Ono, C., Itoh, S., Uesaka, H., Tatsugi, Y., Yatsunami, K., Inaba, T., Mitsuhashi, M., Kohara, S., Masago, H., Kaneuchi, F., Jin, C., & Ono, S. 2001, Identification of chemical structures from infrared-spectra by using neural networks, *Applied Spectroscopy*, vol. 55, no. 10, pp. 1394–1403.

Tandler, P. J., Butcher, J. A., Tao, H., & Harrington, P. D. 1995, Analysis of plastic recycling products by expert-systems, *Analytica Chimica Acta*, vol. 312, no. 3, pp. 231–244.

Tas, A. C. & Vandergreef, J. 1994, Mass-spectrometric profiling and pattern-recognition, *Mass Spectrometry Reviews*, vol. 13, no. 2, pp. 155–181.

Taurino, A. M., Distante, C., Siciliano, P., & Vasanelli, L. 2003, Quantitative and qualitative analysis of VOCs mixtures by means of a microsensor array and different evaluation methods, *Sensors and Actuators, B-Chemical,* vol. 93, no. 1–3, pp. 117–125.

Taylor, J., Goodacre, R., Wade, W. G., Rowland, J. J., & Kell, D. B. 1998, The deconvolution of pyrolysis mass spectra using genetic programming: application to the identification of some Eubacterium species, *FEMS Microbiology Letters*, vol. 160, no. 2, pp. 237–246.

Taylor, J., King, R. D., Altmann, T., & Fiehn, O. 2002, Application of metabolomics to plant genotype discrimination using statistics and machine learning, *Bioinformatics*, vol. 18, pp. S241–S248.

Tchistiakov, V., Ruckebusch, C., Duponchel, L., Huvenne, J.-P., & Legrand, P. 2000, Neural network modeling for very small spectral data sets: reduction of the spectra and hierarchical approach, *Chemometrics and Intelligent Laboratory Systems*, vol. 54, no. 2, pp. 93–106.

Teissier, P., Perret, B., Latrille, E., Barillere, J. M., & Corrieu, G. 1996, Yeast concentration estimation and prediction with static and dynamic neural network models in batch cultures, *Bioprocess Engineering*, vol. 14, no. 5, pp. 231–235.

Tenhagen, A., Feuring, T., & Lippe, W. M. 1998, Testing the identity of hashish samples with ICP-AES and NAA and data handling with neural networks. 2. Data verification with the use of artificial neural networks, *Pharmazie*, vol. 53, no. 1, pp. 39–42.

Tetko, I. V. 2002, Neural-network studies. 4. Introduction to associative neural networks, *Journal of Chemical Information and Computer Sciences*, vol. 42, no. 3, pp. 717–728.

Thaler, S. L. 1993, Neural net predicted Raman spectra of the graphite to diamond transition, *Proceedings of the 3rd International Symposium on Diamond Materials*, Honolulu, Hawaii.

Tham, S. Y. & Agatonovic-Kustrin, S. 2002, Application of the artificial neural network in quantitative structure-gradient elution retention relationship of phenylthiocarbamyl amino acids derivatives, *Journal of Pharmaceutical and Biomedical Analysis*, vol. 28, no. 3–4, pp. 581–590.

Thodberg, H. H. 1996, A review of Bayesian neural networks with an application to near infrared spectroscopy, *IEEE Transactions on Neural Networks*, vol. 7, no. 1, pp. 56–72.

Thompson, S., Fueten, F., & Bockus, D. 2001, Mineral identification using artificial neural networks and the rotating polarizer stage, *Computers & Geosciences*, vol. 27, no. 9, pp. 1081–1089.

Thompson, S. E., Foster, N. S., Johnson, T. J. Valentine, N. B., & Amonette, J. E. 2003, Identification of bacterial-spores using statistical-analysis of Fourier-transform infrared photoacoustic-spectroscopy data, *Applied Spectroscopy*, vol. 57, no. 8, pp. 893–899.

Timmins, E. M. & Goodacre, R. 1997, Rapid quantitative analysis of binary mixtures of Escherichia coli strains using pyrolysis mass spectrometry with multivariate calibration and artificial neural networks, *Journal of Applied Microbiology*, vol. 83, no. 2, pp. 208–218.

Timmins, E. M., Howell, S. A., Alsberg, B. K., Noble, W. C., & Goodacre, R. 1998a, Rapid differentiation of closely related Candida species and strains by pyrolysis mass spectrometry and Fourier transform-infrared spectroscopy, *Journal of Clinical Microbiology*, vol. 36, no. 2, pp. 367–374.

Timmins, E. M., Quain, D. E., & Goodacre, R. 1998b, Differentiation of brewing yeast strains by pyrolysis mass spectrometry and Fourier transform infrared spectroscopy, *Yeast*, vol. 14, no. 10, pp. 885–893.

Tolle, K. M., Chen, H., & Chow, H.-H. 2000, Estimating drug/plasma concentration levels by applying neural networks to pharmacokinetic data sets, *Decision Support Systems*, vol. 30, pp. 139–151.

Tominaga, Y. 1999, Comparative study of class data analysis with PCA-LDA, SIMCA, PLS, ANNs, and k-NN, *Chemometrics and Intelligent Laboratory Systems*, vol. 49, no. 1, pp. 105–115.

Tong, C. S. & Cheng, K. C. 1999, Mass spectral search method using the neural network approach, *Chemometrics and Intelligent Laboratory Systems*, vol. 49, no. 2, pp. 135–150.

Tusar, M., Zupan, J., & Gasteiger, J. 1992, Neural networks and modeling in chemistry, *Journal de Chimie Physique et de Physico-Chimie Biologique*, vol. 89, no. 7–8, pp. 1517–1529.

Udelhoven, T. & Schutt, B. 2000a, Capability of feed-forward neural networks for a chemical evaluation of sediments with diffuse reflectance spectroscopy, *Chemometrics and Intelligent Laboratory Systems*, vol. 51, no. 1, pp. 9–22.

Udelhoven, T., Naumann, D., & Schmitt, J. 2000b, Development of a hierarchical-classification system with artificial neural networks and FT-IR spectra for the identification of bacteria, *Applied Spectroscopy*, vol. 54, no. 10, pp. 1471–1479.

Udelhoven, T., Novozhilov, M., & Schmitt, J. 2003, The NeuroDeveloper: a tool for modular neural classification of spectroscopic data, *Chemometrics and Intelligent Laboratory Systems*, vol. 66, no. 2, pp. 219–226.

Urruty, L., Giraudel, J. L., Lek, S., Roudeillac, P., & Montury, M. 2002, Assessment of strawberry aroma through SPME/GC and ANN methods. Classification and discrimination of varieties, *Journal of Agricultural and Food Chemistry*, vol. 50, no. 11, pp. 3129–3136.

Vaananen, T., Koskela, H., Hiltunen, Y., & Ala-Korpela, M. 2002, Application of quantitative artificial neural network analysis to 2D NMR spectra of hydrocarbon mixtures, *Journal of Chemical Information and Computer Sciences*, vol. 42, no. 6, pp. 1343–1346.

Valcarcel, M., Gomez-Hens, A., & Rubio, S. 2001, Selectivity in analytical chemistry revisited, *TRAC-Trends in Analytical Chemistry*, vol. 20, no. 8, pp. 386–393.

Van Can, H. J. L., teBraake, H. A. B., Dubbelman, S., Hellinga, C., Luyben, K. C. A. M., & Heijnen, J. J. 1998, Understanding and applying the extrapolation properties of serial gray-box models, *AIChE Journal*, vol. 44, no. 5, pp. 1071–1089.

Van der Linden, W. E., Bos, M., & Bos, A. 1989, Arrays of electrodes for multicomponent analysis, *Analytical Proceedings*, vol. 26, no. 10, pp. 329–331.

Vargas, G., Havel, J., & Frgalova, K. 1998a, Capillary zone electrophoresis determination of thyreostatic drugs in urine, *Journal of Capillary Electrophoresis*, vol. 5, no. 1–2, pp. 9–12.

Vargas, G., Havel, J., Babackova, L., & Patocka, J. 1998b, Determination of drugs used as anti-Parkinson's disease drugs in urine and serum by capillary electrophoresis, *Journal of Capillary Electrophoresis*, vol. 5, no. 3–4, pp. 153–158.

Vashi, P. R., Cukrowski, I., & Havel, J. 2001, Stability constants of the inclusion complexes of beta-cyclodextrin with various adamantane derivatives. A UV-Vis study, *South African Journal of Chemistry – Suid-Afrikaanse Tydskrif Vir Chemie*, vol. 54, no. 5.

Vazquez, M. J., Lorenzo, R. A., & Cela, R. 2003, The use of an 'electronic nose' device to monitor the ripening process of anchovies, *International Journal of Food Science and Technology*, vol. 38, no. 3, pp. 273–284.

Veltri, R. W., Miller, M. C., & An, G. 2001, Standardization, analytical validation, and quality control of intermediate endpoint biomarkers, *Urology*, vol. 57, no. 4, suppl 1, pp. 164–170.

Vieira, A. & Barradas, N. P. 2000, Neural-network analysis of Rutherford backscattering data, *Nucl.Instrum.Meth.Phys.Res.B*, vol. 170, no. 1–2, pp. 235–238.

Vieira, A. & Barradas, N. P. 2001a, Composition of NiTaC films on Si using neural networks analysis of elastic backscattering data, *Nucl.Instrum.Meth.Phys.Res.B*, vol. 174, no. 3, pp. 367–372.

Vieira, A., Barradas, N. P., & Jeynes, C. 2001b, Error performance analysis of artificial neural networks applied to Rutherford backscattering, *Surface and Interface Analysis*, vol. 31, no. 1, pp. 35–38.

Vieira, A., Barradas, N. P., & Alves, E. 2002, Analysis of sapphire implanted with different elements using artificial neural networks, *Nucl.Instrum.Meth.Phys.Res.B*, vol. 190, May, pp. 241–246.

Visser, T., Luinge, H. J., & Vandermaas, J. H. 1994, Recognition of visual characteristics of infrared-spectra by artificial neural networks and partial least-squares regression, *Analytica Chimica Acta*, vol. 296, no. 2, pp. 141–154.

Vlasov, Y. G., Legin, A. V., Rudnitskaya, A. M., Damico, A., & Di Natale, C. 1997, Chemical-analysis of multicomponent aqueous-solutions using a system of nonselective sensors and artificial neural networks, *Journal of Analytical Chemistry*, vol. 52, no. 11, pp. 1087–1092.

Vlasov, Y. & Legin, A. 1998, Non-selective chemical sensors in analytical chemistry: from 'electronic nose' to 'electronic tongue', *Fresenius Journal of Analytical Chemistry*, vol. 361, no. 3, pp. 255–260.

Volmer, M., Devries, J. C. M., & Goldschmidt, H. M. J. 2001, Infrared-analysis of urinary calculi by a single reflection accessory and a neural-network interpretation algorithm, *Clinical Chemistry*, vol. 47, no. 7, pp. 1287–1296.

Wailzer, B., Klocker, J., Buchbauer, G., Ecker, G., & Wolschann, P. 2001, Prediction of the aroma quality and the threshold values of some pyrazines using artificial neural networks, *Journal of Medicinal Chemistry*, vol. 44, no. 17, pp. 2805–2813.

Walczak, B., & Wegscheider, W. 1994, Calibration of nonlinear analytical systems by a neuro-fuzzy approach, *Chemometrics and Intelligent Laboratory Systems*, vol. 22, no. 2, pp. 199–207.

Wan, C. H. & Harrington, P. D. 1999, Self-configuring radial basis function neural networks for chemical pattern recognition, *Journal of Chemical Information and Computer Sciences*, vol. 39, no. 6, pp. 1049–1056.

Wan, C. H. & Harrington, P. D. 2000, Screening GC-MS data for carbamate pesticides with temperature-constrained-cascade correlation neural networks, *Analytica Chimica Acta*, vol. 408, no. 1–2, pp. 1–12.

Wang, C., Chen, F., & He, X. W. 2002a, Kinetic detection of benzene/chloroform and toluene/chloroform vapors using a single quartz piezoelectric crystal coated with calix(6)arene, *Analytica Chimica Acta*, vol. 464, no. 1, pp. 57–64.

Wang, D., Dowell, F. E., & Lacey, R. E. 1999, Single wheat kernel color classification using neural networks, *Transactions of the ASEA*, vol. 42, no. 1, pp. 233–240.

Wang, D., Ram, M. S., & Dowell, F. E. 2002b, Classification of damaged soybean seeds using near-infrared spectroscopy, *Transactions of the ASEA*, vol. 45, no. 6, pp. 1943–1948.

Wang, H. W., Zhou, Y. Y., Zhao, Y. K., Li, Q. F., Chen, X. G., & Hu, Z. D. 2001a, Optimization of online microwave flow-injection analysis system by artificial neural networks for the determination of Ruthenium, *Analytica Chimica Acta*, vol. 429, no. 2, pp. 207–213.

Wang, J., Guo, P., Li, X., Zhu, J., Reinert, T., Heitmann, J., Spemann, D., Vogt, J., Flagmeyer, R. H., & Butz, T. 2000, Identification of air-pollution sources by single aerosol-particle fingerprints – micro-PIXE spectra, *Nucl.Instrum.Meth.Phys.Res.B*, vol. 161, Mar., pp. 830–835.

Wang, J. H., Jiang, J. H., & Yu, R. Q. 1996, Robust back-propagation algorithm as a chemometric tool to prevent the overfitting to outliers, *Chemometrics and Intelligent Laboratory Systems*, vol. 34, no. 1, pp. 109–115.

Wang, X. Z. & Chen, B. H. 1998, Clustering of infrared spectra of lubricating base oils using adaptive resonance theory, *Journal of Chemical Information and Computer Sciences*, vol. 38, no. 3, pp. 457–462.

Wang, Y. W., Gao, S. L., Gao, Y. H., Liu, S. H., Liu, M. C., Hu, Z. D., & Fan, B. T. 2003, Study of the relationship between the structure and the relative mobility of chlorophenols in different buffers modified by different organic additives by capillary zone electrophoresis, *Analytica Chimica Acta*, vol. 486, no. 2, pp. 191–197.

Weinstein, J. N., Myers, T., Buolamwini, J., Raghavan, K., Vanosdol, W., Licht, J., Viswanadhan, V. N., Kohn, K. W., Rubinstein, L. V., Koutsoukos, A. D., Monks, A., Scudiero, D. A., Anderson, N. L., Zaharevitz, D., Chabner, B. A., Grever, M. R., & Paull, K. D. 1994, Predictive statistics and artificial-intelligence in the US National Cancer Institutes drug discovery program for cancer and aids, *Stem Cells*, vol. 12, no. 1, pp. 13–22.

Welsh, W. J., Lin, W. K., Tersigni, S. H., Collantes, E., Duta, R., Carey, M. S., Zielinski, W. L., Brower, J., Spencer, J. A., & Layloff, T. P. 1996, Pharmaceutical fingerprinting: evaluation of neural networks and chemometric techniques for distinguishing among same product manufacturers, *Analytical Chemistry*, vol. 68, no. 19, pp. 3473–3482.

Wen, L. J., Zhang, L. Y., Zhou, F., Lu, Y., & Yang, P. Y. 2002, Quantitative-determination of Freon gas using an electronic nose, *Analyst*, vol. 127, no. 6, pp. 786–791.

Werbos, P. 1974, Beyond regression: new tools for prediction and analysis in the behavioral sciences, PhD dissertation, Harvard University.

Werbos, P. 1994, *The Roots of Backpropagation: from Ordered Derivatives to Neural Networks and Political Forecasting*, Wiley.

Werther, W., Lohninger, H., Varmuza, K., & Stancl, F. 1994, Classification of mass-spectra – a comparison of yes/no classification methods for the recognition of simple structural-properties, *Chemometrics and Intelligent Laboratory Systems*, vol. 22, no. 1, pp. 63–76.

Wesolowski, M. & Suchacz, B. 2001, Classification of rapeseed and soybean oils by use of unsupervised pattern-recognition methods and neural networks, *Fresenius Journal of Analytical Chemistry*, vol. 371, no. 3, pp. 323–330.

Wesolowski, M. & Suchacz, B. 2002, Neural networks as a tool for quality assessment of soybean and rapeseed oils analyzed by thermogravimetry, *J.Therm.Anal.Calorim.*, vol. 68, no. 3, pp. 893–899.

Wesolowski, M. & Czerwonka, M. 2003, Quality assessment of medicinal cod liver, edible and technical fish oils based on derivative thermogravimetry data, *Thermochimica Acta*, vol. 398, no. 1–2, pp. 175–183.

Wessel, M. D. & Jurs, P. C. 1994, Prediction of reduced ion mobility constants from structural information using multiple linear-regression analysis and computational neural networks, *Analytical Chemistry*, vol. 66, no. 15, pp. 2480–2487.

Wessel, M. D., Sutter, J. M., & Jurs, P. C. 1996, Prediction of reduced ion mobility constants of organic compounds from molecular structure, *Analytical Chemistry*, vol. 68, no. 23, pp. 4237–4243.

Weigel U.-M., & Herges, R. 1992, Automatic interpretation of infrared spectra: recognition of aromatic substitution patterns using neural networks, *Journal of Chemical Information and Computer Sciences,* vol. 32, pp. 723–731.

Wienke, D., & Kateman, G. 1994, Adaptive resonance theory-based artificial neural networks for treatment of open-category problems in chemical-pattern recognition – application to UV-Vis and IR spectroscopy, *Chemometrics and Intelligent Laboratory Systems,* vol. 23, no. 2, pp. 309–329.

Wilcox, S. J., Hawkes, D. L., Hawkes, F. R., & Guwy, A. J. 1995, A neural network, based on bicarbonate monitoring, to control anaerobic digestion, *Water Research,* vol. 29, no. 6, pp. 1465–1470.

Wilson, D. M., Dunman, K., Roppel, T., & Kalim, R. 2000, Rank extraction in tin-oxide sensor arrays, *Sensors and Actuators, B-Chemical,* vol. 62, no. 3, pp. 199–210.

Wilson, D. M., & Roppel, T., Signal processing architectures for chemical sensing microsystems, *Sensor Update.*

Winquist, F., Krantz-Rulcker, C., Wide, P., & Lundstrom, I. 1998, Monitoring of freshness of milk by an electronic tongue on the basis of voltammetry, *Measurement Science Technology,* vol. 9, pp. 1937–1946.

Winson, M. K., Goodacre, R., Timmins, E. M., Jones, A., Alsberg, B. K., Woodward, A. M., Rowland, J. J., & Kell, D. B. 1997, Diffuse reflectance absorbance spectroscopy taking in chemometrics (DRASTIC). A hyperspectral FT-IR-based approach to rapid screening for metabolite overproduction, *Analytica Chimica Acta,* vol. 348, no. 1–3, pp. 273–282.

Wiskur, S. L., Floriano, P. N., Anslyn, E. V., & McDevitt, J. T. 2003, A multicomponent sensing ensemble in solution: differentiation between structurally similar analytes, *Angewandte Chemie, International Edition,* vol. 42, no. 18, pp. 2070–2072.

Wolbach, J. P., Lloyd, D. K., & Wainer, I. W. 2001, Approaches to quantitative structure-enantioselectivity relationship modeling of chiral separations using capillary electrophoresis, *Journal of Chromatography A,* vol. 914, no. 1–2, pp. 299–314.

Wolf, G., Almeida, J. S., Pinheiro, C., Correia, V., Rodrigues, C., Reis, M. A. M., & Crespo, J. G. 2001, Two-dimensional fluorometry coupled with artificial neural networks: a novel method for on-line monitoring of complex biological processes, *Biotechnology and Bioengineering,* vol. 72, no. 3, pp. 297–306.

Woodward, A. M., Jones, A., Zhang, X. Z., Rowland, J., & Kell, D. B. 1996, Rapid and non-invasive quantification of metabolic substrates in biological cell suspensions using non-linear dielectric spectroscopy with multivariate calibration and artificial neural networks. Principles and applications, *Bioelectrochemistry and Bioenergetics,* vol. 40, no. 2, pp. 99–132.

Woodward, A. M., Kaderbhai, N., Kaderbhai, M., Shaw, A., Rowland, J., & Kell, D. B. 2001, Histometrics: improvement of the dynamic range of fluorescently stained proteins resolved in electrophoretic gels using hyperspectral imaging, *Proteomics,* vol. 1, no. 11, pp. 1351–1358.

Workman, J. J. Jr., Mobley, P. R., Kowalski, B. R., & Bro, R. 1996, Review of chemometrics applied to spectroscopy: 1985–95, Part I, *Applied Spectroscopy Reviews,* vol. 31, no. 1–2, pp. 73–124.

Wu, W., Walczak, B., Massart, D. L., Heueding, S., Erni, F., Last, I. R., & Prebble, K. A. 1995, Artificial neural networks in classification of NIR spectral data: design of the training set, *Chemometrics and Intelligent Laboratory Systems,* vol. 33, pp. 35–46.

Wunderlich, B. 1992, New directions in thermal analysis, *Thermochimica Acta,* vol. 212, pp. 131–141.

Wythoff, B. J., Levine, S. P., & Tomellini, S. A. 1990, Spectral peak verification and recognition using a multilayered neural network, *Analytical Chemistry,* vol. 62, no. 24, pp. 2702–2709.

Xiang, L., Fan, G. Q., Li, J. H., Kang, H., Yan, Y. L., Zheng, J. H., & Guo, D. 2002, The application of an artificial neural-network in the identification of medicinal rhubarbs by near-infrared spectroscopy, *Phytochemistry Analysis,* vol. 13, no. 5, pp. 272–276.

Xiang, Y. H., Liu, M. C., Zhang, X. Y., Zhang, R. S., Hu, Z. D., Fan, B. T., Doucet, J. P., & Panaye, A. 2002, Quantitative prediction of liquid chromatography retention of n-benzylideneanilines based on quantum chemical parameters and radial basis function neural network, *Journal of Chemical Information and Computer Sciences,* vol. 42, pp. 592.

Xie, Y. L., Baeza-Baeza, J. J., Torres-Lapasio, J. R., Garcia-Alvarez-Coque, M. C., & Ramisramos, G. 1995, Modeling and prediction of retention in high-performance liquid-chromatography by using neural networks, *Chromatographia,* vol. 41, no. 7–8, pp. 435–444.

Xing, W.L. & He, X. W. 1997, Prediction of the selectivity coefficients of a berberine selective electrode using artificial neural networks, *Analytica Chimica Acta,* vol. 349, no. 1–3, pp. 283–286.

Yan, A. X., Zhang, R. S., Liu, M. C., Hu, Z. D., Hooper, M. A., & Zhao, Z. F. 1998, Large artificial neural networks applied to the prediction of retention indices of acyclic and cyclic alkanes, alkenes, alcohols, esters, ketones and ethers, *Computers & Chemistry,* vol. 22, no. 5, pp. 405–412.

Yan, A. X., Jiao, G. M., Hu, Z. D., & Fan, B. T. 2000a, Use of artificial neural networks to predict the gas chromatographic retention index data of alkylbenzenes on carbowax-20M, *Computers & Chemistry*, vol. 24, no. 2, pp. 171–179.

Yan, A. X. & Hu, Z. D. 2001, Linear and non-linear modeling for the investigation of gas chromatography retention indices of alkylbenzenes on Cit.A-4, SE-30 and Carbowax 20M, *Analytica Chimica Acta*, vol. 433, no. 1, pp. 145–154.

Yan, W. P., Diao, C. F., Tang, Z. N., & Li, X. 2000b, The study of gas sensor array signal-processing with improved BP algorithm, *Sensors and Actuators B-Chemical*, vol. 66, no. 1–3, pp. 283–285.

Yang, H., Ring, Z., Briker, Y., McLean, N., Friesen, W., & Fairbridge, C. 2002a, Neural network prediction of cetane number and density of diesel fuel from its chemical composition determined by LC and GC-MS, *Fuel*, vol. 81, no. 1, pp. 65–74.

Yang, H. S. & Griffiths, P. R. 1999a, Application of multilayer feed forward neural networks to automated compound identification in low-resolution open-path FT-IR spectrometry, *Analytical Chemistry*, vol. 71, no. 3, pp. 751–761.

Yang, H. S. & Griffiths, P. R. 1999b, Encoding FT-IR spectra in a Hopfield network and its application to compound identification in open-path FT IR measurements, *Analytical Chemistry*, vol. 71, no. 16, pp. 3356–3364.

Yang, H. S., Lewis, I. R., & Griffiths, P. R. 1999c, Raman-spectrometry and neural networks for the classification of wood types. 2. Kohonen self-organizing map, *Spectrochimica Acta part A – Molecular and Biomolecular Spectroscopy*, vol. 55, no. 14, pp. 2783–2791.

Yang, J., Xu, G. W., Kong, H. W., Zheng, W. F., Pang, T., & Yang, Q. 2002b, Artificial neural network classification based on high-performance liquid chromatography of urinary and serum nucleosides for the clinical diagnosis of cancer, *Journal of Chromatography B – Analytical Technologies in the Biomedical and Life Sciences*, vol. 780, no. 1, pp. 27–33.

Yang, H., & Irudayaraj, J. 2003a, Rapid detection of foodborne microorganisms on food surface using Fourier-transform Raman-spectroscopy, *Journal of Molecular Structure*, vol. 646, no. 1–3, pp. 35–43.

Yang, H., Griffiths, P. R., & Tate, J. D. 2003b, Comparison of partial least squares regression and multilayer neural networks for quantification of nonlinear systems and application to gas phase Fourier transform infrared spectra, *Analytica Chimica Acta*, vol. 489, no. 2, pp. 125–136.

Yang, Y. M., Yang, P. Y., & Wang, X. R. 2000, Electronic nose based on SAWS array and its odor identification capability, *Sensors and Actuators B-Chemical*, vol. 66, no. 1–3, pp. 167–170.

Yang, X.-R., Luo, G. Guo, Y., & Ling, Y.-H. 2001, Multisensor poly information fusion technology and its application, *Proceedings of SPIE International Conference on Sensor Technology*, pp. 449–454.

Yea, B., Osaki, T., Sugahara, K., & Konishi, R. 1997, The concentration-estimation of inflammable gases with a semiconductor gas sensor utilizing neural networks and fuzzy inference, *Sensors and Actuators B-Chemical*, vol. 41, no. 1–3, pp. 121–129.

Yea, B., Osaki, T., Sugahara, K., & Konishi, R. 1999, Improvement of concentration-estimation algorithm for imflammable gases utilizing fuzzy rule-based neural networks, *Sensors and Actuators B-Chemical*, vol. 56, pp. 181–188.

Yin, C. S., Shen, Y., Liu, S. S., Li, Z. L., & Pan, Z. X. 2000, Application of linear neural-network to simultaneous determination of 4 B-group vitamins, *Chem.J.Chinese.Univ Chinese.*, vol. 21, no. 1, pp. 49–52.

Yin, C. S., Shen, Y., Liu, S. S., Yin, Q. S., Guo, W. M., & Pan, Z. X. 2001a, Simultaneous quantitative UV spectrophotometric determination of multicomponents of amino-acids using linear neural-network, *Computers & Chemistry*, vol. 25, no. 3, pp. 239–243.

Yin, C. S., Guo, W. M., Liu, S. S., Shen, Y., Pan, Z. X., & Wang, L. S. 2001b, Simultaneous determination of vitamin-B complex using wavelet neural-network, *Chinese.J.Chem.*, vol. 19, no. 9, pp. 836–841.

Yin, C. S., Guo, W. M., Lin, T., Liu, S. S., Fu, R. Q., Pan, Z. X., & Wang, L. S. 2001c, Application of wavelet neural-network to the prediction of gas-chromatographic retention indexes of alkanes, *J.Chin. Chem.Soc.*, vol. 48, no. 4, pp. 739–749.

Yu, R. F., Liaw, S. L., Chang, C. N., & Cheng, W. Y. 1998, Applying real-time control to enhance the performance of nitrogen removal in the continuous-flow SBR system, *Water Science and Technology*, vol. 38, no. 3, pp. 271–280.

Zakaria, P., Macka, M., & Haddad, P. R. 2003, Mixed-mode electrokinetic chromatography of aromatic bases with two pseudo-stationary phases and pH control, *Journal of Chromatography A*, vol. 997, no. 1–2, pp. 207–218.

Zampronio, C. G., Rohwedder, J. J. R., & Poppi, R. J. 2002, Artificial neural networks applied to potentiometric acid-base flow-injection titrations, *Chemometrics and Intelligent Laboratory Systems*, vol. 62, no. 1, pp. 17–24.

Zeng, Y.-B., Xu, H.-P., Liu, H.-T., Wang, K.-T., Chen, X.-G., Hu, Z.-D., & Fan, B.-T. 2001, Application of artificial neural networks in multifactor optimization of an on-line microwave FIA system for catalytic kinetic determination of ruthenium (III), *Talanta*, vol. 54, no. 4, pp. 603–609.

Zhang, Q., Reid, J. F., Litchfield, J. B., Ren, J. L., & Chang, S. W. 1994, A prototype neural-network supervised control-system for Bacillus-Thuringiensis fermentations, *Biotechnology and Bioengineering*, vol. 43, no. 6, pp. 483–489.

Zhang, J. & Chen, Y. 1997a, Food sensory evaluation employing artificial neural networks, *Sensor Review*, vol. 17, no. 2, pp. 150–158.

Zhang, L., Jiang, J.-H., Liu, P., Liang, Y.-Z., & Yu, R.-Q. 1997b, Multivariable nonlinear modelling of fluorescence data by neural network with hidden node pruning algorithm, *Analytica Chimica Acta*, vol. 344, no. 1, pp. 29–39.

Zhang, L., Liang, Y.-Z., Jiang, J.-H., Yu, R.-Q., & Fang, K.-T. 1998, Uniform design applied to nonlinear multivariable calibration by ANN, *Analytica Chimica Acta*, vol. 370, pp. 65–77.

Zhang, R. S., Yan, A. X., Liu, M. C., Liu, H., & Hu, Z. D. 1999, Application of artificial neural networks for prediction of the retention indexes of alkylbenzenes, *Chemometrics and Intelligent Laboratory Systems,* vol. 45, no. 1–2, pp. 113–120.

Zhang, X. Q., Liu, H., Zheng, J. B., & Kao, H. 2002a, The application of signal processing in resolving overlapping chemical bands, *Progress in Chemistry*, vol. 14, no. 3, pp. 174–181.

Zhang, X. Y., Qi, J. H., Zhang, R. S., Liu, M. C., Hu, Z. D., Xue, H. F., & Fan, B. T. 2001, Prediction of programmed-temperature retention values of naphthas by wavelet neural networks, *Computers & Chemistry*, vol. 25, no. 2, pp. 125–133.

Zhang, Y., Liu, J. H., Zhang, Y. H., & Tang, X. J. 2002b, Cross-sensitivity reduction of gas sensors using genetic neural-network, *Optical Engineering*, vol. 41, no. 3, pp. 615–625.

Zhang, Z.-X., & Ma, X.-G. 2002c, Methods for correction of spectral interferences in inductively coupled plasma atomic emission spectrometry, *Current Topics in Analytical Chemistry*, vol. 3, pp. 105–123.

Zhao, R. H., Yue, B. F., Ni, J. Y., Zhou, H. F., & Zhang, Y. K. 1999, Application of an artificial neural network in chromatography – retention behavior prediction and pattern recognition, *Chemometrics and Intelligent Laboratory Systems*, vol. 45, no. 1–2, pp. 163–170.

Zheng, P., Harrington, P. de B., & Davis, D. M. 1996, Quantitative analysis of volatile organic compounds using ion mobility spectrometry and cascade correlation neural networks, *Chemometrics and Intelligent Laboratory Systems*.

Zhou, Q., Minhas, R., Ahmad, A., & Wheat, T. A. 2000a, Intelligent gas-sensing systems, *Proceedings of the International Symposium on Environment Conscious Materials: Ecomaterials*, Ottawa, pp. 521–535.

Zhou, W. C. & Mulchandani, A. 1995, Recent advances in bioprocess monitoring and control, *Biosensor and Chemical Sensor Technology*, vol. 613, pp. 88–98.

Zhou, Y.-Y., Wang, H.-W., Sun, G., Fan, Y.-Q., Chen, X.-G., Hu, Z. 2000b, Application of artificial neural networks in multifactor optimization of an FIA system for the determination of aluminum, *Fresenius Journal of Analytical Chemistry*, vol. 366, no. 1, pp. 17–21.

Zupan, J. & Gasteiger, J. 1991, Neural networks – a new method for solving chemical problems or just a passing phase, *Analytica Chimica Acta*, vol. 248, no. 1, pp. 1–30.

Zupan, J. & Gasteiger, J. 1993, *Neural Networks for Chemists – An Introduction*, VCH, Weinheim.

Zupan, J., Novic, M., Li, X. Z., & Gasteiger, J. 1994, Classification of multicomponent analytical data of olive oils using different neural networks, *Analytica Chimica Acta*, vol. 292, no. 3, pp. 219–234.

Zupan, J., Novic, M., & Gasteiger, J. 1995, Neural networks with counter-propagation learning-strategy used for modeling, *Chemometrics and Intelligent Laboratory Systems*, vol. 27, no. 2, pp. 175–187.

Zupan, J., Novic, M., & Ruisanchez, I. 1997, Kohonen and counterpropagation artificial neural networks in analytical chemistry, *Chemometrics and Intelligent Laboratory Systems*, vol. 38, no. 1, pp. 1–23.

Zupan, J. & Gasteiger, J. 1999, *Neural Networks in Chemistry and Drug Design*, 2nd edition, Wiley-VCH, Weinheim.

Zupan, J. 2002, 2D mapping of large quantities of multi-variate data, *Croatica Chemica Acta*, vol. 75, no. 2, pp. 503–515.

Appendix A: A brief description of the PCA-CG algorithm

(Guterman, H., 1994, *Neural Parallel Sci. Comput.*, vol. 2, pp. 43–53)

Step 1: Form joint input-output data vector $\mathbf{X} = x_p \cup y_p$, making N_p rows of matrix $\mathbf{X}$ represent the entire data set. The columns of $\mathbf{X}$ are scaled in one of two ways: (i) each column is divided by its standard deviation, or (ii) each column is divided by the range of the data. In the present work, the data were scaled by the standard deviation and the zero-centered mean.

Step 2: Calculate $\Sigma_x[a \quad x \quad a]$ as

$$\Sigma_x = E\{(\mathbf{X} - E\{\mathbf{X}\}^T(\mathbf{X} - E\{\mathbf{X}\}\} \tag{A1}$$

Step 3: Determine the eigenvectors and eigenvalues of Σ_x. Select eigenvectors $\phi_1, \ldots, \phi_r$ corresponding to the largest eigenvalues $\lambda_1, \ldots, \lambda_r$ necessary for reconstructing $\mathbf{X}$ with a chosen information content ξ:

$$\mu_i = \frac{\lambda_i}{\text{tr}(\Sigma_x)} = \frac{\lambda_i}{\sum_{i=1}^{a} \lambda_i} \tag{A2}$$

Then, assuming that λ_i and ϕ_i are ordered, the number of neurons in the hidden layer, r, would be equal to the number of dimensions necessary to reconstruct the original information with a ξ degree of fidelity,

$$\sum_{i=1}^{r} \mu_i \geq \xi \tag{A3}$$

There are n inputs and m outputs, and $a = m + n$. The r index represents the number of nodes in the hidden layer.

Step 4: Compute the initial input to the hidden weight matrix, $\mathbf{W}_H$ as follows (last column is the bias values):

$$\mathbf{W}_H = \begin{bmatrix} \phi_{11} & \cdots & \phi_{n1} & h_1 \\ \phi_{12} & \cdots & \phi_{n2} & h_2 \\ \cdots & \cdots & \cdots & \cdots \\ \phi_{1r} & \cdots & \phi_{nr} & h_r \end{bmatrix} \tag{A4}$$

$$h_i = \sum_{j=n+1}^{a} \phi_{ij}^T E\{X_j\} \tag{A5}$$

Step 5: Compute the initial hidden to output weight matrix, $\mathbf{W}_O$, as follows (last column is the bias values):

$$\mathbf{W}_o = \begin{bmatrix} \phi_{(n+1)1} & \cdots & \phi_{(n+1)r} & u_{n+1} \\ \phi_{(n+2)1} & \cdots & \phi_{(n+2)r} & u_{n+2} \\ \cdots & \cdots & \cdots & \cdots \\ \phi_{a1} & \cdots & \phi_{ar} & u_a \end{bmatrix} \tag{A6}$$

$$\mathbf{U}_{\text{bias}} = \sum_{i=r+1}^{a} \boldsymbol{\phi}_1^{\text{T}} \mathbf{E}\{\mathbf{X}\} \boldsymbol{\phi}_i = [u_1, \ u_2, \ \ldots, \ u_a]^{\text{T}} \tag{A7}$$

Step 6: A conjugate gradient method (CG) similar to the one proposed by Leonard, J. & Kramer, M., 1990, *Comput. Chem. Eng.*, vol. 14, pp. 337–43, is employed for searching the optimum weights.

Appendix B: Network reduction algorithm

(Boger, Z. & Guterman, H., 1997, *Proc. IEEE Int. Conf. on Systems Man. and Cybernetics, SMC'97*, Orlando, Florida, pp. 3030–35).

Basic statistics (Vandaele 1983) provide the variance of a linear combination of random variables **x**, given by:

$$y = \sum_{i=1}^{n} a_i x_i = \mathbf{A}\mathbf{X} \tag{B1}$$

$$\text{var}(y) = \sum_{i=1}^{n} \sum_{j=1}^{n} a_i a_j r_{ij} = \mathbf{A}\mathbf{R}\mathbf{A}^{\mathrm{T}} \tag{B2}$$

where **R** is an $n \times n$ covariance matrix i.e. $\mathbf{R} = \text{cov}(\mathbf{x})$. If the variable set **x** is uncorrelated, the calculation can be further simplified to:

$$\text{var}(y) = \sum_{i=1}^{n} a_i^2 r_{ii} \tag{B3}$$

However, in normal industrial plant data the variables are usually correlated in some way, so Equation (B2) is used in the present procedure. Defining the matrix $\mathbf{W}_{\mathrm{H}}$ to be the input-hidden layer weight set of the original (trained but unreduced) NN, and taking $\mathbf{x}_{\mathrm{p}}$ to represent the network inputs for the pth training example, then the input to the jth hidden layer node for the pth example is:

$$y_{jp} = \sum_{i=1}^{n} (\mathbf{W}_{\mathrm{H}})_{ji} x_{ip} \tag{B4}$$

where n is the number of network inputs. Following Equation (B2), the variance $(V_{\mathrm{H}})_j$ of the input to hidden layer node j over the full set of training examples is:

$$(V_{\mathrm{H}})_j = \text{var}(y_j) = (\mathbf{W}_{\mathrm{H}})_j \mathbf{R}(\mathbf{W}_{\mathrm{H}})_j^{\mathrm{T}} \tag{B5}$$

where $(\mathbf{W}_{\mathrm{H}})_j$ represents the jth row of $\mathbf{W}_{\mathrm{H}}$ and **R** is the covariance matrix for the network inputs **x**, estimated from the training set. From Equation (B5), the relative variance of the input to hidden node j can be calculated as:

$$(V_{\mathrm{H}})_j^{\text{rel}} = \frac{(V_{\mathrm{H}})_j}{\sum_{k=1}^{n_h} (V_{\mathrm{H}})_k} \tag{B6}$$

where n_{h} is the number of hidden nodes. Equations (B5)–(B6) are the essential equations for determining the statistical relevance of a hidden node.

Moving to the input relevance analysis, $\mathbf{R}_i$, the ith row of **R**, is associated with the ith input variable. From Equation (B5), the contribution of input i to the variance $(V_{\mathrm{H}})_j$ is:

$$(V_1)_{ji} = (\mathbf{W}_{\mathrm{H}})_{ji} \mathbf{R}_i (\mathbf{W}_{\mathrm{H}})_j^{\mathrm{T}} \tag{B7}$$

and the contribution of input i to the total variance of the hidden layer inputs is:

$$(V_1)_i = \sum_{j=1}^{n_h} (\mathbf{W}_H)_{ji} \mathbf{R}_i (\mathbf{W}_H)_j^T \tag{B8}$$

An equivalent, more concise form of Equation (B8) is:

$$(V_1)_i = (\mathbf{W}_H)_i^T \mathbf{W}_H \mathbf{R}_i^T \tag{B9}$$

where $(\mathbf{W}_H)_i^T$ is the ith row of the transpose of $\mathbf{W}_H$, i.e. the ith column of $\mathbf{W}_H$ expressed as a row vector. The relative contribution of input i to the variance of the hidden layer inputs can now be calculated:

$$(V_1)_i^{\text{rel}} = \frac{(V_1)_i}{\sum\limits_{k=1}^{n} (V_1)_k} \tag{B10}$$

Equations (B9–B10) are the equations used to determine the statistical relevance of network inputs. The V_H^{rel} values serve as a check on the significance of particular hidden layer nodes. Nodes with low V_H^{rel} values are good candidates for elimination, and if trial removal yields an improvement in AIC and/or FPE, these nodes can be permanently removed. Calculation of the adjustment of the bias for an output node k compensating for the removal of the hidden node j is based on the expected value of o_j, $\mathbf{E}(o_j)$:

$$\phi_k' = \phi_k + (\mathbf{W}_o)_{kj} \mathbf{E}(o_j) \tag{B11}$$

where $\mathbf{W}_O$ represents the hidden-output layer weight set. A reasonable approximation of Equation (B11) is given by:

$$\phi_k' = \phi_k + (\mathbf{W}_o)_{kj} f\left(\sum_{i=1}^{n} (\mathbf{W}_o)_{ji} \mathbf{E}(x_i) + \phi_j \right) \tag{B12}$$

Since Equation (B12) is approximate, additional training by PCA-CG is used to re-tune the network after hidden nodes are removed. Input variables with low V_I^{rel} values do not contribute much information to the network and can be eliminated. Analogous to Equation (B12), the bias adjustment for hidden layer node j to compensate for removal of input i is:

$$\phi_j' = \phi_j + (\mathbf{W}_H)_{ji} \mathbf{E}(x_i) \tag{B13}$$

Procedurally, one begins by training the NN with the whole data set, with a reasonable value of the information content used for estimating the number of hidden nodes, for example, 70%. After training the ANN with the PCA-CG algorithm, the best candidates for removal according to Equations (B6) and (B10) are identified. The group of top candidates is removed, and the network is retrained using PCA-CG. On the retraining step, different information content can be used in selection of the hidden layer architecture, according to the relative variance values of the hidden nodes. Our experience shows that optimum results are obtained when there is only one hidden node with a relative variance smaller than 10%, so a higher information content is used when the least significant hidden node has relative variance higher than 10%, and a smaller information content is used when more than one hidden node have a relative variance smaller than 10%. After retraining, various criteria can be checked, such as the Akaike information theoretic content (AIC) or the final prediction error (FPE) (Ljung 1987). Evaluation of the criteria requires only a single forward pass through the training set to determine the prediction error. As long as the FPE and AIC decline, more inputs and/or hidden nodes should be removed (if the AIC has a negative value, it means that the model is over-fitted). When the network is in the range of the minimum of FPE and AIC (the minima many occur at different points), the same analysis of variance procedure may be performed on the reduced net as a further check that the optimal net was obtained.

References

Ljung, L., 1987 *System Identification – Theory for the User*, Prentice-Hall.
Vandaele, A., 1983 *Applied Time Series and Box – Jenkins Models*, Academic Press.

Appendix C: Prediction of diesel fuel cetane number

(Reprinted from Boger, Z., 2003, *Analytica Chimica Acta*, vol. 490, pp. 31–40, with permission from Elsevier)

The NIR spectra of diesel fuels, together with five properties that were measured at the Southwest Research Institute are available on the Eigenvector Research Corporation site page (http://www.evriware.com/Data/SWRI/index.html). As an example, the cetane number (equivalent to the octane number in gasoline) was predicted by the ANN model using the 401 wavelengths-wide NIR spectra as inputs. One hundred and twenty-four examples were used as the training set, while 112 examples were used as the validation set. The spectral data were zero-centered and unit scaled, while the cetane output values were further compressed in to the (0.1–0.9) range. The initial 401-6-1 ANN was trained and its input dimension automatically reduced. The ANN models were trained to the minimal training set root mean square prediction (RMSP) error, with no early stopping of the training. The results are shown in Fig. C.1.

It can be seen that both the final *training* and validation RMSP errors, with 23 wavelengths as inputs, are lower than the initial errors with the full 401 wide

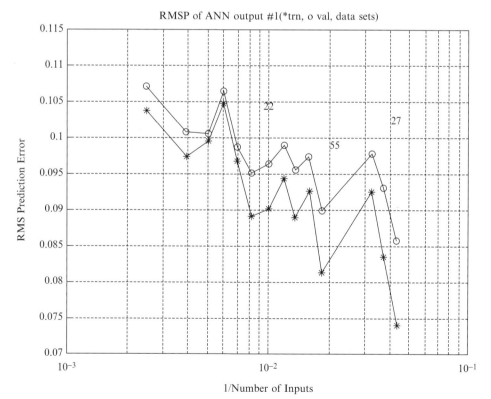

Fig. C.1 The reduction of the cetane number ANN model RMSP errors as a function of the number of wavelength. (*) training, (o) validation sets.

spectra. The jagged RMSP error history may be the result of the ANN trapped in a local minimum, that was not avoided or escaped from, but it is interesting to note that the input relevance algorithm continues to perform well. The selected wavelengths are shown as circles on the spectra of the last validation example in Fig. C.2.

Specific peaks and peak shoulders were selected to give the minimal RMSP errors. The causal index of the minimal ANN model was calculated and it seems that the compounds that produce the peaks at the 121–122, 256, and 323–328 regions are those that lower the cetane number, while those compounds in the 148–173 region are those that increase the cetane number. It also means that the compounds responsible for the rest of the spectra have little effect on the cetane number, or that they are correlated with the selected inputs and thus are redundant.

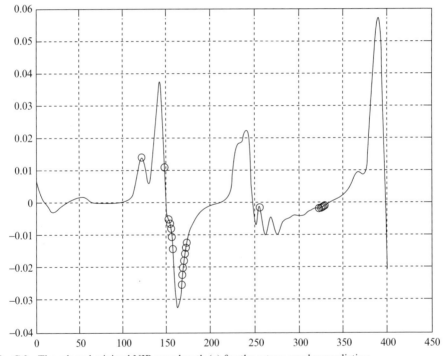

Fig. C.2   The selected minimal NIR wavelength (○) for the cetane number prediction.

Appendix D: Meanings of acronyms

Acronym	Full name
AA-ANN	Auto Associative ANN
AE	Atomic Emission
AES	Atomic Emission Spectroscopy
AI	Artificial Intelligence
ANN	Artificial Neural Networks
ART	Adaptive Resonance Theory
ATOFMS	Aerosol Time-Of-Flight Mass Spectrometry
CC	Counter Correlation
CG	Conjugate Gradient
CI	Causal Index
COD	Chemical Oxygen Demand
CWA	Chemical Warfare Agent
EDX	Energy Dispersive X-Ray
EDXRF	Energy-Dispersive X-Ray Fluorescence
EN	Electronic Nose
ES	Expert Systems
ESI	Electro Spray Ionization
FAAS	Flame Atomic Absorption Spectrometry
FFT	Fast Fourier Transform
FIA	Flow Injection Analysis
FID	Flame Ionization Detector
FL	Fuzzy Logic
FT	Fourier Transform
FTIR	Fourier Transform Infra Red
GA	Genetic Algorithm
GC	Gas Chromatograph
HPLC	High Performance Liquid Chromatograph
IC	Ion Chromatograph
ICP	Inductively Coupled Plasma
IMS	Ion Mobility Spectroscopy
IR	Infra Red
IX	Ion Exchanger
LC	Liquid Chromatograph
LM	Marquand-Leuvenberg
LVQ	Learning Vector Quantization
MAB	Metastable Atom Bombardment
MLADI	Matrix Laser Assisted Desorption Ionization
MLP	Multi Layer Perceptrons
MOS	Metal Oxide Sensor
MRI	Magnetic Resonance Imaging
MS	Mass Spectrometer
NIR	Near Infra Red
NMR	Nuclear Magnetic Resonance
ORP	Oxidation Reduction Potential
PAH	Polycyclic Aromatic Hydrocarbons
PAS	Photo Acoustic Spectroscopy
PCA	Principal Component Analysis
PCB	Poly Chloro Biphenyl

PLS	Partial Least Square
PNN	Probabilistic Neural Networks
Py	Pyrolysis
PZ	Piezoelectric
QA	Quality Assurance
QMB	Quartz Micro Balance
QSAR	Quantitative Structure Activity Relationship
RBF	Radial Based Function
RI	Retention Index
ROC	Receiver Operator Characteristic
RPLC	Reversed Phase Liquid Chromatograph
SAW	Surface Acoustic Wave
SIMCA	Soft Independent Modeling of Class Association
SIMS	Secondary Ion Mass Spectrometry
SOM	Self Organized Map
SVD	Singular Value Decomposition
SVM	Support Vector Machine
TOF	Time of Flight
UV	Ultra Violet
Vis	Visible
VOC	Volatile Organic Compounds
XRD	X-Ray Diffraction
XRF	X-Ray Fluorescence

Index